PROTEIN METABOLISM and BIOLOGICAL FUNCTION

James Boyd Allison, January 28, 1901–September 25, 1964

PROTEIN METABOLISM and BIOLOGICAL FUNCTION

Edited by
C. PAUL BIANCHI and
RUSSELL HILF

RUTGERS UNIVERSITY PRESS *New Brunswick, New Jersey*

Library of Congress Catalog Card Number: 72–96030
SBN: 8135-0617-4
Manufactured in the United States of America by Quinn & Boden Company, Inc., Rahway, New Jersey

Dedication

This volume has been compiled and dedicated as a memorial for Dr. James Boyd Allison. It seemed to us to be an appropriate way to evidence our esteem for the man. We sought as contributors those who had known Jim, some who had studied under him as graduate students, others who had shared research interests and experiences with him. It proved to be an easy task to gather together the manuscripts presented here, and it is no doubt true that there are others who should also have been included. The fact is that Dr. Allison's influence, professional and in many other respects, was felt by a very large number of persons. He was a friend and mentor to many; far beyond the numbers he trained as graduate students or taught as undergraduates, for he always shared generously his knowledge and experience. Despite an extremely busy schedule, he found time to advise and to encourage—the effect was to inspire—those who sought his counsel. It should be pointed out that in addition to his role as a scientist and a teacher, he was also an administrator, and he was at all times concerned with the whole of Rutgers University, with which he was associated for so many years. Beyond that and in his view, we feel, he was first and foremost a husband and father. We imply that his talents were many and they were. We propose that one of the most noteworthy was his feeling for his fellow man. In short, we count it a privilege to have known him.

J.H.B

Acknowledgment

The editors wish to express their sincere appreciation to Mrs. James B. Allison, for the photography of Dr. Allison; to Mrs. Judith Holman Bianchi, former Administrative Assitant to Dr. Allison, for the Dedication Section; to the various Journals and Publishers who have granted permission for reproduction of material; and to all of the contributors for their continued cooperation to meet deadlines in the face of their busy schedules. We wish to personally thank the staff of the Rutgers University press for their prompt and expert advice in many matters and for their support in making this volume possible.

C.P.B.
R.H.

June 1969

Preface

The first half of the twentieth century witnessed a remarkable development of the subject of nutrition from a field dominated by a knowledge of good cooking from the gourmet point of view, to a quantitative science which, by utilizing the tools of chemistry and physiology, identified the critical components of foods. Within a relatively short period of time all of the vitamins were isolated and their structure determined, and the essential amino acids were differentiated from those which were dispensable. The importance of certain unsaturated fatty acids and trace metals in the diet was also demonstrated during this period. At this point it appeared that all the basic knowledge of nutrition had been elucidated and the major task ahead was its application to the benefit of living organisms, especially man, in sickness and in health.

Instead of reaching a plateau at this stage of its development, the science of nutrition rounded a corner and became deeply involved with intermediary metabolism and the mechanism of action of dietary components. Attempts were made to understand the nutritional requirements of the individual cell in terms of the metabolic role played by each indispensable and nonessential component which must penetrate the plasma membrane and enter the cytoplasm. Intensive studies, particularly in physiology and biochemistry, performed in many countries of the world over several decades finally led to the general conclusion that most of the dietary essentials were required for the synthesis and biological activity of proteins which were endowed with structural, contractile, hormonal, enzymatic, and other functions. Amino acids essential for protein synthesis were required for distribution, together with nonessential ones, in proper sequence along the polypeptide chain. The precise ordering of the amino acids is dictated by a different gene (DNA) for each of approximately 2000 varieties of proteins in a single living organism. Most of the proteins are enzymes, the majority of which are catalytically inert unless they have bound to the protein moiety as a trace metal or a coenzyme containing a vitamin. The convergence on the problems of the protein molecule by the nutritionist on the one hand and the enzymologist on the other, has made possible our understanding of the role of dietary essentials in biosynthesis and intermediary metabolism and has enabled molecular biology, medicine, and agriculture to develop at a remarkable pace. It has provided the necessary theoretical underpinning for the spectacular progress which has been made in the application of the life sciences to the needs of mankind.

As an exception to the basic principle described above, certain unsaturated fatty acids (and perhaps other organic molecules required in trace amounts) which are not directly required for protein synthesis, may play an indirect role in the regulation of protein activity. For example, linolenate has recently been shown to be a precursor of the hormone, prostaglandin, which can stimulate the contraction of smooth muscle as well as have other regulatory functions. Now that all of the major dietary essentials have been identified and their metabolic roles at least partially elucidated, it has become possible to utilize this knowledge in the study of many fields of vital importance to our understanding of living organisms.

It was in the biochemical approach to nutrition and intermediary metabolism that Dr. James B. Allison was involved for much of his scientific career. When I came to Rutgers in 1931 as a graduate student, he had already begun a fruitful collaboration with the physiologist, William H. Cole, in which they developed concepts of chemical stimulation using several series of organic compounds and animals such as paramecia, barnacles, fish and frogs. This collaboration was so successful that it influenced other life scientists at Rutgers and soon microbiologists, botanists, zoologists, endocrinologists and pathologists were involved in cooperative research with biochemists and physiologists. This association was formalized in 1936 with the establishment of the Bureau of Biological Research. While a member of the Bureau, as well as the Department of Physiology and Biochemistry, Dr. Allison's major research interests were in the area of nutritional biochemistry, with an emphasis on the highly quantitative approach to mammalian nutrition. He and his associates were able to relate specific changes in the diet to the variations in activity of certain enzymes in the liver while other enzymes remained intact. He was particularly interested in the relationship of dietary components to disease, especially cancer, and made many basic contributions to an understanding of the interrelation of nutrition to intermediary metabolism of diseased and healthy animals.

Of even greater importance than Allison's abilities as an administrator and investigator, were his skills as a teacher. His influence went far beyond the classroom and at the laboratory bench he was especially effective in explaining and discussing research not only with students but also with staff and faculty as well. Never flamboyant but shy and modest to a fault, his quiet, friendly, sympathetic, dedicated and always helpful ways made him one of the most loved and respected men on the Rutgers faculty. It is therefore fitting that this volume, containing chapters by former students and associates on topics often pursued in the Bureau of Biological Research, be dedicated to his memory. Topics discussed range from development, hormones, and immunology to enzymes, physiology, muscle, and disease. Although these subjects are sometimes far removed from nutritional biochemistry in the narrow sense, they certainly lie within the mainstream of Allison's broad interests in physiology and biochemistry. All of the authors are happy to use this publication as a means of expressing their deep love and respect for this great scientist.

IRWIN W. SIZER

Massachusetts Institute of Technology
February 1, 1968

Contents

PREFACE—*Irwin W. Sizer* ix

I NUTRITION AND DEVELOPMENT

Behavioral Abnormalities Resulting from Early Malnutrition in Rats and Pigs—*Richard H. Barnes* 3

The Effect of Severe Protein Deficiency on the Human Infant —*J. S. Garrow* 28

II ENDOCRINOLOGY

Hormones and Growth in Mammals—*James H. Leathem* 51

Endocrine and Nutritional Interrelationships—*Leonard J. Lerner* 61

III IMMUNOLOGY

Immunoelectrophoretic Analysis of Cytoplasmic Protein Synthesis in the Liver—*Curtis A. Williams* 75

A Review of the Nature and Significance of Hemagglutinins of Selected Invertebrates—*Elias Cohen* 87

IV PHYSIOLOGY

Some Effects of Physiological State on Various Metabolic Systems—*Harold L. Rosenthal* 97

Role of Calcium in the Regulation of Muscle Contraction—*C. Paul Bianchi* 111

Relationship Between Protein Metabolism in Muscle Tissue and the Concept of "Protein Reserves"—*R. W. Wannemacher and W. K. Cooper* 121

V BIOCHEMISTRY

The Phosphorylation of Thymidine as a Factor Controlling Cellular Proliferation—*Jay S. Roth* 141

Implications of the Kinetic Behavior of a Soluble Cytochrome Oxidase Preparation—*A. J. Davison and W. W. Wainio* 220

VI VIRAL AND NEOPLASTIC DISEASE

A Model for the Study of the Effect of Viral Stress on Protein Metabolism—*Robert L. Squibb* 233

Investigations of Experimental Mammary Tumors—*Russell Hilf* 251

Research on Tumor-Host Relationship and Chemotherapy of Canine Malignancies—*John R. McCoy* 262

James Boyd Allison—An Appreciation—*James R. McCoy* 283

INDEX 289

PROTEIN METABOLISM and BIOLOGICAL FUNCTION

I. Nutrition and Development

Behavioral Abnormalities Resulting from Early Malnutrition in Rats and Pigs *

RICHARD H. BARNES †
Graduate School of Nutrition
Cornell University
Ithaca, New York

The concept that severe protein-calorie malnutrition in the early life of children might lead to permanent abnormalities in behavior, particularly a retardation in learning ability, is relatively new. Nevertheless, the implications of this concept relating to the vast problems associated with the development and modernization of large populations of underprivileged people with whom malnutrition is endemic have encouraged groups of nutritionists and psychologists to initiate large research programs to examine the validity of this concept by studying malnourished children from all parts of the world. The recognition that social and sensory deprivations in the young child result in long-lasting behavioral abnormalities and that in population groups suffering from malnutrition, social deprivations usually accompany nutritional deprivations has made the problem of validating the concept extremely difficult. As a consequence, a definitive answer regarding the specific contribu-

* This research was supported in part by funds provided through the State University of New York and Public Health Service Research Grant No. HD-02581 from the National Institute of Child Health and Human Development.

† An account of investigations conducted in the author's laboratories during recent years through the collaboration of Wilson G. Pond, A. Ulric Moore, Susan R. Cunnold, Ian M. Reid, S. Fraňková, and L. Krook representing the College of Agriculture, the Veterinary College, and the Graduate School of Nutrition, Cornell University, Ithaca, N. Y.

tion of malnutrition has not yet been achieved even though strong suggestive evidence of a causal relationship between malnutrition and long-lasting behavioral changes has been provided in several studies. We must await the results of carefully controlled field studies, which unfortunately will probably have to employ extremely complex experimental designs, before this cause and effect relationship can be stipulated for the human being. However, much information can be gained by using experimental animals in which social and sensory factors can be controlled and one can introduce important nutritional variables such as the extent of malnutrition, the type of nutrient deficiency and the period during the development of the organism when malnutrition is imposed. Furthermore, one can study changes in the structural and biochemical makeup of nervous tissue, the role of endocrine factors and the effect that certain drugs may have in altering abnormal behavior both during the period of its development and following its long-lasting establishment. One can also identify different forms of abnormal behavior resulting from early malnutrition and attempt to relate these to the various factors that have been mentioned. Our laboratories, as well as many others around the world, have been studying certain of these problems and it will be the purpose of this dissertation to regroup and restate the experimental findings we have obtained and which have been reported in the literature, and also to include some unpublished findings that may be of interest.

Protein-Calorie Malnutrition in the Baby Pig

The baby pig is a valuable experimental animal for the study of severe protein-calorie malnutrition. Platt and his associates [1] found the young pig would consume large quantities of carbohydrate when added to a very low protein containing diet and that several of the characteristic signs of kwashiorkor would develop. Other experimental animals, such as the young rat, develop severe anorexia when fed low protein diets so that it is essentially impossible to induce a sufficiently high caloric intake by *ad libitum* feeding so as to produce signs of fatty livers and grossly observable edema. In our laboratories we were unable to obtain high caloric intakes in baby pigs with extra carbohydrate because of anorexia when the animals were being fed a diet containing approximately 5% protein. On the other hand, increasing the caloric density of the diet by adding 24% corn oil to the 5% protein diet did cause baby pigs to develop extremely low serum albumin levels, fatty livers, and grossly observable edema.[2] Autopsies performed after the pigs had received this diet for 8 weeks showed a number of tissue changes, but for purposes of the present discussion it was of particular interest that by microscopic examination these animals appeared to have decreased numbers of neurones in the cerebral cortex and cerebellum. Winick and Noble[3] have shown that baby rats that were malnourished during the suckling period by giving the lactating dam large litters to nurse have reduced numbers of cells in the brain. They utilized the technique of Enesco and Leblond[4] which related DNA to cell numbers, a procedure that is undoubtedly far more reliable than the visual technique used in our earlier studies. This substantiation of the effect of early malnutrition gives strong support to the conclusion that the type of malnutrition induced in pigs does result in an impairment in normal brain development.

The protein-calorie malnutrition that developed in our early studies was complicated by several factors that were not appreciated at that time. Later work showed that the combination of feeding a diet that was very low in protein and high in unsaturated fat (24%) resulted in deficiencies in selenium and vitamin E which were not evident in animals fed diets low in protein but also low in fat (3%). In addition, the malnourished condition of the animals made them susceptible to monilial infections that were particularly evident in the mouth, esophagus and upper regions of the stomach. These factors were corrected by increasing dietary levels of selenium and vitamin E and by the inclusion of the antifungal antibiotic, mycostat.* Most of the behavioral studies that will be discussed here were conducted with pigs that had been provided with these dietary fortifications. Two other forms of dietary restriction have been studied. One of these was to feed pigs the low protein diet, but with a low fat level also, and the second was to feed a diet of normal composition in reduced amount so as to prevent any increase in body weight during the experimental period. In all studies, the pigs were fed the experimental diets from the third to the eleventh week of life and in most experiments, the pigs were nutritionally rehabilitated by feeding a commercial swine ration for an additional 3 to 6 months. A description of the major nutrients in the experimental diets is given in Table I. Typical growth responses which have been described previously, showed permanent stunting of the three groups that were nutritionally deprived in early life.[5]

* Unpublished results.

Table I. The Major Components of the Experimental Diets Used for Pigs – Minerals and Vitamins Were Added in Adequate Amounts *

	1 Low Protein, High Calorie	2 Low Protein, Low Calorie	3 Control "Restricted"	4 Control *"ad lib."*
Protein (casein) (g/100 g food)	3	3	24	24
Carbohydrate (glucose) (g/100 g food)	73	94	73	73
Fat (corn oil) (g/100 g food)	24	3	3	3
Calories/100 g food	520	415	415	415
Calories from protein (%)	2.3	2.9	23	23
Food administered per day	*ad lib.*	*ad lib.*	"pair-fed" with 2	2–3% b.w.

* Data taken from Barnes *et al.*[23]

Protein-Calorie Malnutrition in Baby Rats

We have employed two techniques for the induction of severe protein-calorie malnutrition in the rat. In spite of the drastic steps taken the gross signs that develop are not so much those of kwashiorkor, with fatty livers and grossly observable edema, but rather appear to simulate marasmus with the extreme atrophy of body tissues. In our early studies, we used the large litter versus small litter technique [6,7] in order to restrict milk consumption during the first 3 weeks of life. Difference in litter size was not as great as others have employed and involved either 14 or 16 pups versus 8 per lactating dam. This was a sufficient restriction during the nursing period so that the rats in the large litters were about 25% smaller in body size than the controls, at 3 weeks of age. At this age, the normal weaning time in our laboratory, some of the rats were fed a purified diet containing only 4% protein for an additional 8 weeks, and others were fed the normal high protein diet at weaning. Those rats restricted for an additional 8 weeks following weaning were fed the normal control diet at that time to establish a complete nutritional rehabilitation.

The technique of large litter versus small litter nursing is extremely difficult to control from experiment to experiment. We have had several trials in which the body weight of rats was approximately 25% different at 3 weeks of age, but after weaning all the rats and feeding a normal control diet, no difference in ultimate body weight between the groups was observed. This is contrary to previous observations in our laboratory and also to the reported results of many investigators who have found that restriction during the nursing period results in permanent stunting of body size. This, as well as other difficulties with this procedure, led us to use another technique which is far more reproducible and which also results in greater dietary restriction in the nursing rat. The procedure is simply to change the diet of the lactating dam on the second day following parturition to one containing 12% casein (10% protein) and this diet is continued to the normal weaning age. All litters are adjusted to a uniform number which in our experiments has been 8, and at 3 weeks of age the restricted rats are uniformly close to 50% of the body weight of the normal animals. To establish a more severe protein-calorie deficiency, the pups from the restricted dams, in some studies, have been weaned to a diet containing 5% casein and maintained with this diet for 4 weeks. One dietary modification we have used has been to feed the previously undernursed rats a diet of normal composition in a restricted amount so that these rats did not gain weight for the 4-week period and thus were comparable, insofar as growth was concerned, with the rats fed the 5% protein diet. At the end of the 4-week postweaning restriction, all rats were rehabilitated by feeding a diet of normal composition *ad libitum.* Milk composition has not been determined in our laboratory but presumably the effect of feeding a 10% protein diet to the lactating female is primarily to reduce the amount of milk produced. If there were any changes in the composition of the milk resulting from this regimen, it is assumed that the change would be a decrease in fat content without there being any appreciable change in the amount or the type of protein. A description of the dietary regimens involved in the large litter, small litter studies and growth curves illustrating the effects of the different treatments have been reported previously.[8]

State of Excitability in Stressful Conditions

The first behavioral abnormality noted in our studies in which an animal had been subjected to severe malnutrition in early life and then rehabilitated with a normal diet, was the high state of excitability or emotionality exhibited when an animal is subjected to a stressful situation.

The significance of our early observation was not appreciated at the time it was made and it is believed now that this behavioral characteristic of high excitability may be one of the most important and most characteristic behavioral manifestations of early malnutrition. It was originally observed in pigs that had been subjected to either the low protein, low fat or low protein, high fat diet during the third to eleventh week of life. The test procedure in which it was observed was the extinction of a previously established classic Pavlovian conditioned response. The animals that had been rehabilitated to a weight of approximately 100 kg were conditioned to an electric shock given 10 seconds after the introduction of a conditional stimulus which was a change in the intensity of room light. Actually four groups of pigs were studied, for in addition to the two protein-malnourished groups there was a group that had been fed a restricted amount of a good diet so as to maintain the same body size as the malnourished animals during the 8-week period of experimental nutrition deprivation, and a fourth group of control animals that were fed the diet of normal composition *ad libitum* from the beginning of the experiment. The results which have been published [5,9] showed all groups of animals achieving the conditioned response at essentially the same rate. After conditioning had been established an attempt was made to extinguish the response by subjecting the animals to the conditional stimulus, the change in light intensity, without accompanying this stimulus with an electric shock. The normal control animals and those that had received a restricted amount of the normal diet, readily extinguished the response so that they did not react to the change in light intensity after a few trials. The protein-malnourished pigs, however, continued to respond time after time to the changing light intensity even though this was not followed by a shock. This relative inability to extinguish a previously established conditioned response has been observed many times in our pig experiments. What was not noticed in the early experiments was that the previously malnourished pigs were considerably more excited during the inter-trial period and during the conditional stimulus than the pigs in the other two groups. In anticipation of receiving a shock, these malnourished pigs vocalized, defecated, and urinated with considerably greater frequency than observed in the others. This excited behavior continued during the extinguishing trials and was undoubtedly the explanation for the relative inability of these animals to lose their conditioned response.

The effect of early malnutrition on the state of excitability of the adult rat was also missed in our early investigations. In fact we reported [8] that the normal control animals appeared to be more highly excited in a water maze. This subjective observation proved to be nonrepeatable for in all successive studies there was a general impression that the previously malnourished animals were more excited when placed in the cold water of the swimming water maze. The state of excitability of the previously malnourished animals remained an uncertainty in our laboratories despite numerous attempts to evaluate it more accurately, until an experimental psychologist, Dr. Slávka Fraňková from the Institute of Human Nutrition in Prague, Czechoslovakia, came to spend a year in our laboratories. Dr. Fraňková used a conditioned avoidance procedure whereby rats placed in a box with a grid floor through which they could be shocked were conditioned to respond to a constant frequency sound which was followed in 10 seconds by an electric shock. A wire mesh screen was suspended at one end of the box and if the rat jumped upon this screen its weight would pull the screen down and break a circuit so that the conditional stimulus was interrupted and the rat was not shocked. The rats used in this study were subjected to one of three different dietary treatments. They were either normally nursed and normally fed postweaning, or secondly, they were restricted during nursing by giving the mothers a low protein diet from the second day following parturition and then when the animals reached 3 weeks of age they were fed a 5% protein diet for 4 weeks, or thirdly, they were restricted during the nursing period by feeding a low protein diet to the lactating female and then following weaning were fed for 4 weeks with a diet of normal composition but in a restricted amount so that they did not grow. Nutritional rehabilitation was established following these experimental regimes and the animals were studied with the conditioned avoidance procedure when they were approximately 6 months of age. All three groups of rats learned to avoid the shock at essentially the same rate. They were continued to be exposed to the cycle of events which involved test periods every other day, each test consisting of 7 cycles of a 60-second inter-trial interval in which the animal was not exposed to any stimulus, followed by a 10-second conditional stimulus in the form of a constant frequency sound culminating in an electric shock. It was found that previously malnourished animals made ineffectual jumps on the wire screen during the 60-second inter-trial interval with increasing frequency. This is shown in Figure 1 in which both the frequency of jumping on the screen and the total amount of time that the animals spent on the screen were greatest for the animals subjected to the postweaning protein-deficient diet, and less marked, but still more excitable than the controls, were the animals that were restricted during the nursing period and fed a diet of normal composition but in restricted amount during the postweaning period. In this series of experiments it was also shown that once the conditioned avoidance had been established and the animals were then subjected to a type of extinction procedure in which no shock was administered, the previously malnourished animals continued to respond; in other words, they did not extinguish the response, in contrast with a considerably more rapid extinction of the response by the control animals and the

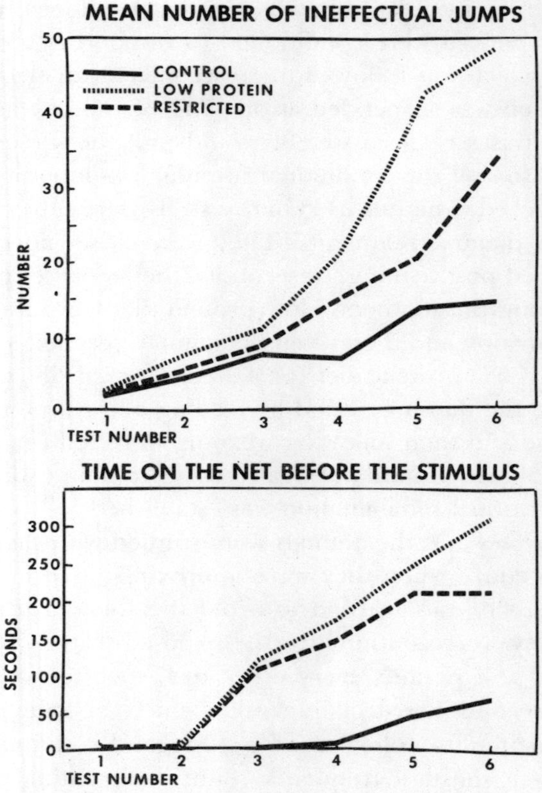

Figure 1. Example of the elevated excitability of previously malnourished rats when given repeated series of tests in a conditioned-avoidance situation. Jumps refer to the rat jumping from the wire grid floor of the box to a screen net hanging from one end of the box. They were classed as ineffectual jumps if they were during the inter-trial period, that is, not during exposure to the conditional stimulus.

animals that had been given the normal diet in restricted amounts. The entire pattern of this series of experiments is strikingly similar to the behavioral manifestations of the pigs just described. Other observations made in our laboratories both with pigs and rats have involved an elevated state of excitability as a component of some more complex behavioral manifestation being measured. These will be described in later sections of this report.

Spontaneous Exploratory Activity

Several years ago, Dr. Joseph Lát from the Institute of Physiology in Prague, conducted some behavioral studies on rats raised in large and small litters in the laboratories of McCance and Widdowson.[10] Lát had developed a technique for measuring the spontaneous exploratory activity of rats (which he has frequently used in the expression "nonspecific excitability level" or NEL) which differs in some major respects from the well-known open-field behavioral tests for small animals. Lát found that rats raised during the first 3 weeks of life in large litters exhibited less spontaneous exploratory activity than the control animals that were raised in small litters. This difference was also observable many months after nutritional rehabilitation had been instituted. These studies regarding early malnutrition and spontaneous activity were extended in our laboratories by Dr. Slávka Fraňková. The spontaneous activity characteristics that are measured by the procedures originated by Lát are primarily five behavioral patterns exhibited by a rat when it is placed in an environment different from that in which it has been living. This environment is a Plexiglas box with the floor marked in squares, which is set behind a half-silvered, one-way mirror. The rat is taken from its home—the cage in which it has lived all its life—and is placed in this new environment and observed for 10 minutes. The characteristics that are observed are the frequency and duration of standing up on the hind legs, number of squares traversed by walking around the box, number and duration of grooming reactions, and the frequency and duration of complete inactivity. At 10 days of age when the eyes are open, rats can be tested satisfactorily by this procedure and at this early age a decrease in spontaneous activity of animals that are being undernourished by their mothers is seen. At 3 weeks of age the previously deprived animals were divided into two groups and fed either a 5% protein diet or a normal diet but in restricted quantity. These two groups were compared with a third group normally nourished from birth, in the test of spontaneous exploratory activity about the fifth week of life while the deficient regimens were still being fed. The deficient animals were much more active than the controls, with the animals fed the restricted diet of normal composition having the highest spontaneous activity. This manifestation of activity is undoubtedly different from that observed during the nursing period and in the completely rehabilitated animal, which will be described presently. During the postweaning period of the low protein or restricted feeding the animals were highly active in their home cages as well as in the new environment. If placed in revolving cages so as to measure their general activity the deficient animals showed considerably higher numbers of revolutions per day than the normal controls. This latter observation has been well documented in the literature by many laboratories.[11–13] This generally elevated level of activity obviously overrides any manifestation of spontaneous exploratory activity due solely to the stimulation of a new environment. When the animals are rehabilitated and tested for spontaneous exploratory activity at about 6 months of age the rats restricted during nursing and then fed a low protein diet for 4 weeks showed the lowest spontaneous activity. Rats restricted during the nursing period and then fed a restricted quantity of a diet of normal composition had an intermediate level of spontaneous activity. The highest level of activity was exhibited by the controls.

The relative lack of interest in the surroundings of an adult pig that was malnourished as a baby has been observed but in a subjective manner that could not be quantitated. This behavioral characteristic will be described in a later part of this report dealing with learning behavior. At this point let it suffice to say that in both rats and pigs there is evidence that early malnutrition results in a long-lasting decrease in the animal's exploratory activity or perhaps in more descriptive terms, their curiosity, when placed in a new environment.

Food Drive and Food Consumption

Several investigators have noted that if rats are deprived of food early in life they will thereafter have an abnormal drive for food. This has been described frequently as an increased tendency of adults to hoard food (termed "hoarding") and also to consume food more rapidly.[6,14–16] We have reported an increased drive for food in both rats and pigs that had been malnourished in early life.[17] Unfortunately in this earlier report it was not recognized that if the condition of malnutrition is sufficiently severe the animal may show not only abnormal behavior toward food but also it will have an increased requirement for food. In these earlier studies, rats that had been raised in large or small litters at 3 weeks of age were transferred to a diet of normal composition and thus nutritionally rehabilitated. Food consumption under special conditions was measured when their growth weight had essentially reached a plateau at about 4 months of age. The technique was to change the rats from *ad libitum* feeding to feeding one hour per day and during this one hour to observe the general attitude of the animals and to measure the amount of food consumed and the amount of food spilled. In Figure 2 is shown food consumption and spillage over the first 3 weeks of this one-hour-per-day feeding regime. The previously undernursed rats ate more food during this first 3-week period and also spilled more food than the normal controls. In addition, the excited behavior of the animals when food cups were either put in or taken out of the cage was

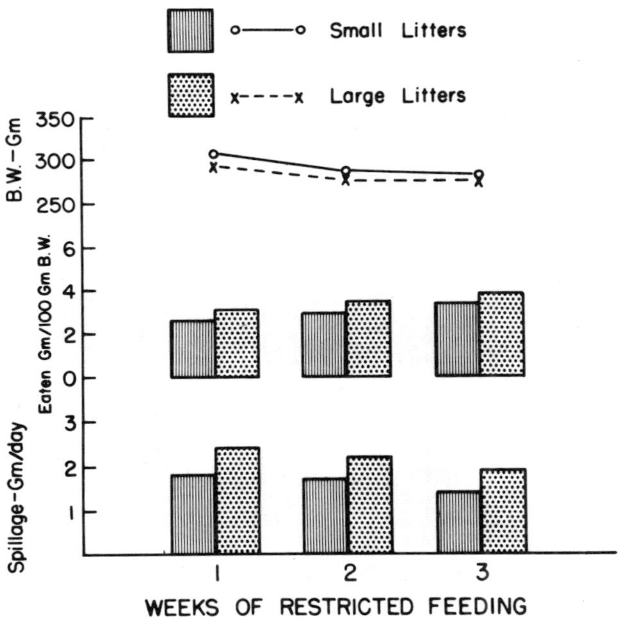

Figure 2. Food consumption pattern of rats that had been malnourished by nursing in large litters. After reaching the adult plateau of body weight through *ad libitum* consumption of an adequate diet they were switched to a one-hour-per-day feeding schedule. Data are for the first 3 weeks on this restricted feeding schedule. Reproduced from Barnes *et al.*[17]

quite marked as compared with the controls. One of the most unexpected features of this particular experiment was that during the period of time when food consumption was being measured there was little difference in the average body weights for the two groups. Nevertheless, as in all of the early experiments of this type carried out in our laboratories, food consumption was expressed on the basis of the amount per 100 g of body weight. The one-hour-per-day feeding was continued and in Figure 3 are shown the results obtained over the 7 to 10 weeks of restricted feeding, but with another group of animals. It is seen that the food intake was equalized when expressed on a body weight basis but the food spillage continued to be high for the previously undernursed rats. Up to this time, on the basis of *ad libitum* food consumption data, it had been our supposition that an early nursing deficit caused by large litters did not affect the food requirement of the adult rat. In more recent experiments where the degree of malnutrition during early life has been

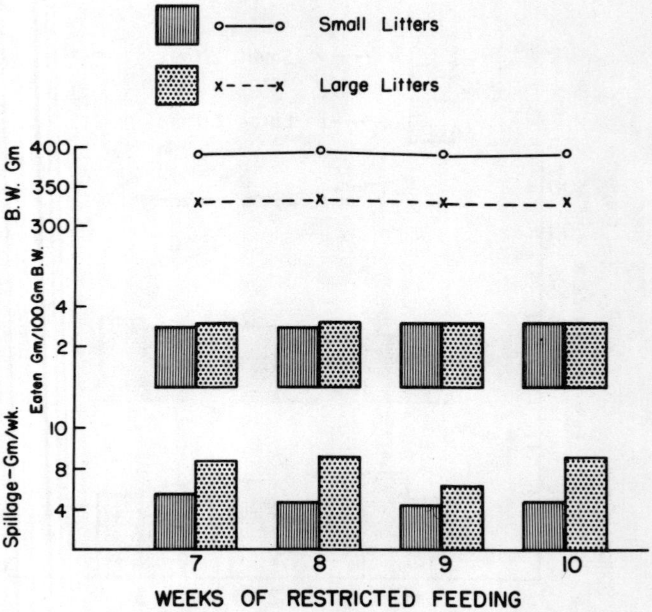

Figure 3. Food consumption pattern of rats that had been malnourished by nursing in large litters. After reaching the adult plateau of body weight through *ad libitum* consumption of an adequate diet they were switched to a one-hour-per-day feeding schedule. Data are for a different group of rats than those shown in Figure 2 and represent the seventh through tenth week of restricted feeding. Reproduced from Barnes *et al.*[17]

considerably greater, we have now clearly established the fact that these animals have a greater food requirement. It seems possible that in our earlier experiments previously malnourished animals did have a higher food requirement but quantitatively it was too small to be picked up by the usual food consumption measurements.

The results of a more definitive study are shown in Table II. In this experiment the rats had been underfed during nursing as a consequence

Table II. Daily Food Consumption and Food Spillage for Adult Male Rats Eating *ad libitum* During One Week Observation

Early Diet Treatment *	Body Weight †	g per 100 g b.w.	Percent Increase Over Control	g per $b.w.^{0.75}$	Percent Increase Over Control
		Food Consumption			
Group 1	447 (24)	4.4		0.202	
Group 2	367 (16)	4.9	11	0.215	6
Group 3	337 (24)	5.3	20	0.227	12
		Food Spillage			
Group 1		1.3		0.060	
Group 2		1.1	−15	0.048	−20
Group 3		1.9	46	0.081	33

* Group 1, control diet throughout; group 2, preweaning protein deprivation and postweaning normal diet in restricted quantity; group 3, preweaning protein deprivation and postweaning protein deprivation.

† Numbers in parentheses represent number of rats.

of feeding the lactating female a 10% protein diet and at weaning, half of the animals thus treated were fed a 5% protein diet and the other half fed a diet of normal protein content restricted in amount so that they did not gain weight during a 4-week period. Approximately 4 months after nutritional rehabilitation or when the animals were roughly 6 months of age, *ad libitum* food consumption was measured. At this time the rats differed considerably in body weight, with the postweaning restricted animals being intermediate between the controls and those fed the low protein diet postweaning. When food consumption was expressed on the basis of grams of food intake per 100 g of body weight there were large differences between the previously malnourished and the control animals. Since food intake is known to be related to metabolic rate, which is more correctly related to body surface area than body weight, food intake was recalculated on the basis of grams of food intake per body weight to the 0.75 power (b.w.$^{0.75}$). With this calculation the difference between the groups is somewhat smaller than with the original body weight calculation. It is anticipated that this apparent difference in food intake would disappear if we could measure the body composition of these animals and express food intake on the basis of lean body mass or more appropriately, total protoplasmic mass. It is probable that the stunted rats that were previously malnourished would have a lower fat content than the normally fed rats. Shock *et al.*[18] have shown that if a correction for body composition is made in aging humans, basal metabolic rate remains constant with age and it will remain for experimentation to determine whether this same type of calculation corrects for difference in food consumption in the previously malnourished rat. Lee and Chow [19] have shown that the offspring of female rats that were partially restricted in food intake during gestation and lactation have a higher food intake than normal control animals after nutritional rehabilitation. Their results are expressed on the basis of gain in weight per gram of food intake during a 10-day observation period. Differences were so great that corrections for body size and composition could not resolve the differences. On the other hand, gains as high as 0.78 g per day for a 20-month-old rat weighing 488 g were reported. There may have been some unusual circumstances associated with the 10-day observation period that resulted in such large growth increments at this age.

Whether or not a strictly behavioral component is involved in food spillage which relates to the higher stress-induced excitability of the previously malnourished animals will have to be determined by further experimentation. Regardless of the explanation for this increased food spillage, there is no question that the early-deprived animals do have an altered attitude toward food. Previously deprived animals that are on a one-hour-per-day feeding schedule are more aggressive in their eating. When the food cups are placed in the cage they immediately mount the cup with the front paws and start eating, remaining in this position throughout the one-hour-period that the cup is left in the cage. The normal controls eat in a more sporadic fashion, with frequent periods when they leave the cup and by the end of the hour are usually away from the cup in a sleeping posture due, presumably, to the engorgement of their stomachs. From the standpoint of behavioral testing this attitude toward food must be of considerable importance. Many tests of learning behavior involve a food reinforcement and in such tests the animals are underfed before the test procedure so as to increase their drive. The results of learning behavior tests involving food reinforcement are undoubtedly affected by differences in food drive resulting from the early nutritional history of the animal.

Some preliminary studies of food intake patterns in pigs that had been previously malnourished have been reported from our laboratory.[17] These studies show that pigs fed over a period of several weeks an amount of food less than they would normally ingest by *ad libitum* feeding would eat tremendous quantities of food when suddenly switched to an *ad libitum* schedule. Using an automatic device for recording the frequency and amount of food consumed 24 hours a day, it was shown that animals that had been previously malnourished continued to eat throughout the day and night. During the night (7:00 P.M. to 7:00 A.M.) the pig would approach the feeder less frequently than it would in the daytime, but nevertheless it would eat off and on throughout the night. In contrast, the normal control pigs consumed their full allotment of feed during the day and only rarely would work the automatic feed dispensing device during the night hours.

One consistent observation with the pigs that had been previously malnourished was that they frequently worked the device to dispense food to such an extent that far more food accumulated in the feeding pan than they could consume at one time. This excess food was scattered out of the feeder onto the floor of the pen, resulting in far greater wastage than in the case of the normal control animals. More recent studies in our laboratories have shown that the stunted pigs that were previously malnourished do consume more food per unit body size than the normal controls, although we have not been able to determine whether this can be accounted for by corrections for body surface area and body composition. Nevertheless, the same line of reasoning that has just been applied for the rat leads us to the conviction that the increased drive for food must influence the outcome of learning behavior studies with the pig in which a food reinforcement is used.

Learning Behavior

It is believed that tests intended to measure learning behavior should be interpreted on the basis of behavioral characteristics that typify the animal that has been malnourished in early life. The elevated state of excitability under stressful conditions, the decreased spontaneous exploratory activity and the exaggerated food drive, especially under conditions of hunger, obviously can all influence the performance of an animal in tests designed to measure learning behavior. It appears that the food drive is particularly critical in the interpretation of results of learning tests, for this drive may be due in large measure to an apparent increase in caloric requirement and may not reflect accurately altered learning behavior due to higher nervous system malfunction caused directly by the malnutrition induced during the early development of the animal. Cowley and his associates [20,21] have observed decreased learning behavior in rats that have been subjected to long-term feeding of a diet containing a decreased protein content. In these studies the role of diet alteration is not clear since the experimental diet was fed throughout gestation and lactation and was also being fed at the time the behavioral studies were conducted. In these studies the Hebb-Williams maze with a food reinforcement was used. In more recent studies Cowley [22] has given some evidence of behavioral abnormalities under different dietary conditions. After a generation of feeding the low protein diet the young animals were rehabilitated and appeared to have altered behavior which was observed also in the young of the next generation even though a normal diet was fed throughout gestation and lactation and was continued with the offspring which continued to show altered behavior. Food reinforcement was involved in the second generation rehabilitated animals. In studies from our laboratories the first clear demonstration of altered learning behavior in rats that had been rehabilitated following a combination of large-litter nursing and 8 weeks of a low protein diet following weaning was found in tests using a Y-shape water maze.[8] More extensive experience with the water maze has convinced us that this technique has serious shortcomings for the measurement of learning behavior under the conditions of our procedures for induction of malnutrition and rehabilitation. Difference in learning can be measured but it requires severe malnutrition over a long period in the early life of the animal. To learn the discrimination which is in the form of a light fixed in one of the arms of the Y which contains the escape ramp, takes a long period of trials, about 30 days of two trials each day. All animals become excited after continued exposure to this test so that any difference in the state of excitability due to malnutrition becomes masked. The nature of the discrimination is such that a random series of numbers must be used in order to move the escape ramp from one arm to the other in a completely random order. This means that two or more successive trials may involve the ramp in the same arm of the maze and a very strong, natural reaction of rats is to establish a positional preference. This randomization of the arm in which the rat can achieve escape obviously makes the discrimination problem very difficult to learn. It is not a test that permits the measurement of behavioral components of learning and it also requires such a long period of time to learn that it is not practical to use for the evaluation of a large number of nutritional and biochemical variables that might be related to learning behavior.

Bar pressing in a conventional Skinner box has been studied with rats with relatively little success. The only demonstrable differences seen in the rats malnourished in early life as compared with normal controls was in the first few days when the animals were learning to associate bar pressing with a food reward. These results, given in Table III, show that with males the normal control rats learn bar-pressing more quickly than those that had been malnourished by the three procedures, that is, restriction pre- plus postweaning, preweaning alone, or postweaning alone. Female rats showed no differences and this will be commented on later. The interesting point here is that all three forms of malnutrition, including the postweaning treatment alone, caused an apparent learning deficit. These results may have been influenced by the enhanced food drive caused by all forms of early restriction. However, the condition of the test, which involves placing a rat in a new environment and the known decrease in spontaneous exploratory activity that results from restriction in the preweaning period and possibly also for the postweaning period, undoubtedly influenced the speed with which the response was learned.

Table III. Number of Days Taken by Adult Rats Subjected to Nutritional Deprivation in Early Life to Obtain 300 Rewards in All *

	Males		Females	
Form of Restriction	Group	Days to Obtain 300 Rewards in All	Group	Days to Obtain 300 Rewards in All
Pre- + postweaning	1	3.5	5	3.8
Preweaning	2	3.6	6	3.6
Postweaning	3	3.4	7	4.0
Normal control	4	3.0	8	3.7

* Data taken from Barnes *et al.*[17]

It may be of some significance that further testing of these animals in the Skinner box after a high fixed ratio response was achieved did not show any differences between the groups. These tests involved variable ratio and variable interval reinforcement programs and the delayed response procedure, known as DRL, in which the animal must learn not to press the bar for a given interval in order to obtain a food reward. These are well-established operant procedures for the measurement of learning in the rat but in our hands there has been no instance in which we have observed differences between control and early malnourished animals.

Pigs approximately 9 months old that had been malnourished by feeding a low protein, high fat diet in early life, together with appropriate controls, were observed when exposed for the first time to a modified Skinner box in which the lifting of a bar with the snout was associated with the delivery of a small portion of food. A ratio of one food reward per five responses was used. The pigs were left in the modified Skinner box for half an hour on successive days and were considered to have reached a criterion when on two consecutive days a minimum of 20 food reinforcements was obtained. The results for ten previously malnourished and six normal control pigs are shown in Table IV. The controls learned the relationship between bar-lifting and food reinforcement much more rapidly than the previously malnourished pigs. The observ-

Table IV. Response Rate and Emotionality (Defecation) of Pigs When First Exposed to Bar Lifting—Food Reinforcement at a Fixed Ratio of 1:1 *

	Controls (6)		Previously Malnourished (10)	
Trial Number	Pigs Defecating %	Pigs Achieving Minimum of 20 Responses %	Pigs Defecating %	Pigs Achieving Minimum of 20 Responses %
1	67	83	100	30
2	33	100	80	50
3	x	100	x	60
4			x	80
5			x	100
6			x	100

* Data taken from Barnes *et al.*[17]

able behavioral characteristics of the animals were of particular significance. The normal control pigs, when first placed in the experimental box, extensively explored the box, the food pan, and the bar which was to be lifted by the snout. After an experimental lifting of the bar and consequent acquisition of a food reinforcement, the pigs seemed very quickly to recognize a relationship between the lifting of the bar and the reward of food. The previously malnourished pigs acted in a completely different fashion. When first placed in the special pen a small amount of food was put into the feed pan. The animals went immediately to this location, ate up the food, and then continued to sniff around the area of the food pan trying to find more food. These pigs appeared to show relatively little interest in their surroundings and were very slow in exploring the other areas of the box and as a result required much more time in locating the bar device. After a few random movements of the bar which resulted in dispensing a small quantity of food, the animals voraciously ate the food and then repeated the previous performance of spending much time sniffing around the empty feed pan, and only after a relatively long delay returning to the bar. Random operation of the bar which resulted in only occasional dispensing of food appeared to frustrate the pigs and this resulted in frequent defecation and urination. The general appearance and actions of the previously malnourished pigs indicated a high state of excitability and emotionality. This pattern of behavior which was characteristic of the previously malnourished animals, clearly illustrated the three primary behavior patterns described previously, namely, an elevated state of excitability under stress, a decreased spontaneous exploratory activity and a heightened food drive. The defecation and urination performance is shown in Table IV, but the other manifestations of behavior can be reported only in descriptive terms.

High rates of response were achieved by all pigs, regardless of previous nutritional treatment. In early studies only sows were used since the 9- to 12-month-old boar becomes difficult to handle and can be destructive of caging facilities. There appeared to be no difference between treatment groups of these sows when given different bar-lifting programs such as increasing fixed ratios, variable intervals, and DRL schedules. In a more recent study, not completed at the time of writing this dissertation, groups of male pigs have been carried through the procedures of early malnutrition described above and then rehabilitated. At about 6 months of age the two groups previously subjected to severe protein deficiency appear to be inferior in establishing high response rates even after prolonged training in bar lifting. This difference from earlier experiences in our laboratories may be related to the consistent observation of a marked sex difference in behavior due to early malnutrition.

Conditioned avoidance studies were carried out in pigs during the eighth and final week in which they were being subjected to the experimental diets described in Table I. At the time the tests were run, pigs receiving the two low protein diets were extremely apathetic and their voluntary food consumption had dropped to a very low level. The animals that had been fed the diet of normal composition but in restricted quantity were more active although severely stunted in body size. Behaviorally they appeared much more similar to the controls than the protein-calorie malnourished animals. The results of a shuttle box type of conditioned avoidance experiment which have been published previously[23] are shown in Table V. It appears from these figures that the normal control animals were not as successful in avoiding the shock as

Table V. Conditioned Avoidance – Sixth to Eighth Week of Deprivation *

Group	Diet	No. Pigs	Response per Pig in 30 Trials		
			Avoidance	Escape	Nonescape
1	L. prot. H. fat	8	1.3	19.6	9.1
2	L. prot. L. fat	4	3.0	11.7	15.3
3	Control-restricted	4	12.5	16.0	1.5
4	Control-*ad lib.*	6	5.8	22.8	1.3

* Data taken from Barnes *et al.*[23]

were the animals receiving restricted amounts of the control diet. It is believed that this is an artifact due to the relatively large size of the control pigs in relationship to the size of the shuttle box that was being used. The results in terms of nonavoidance and escape show clearly that the animals receiving the control diet *ad libitum* and the control diet in restricted quantity were similar and their performance was superior to that of the two groups being subjected to protein-calorie malnutrition. Only recently has it been possible to conduct satisfactory conditioned avoidance studies in adult pigs that had been rehabilitated from the nutritional deprivations described. A circular arrangement of compartments from which the pig must escape in order to avoid a shock after receiving a signal has been constructed. This makes it possible for the animal to move in the same direction in avoiding the painful conditional stimulus and thus obviates the confusing situation in which the pig must jump over a hurdle into an area in which it was previously shocked, in order to avoid a shock. Avoidance in the circular series of compartments was very quickly learned by all animals but there was a clear-cut decrease in the rate of learning by previously protein-calorie malnourished pigs as compared with the controls and the group fed the control diet in restricted quantity. In this test the behavioral manifestations of a high state of excitability under stress and decreased spontaneous exploratory activity was very clearly seen in the two groups of animals previously subjected to protein-calorie malnutrition.

Sex Differences

In all the studies with rats that have been described a clearly demonstrable sex difference has been observed. Consistently the male rats exhibited larger differences in behavior resulting from early malnutrition than has been the case with the females. The results of the light-discrimination studies and the position reversal studies in the Y water maze show certain clear-cut differences between groups of the males and no significant difference in the females. In studies of food drive and food spillage greater differences have been observed in the males than the females. Spontaneous exploratory activity differences are consistently greater in the males that have been previously malnourished and in comparisons of previously malnourished and control females. In our pig studies most experiments have been performed with female animals. The reason for this has been that following rehabilitation, particularly with the controls, boars become large and difficult to manage and are more destructive of equipment than sows. However, in recent experiments males have been used and it is our impression that we have obtained much more marked differences due to previous malnutrition than we had ever observed before in female animals subjected to the same nutritional conditions. In Skinner box performance with pigs, higher learning rates are currently being observed with controls and animals previously fed restricted quantities of a normal diet than animals that had been subjected to early protein-calorie malnutrition. As mentioned above, these current studies involve male pigs only. In previous Skinner box programing with females no differences due to prior nutritional deprivation were seen except for the first few days when the association between bar lifting and food reward was being established. It is our belief that the behavioral differences noted between males and females subjected to early malnutrition are of a quantitative nature only. The females undoubtedly are affected but appear to be considerably more stable insofar as excitability, emotionality, and curiosity are concerned, although all of these behavioral characteristics appear to be at a lower plane than is the case with males. Sex differences appear also in learning performance, for as reported in the water maze discrimination studies, females performed with significantly more errors than the males. Superior performance of control male pigs in operant testing has just been described. More clear-cut effects of malnutrition on behavior in male than in female rats have been noted previously by Cowley and Griesel.[20] This will be an interesting avenue of study for some future investigations, but for the present if clear-cut effects of early malnutrition on behavioral characteristics are to be measured in experimental animals, male animals only should be used as subjects.

Summation

In this narration of our studies over the past few years in which we have attempted to relate early malnutrition to behavioral development, relatively few questions have been answered. The answers that have become available are exciting in their implications relating early nutrition to behavioral development in the human being and they also offer encouragement to psychologists and nutritionists to combine forces so as to explore in more detail this very fruitful field of investigation.

It has been shown in repeated studies that many months after complete rehabilitation with a normal diet there is an abnormal behavioral pattern in the animal subjected to severe protein-calorie malnutrition in early life. These long-lasting manifestations, which in some degree may be considered permanent, may be described as, 1) an increased level of excitability; 2) a decreased level of spontaneous exploratory activity; and 3) an increased drive-related attitude toward food. These three alterations from normal behavior cause impairment in tests of learning behavior performance. Even though there is considerable evidence of depressed learning ability in the previously malnourished animal, there is no evidence that this is due to an impaired capacity to learn.

The effects of early malnutrition on behavior are considerably greater in male than in female rats and pigs. It is believed that these very marked differences resulting from early malnutrition represent quantitative rather than qualitative changes in behavioral manifestations.

Food consumption is increased in the previously malnourished animal. The increased food consumption can undoubtedly be related to total metabolic rate. If one corrects for the small body size of the previously deprived animal by expressing food intake on the basis of body surface area and if body composition is estimated so as to relate food consumption to active protoplasmic mass, the apparent differences in food intake or food requirement will presumably disappear. There remains, however, some indication of an enhanced drive for food that may not be accounted for on the basis of increased food (calorie) requirement, as evidenced by the animal's behavior toward food when it is hungry and also by the amount of food that is spilled.

Behavioral abnormalities of the previously malnourished animal have been observed in two species, the rat and the pig, which differ widely in their rates of development. The consistency with which behavioral characteristics that are altered from normal due to early malnutrition can be seen in these two species gives us encouragement that these results can, in principle, be extrapolated to man.

The extent of malnutrition and the critical period when it must be imposed in order to cause a permanent alteration in behavioral development have not been determined. The results presented in this dissertation give encouragement that a great deal can be learned about these nutritional parameters. It is believed that the rat is not a good animal to use for the purpose of measuring nutritional variables because of its extremely rapid development and the difficulty in inducing characteristic signs of severe protein-calorie malnutrition. The slower developing pig gives more hope that nutritional factors in early life affecting behavioral development can be studied. Most encouraging in this respect has been the consistent observation that pigs severely restricted in growth during early life by feeding a diet of normal composition but in restricted quantity do not exhibit behavioral characteristics differing greatly from normal control animals. However, protein-calorie malnutrition that is imposed in the pig in early life (by feeding either a low protein, low fat diet or a low protein, high fat diet) which simulates in some respects the human infancy diseases of marasmus and kwashiorkor, respectively, causes easily measured and long lasting behavioral abnormalities in the rehabilitated animal. Furthermore, in all tests performed there has been no difference in the degree of abnormal behavioral development with these two conditions, even though during the period of the experimental deficiency biochemically and morphologically the animals receiving the low protein, high fat diet are considerably more abnormal than the ones receiving the low protein, low fat diet.

Growth is markedly affected by early malnutrition but there is no apparent correlation between growth retardation (stunting) and the development of the abnormal behavioral characteristics. A nutritional condition can be imposed in early life which causes a serious degree of stunting of body size but without any evidence of behavioral abnormality. On the other hand, animals restricted during the nursing period may, on occasion, show relatively little stunting in body development and yet exhibit abnormal behavior.

There is undoubtedly a critical period during development when nutritional deprivations will have their greatest effect. However, in the studies just described in which nutritional variables resulted in different behavioral manifestations, the time of inducing the deprivation remained constant.

The abnormal behavior pattern of previously malnourished animals is easily reproduced and is unusually consistent in its manifestations. The combination of decreased spontaneous exploratory activity and elevated excitability is believed to be pathologic. This type of adult, nutritionally rehabilitated animal should be of considerable value in studying the possible modifying effects of interventions such as sensory stimulation or psychopharmacologic agents, as well as major nutritional modifications.

References

1. Platt, B. S., Heard, C. R. C., and Stewart, R. J. C., in "Mammalian Protein Metabolism," H. N. Munro and J. B. Allison, Eds., Academic Press, New York, Vol. 2, pp. 445–521 (1964).
2. Lowrey, R. S., Pond, W. G., Barnes, R. H., Krook, L., and Loosli, J. K., *J. Nutr.,* **78,** 245 (1962).
3. Winick, M., and Noble, A., *J. Nutr.,* **89,** 300 (1966).
4. Enesco, M., and Leblond, C. P., *J. Embryol. Exp. Morphol.,* **10,** 530 (1962).
5. Barnes, R. H., Moore, A. U., Reid, I. M., and Pond, W. G., *J. Amer. Diet. Assoc.,* **51,** 34 (1967).
6. Seitz, P. F. D., *Am. J. Psychiat.,* **110,** 916 (1954).
7. Kennedy, G. C., *J. Endocrinol.,* **16,** 9 (1957).
8. Barnes, R. H., Cunnold, S. R., Zimmermann, R. R., Simmons, H., MacLeod, R. B., and Krook, L., *J. Nutr.,* **89,** 399 (1966).
9. Barnes, R. H., *Fed. Proc.,* **26,** No. 1, 144 (1967).

10. Lát, J., Widdowson, E. M., and McCance, R. A., *Proc. Roy. Soc. (B)*, **153,** 347 (1960).
11. Anderson, J. E., and Smith, A. H., *J. Comp. Psychol.,* **6,** 336 (1926).
12. Anderson, J. E., and Smith, A. H., *Psychol. Bull.,* **24,** 177 (1929).
13. Bevan, W., Lewis, G. T., Bloom, W. L., and Abers, A. T., *Amer. J. Physiol.,* **163,** 104 (1950).
14. Hunt, J. McV., *J. Abn. Soc. Psychol.,* **36,** 338 (1941).
15. Marx, M. H., *Psychol. Rev.,* **57,** 80 (1950).
16. Marx, M. H., *J. Comp. Physiol. Psychol.,* **45,** 42 (1952).
17. Barnes, R. H., Reid, I. M., Pond, W. G., and Moore, A. U., in "Calorie and Protein Deficiencies," R. A. McCance and E. M. Widdowson, Eds., J. A. Churchill, Ltd., London, 1968.
18. Shock, N. W., Watkin, D. M., Yiengst, M. J., Norris, A. H., Gaffney, G. W., Gregerman, R. I., and Falzone, J. A., *J. Gerontol.,* **18,** 1 (1963).
19. Lee, C-J., and Chow, B. F., *J. Nutr.,* **87,** 439 (1965).
20. Cowley, J. J., and Griesel, R. D., *J. Genetic Psychol.,* **95,** 187 (1959).
21. Cowley, J. J., and Griesel, R. D., *J. Genetic Psychol.,* **103,** 233 (1963).
22. Cowley, J. J., and Griesel, R. D., *Animal Behavior,* **14,** 506 (1966).
23. Barnes, R. H., Moore, A. U., Reid, I. M., and Pond, W. G., in "Malnutrition, Learning and Behavior," N. S. Scrimshaw and J. E. Gordon, Eds., M. I. T. Press, Cambridge, Mass., pp. 203–218, 1968.

The Effects of Severe Protein Deficiency on the Human Infant

J. S. GARROW
Medical Research Council Staff
Department of Obstetrics and Gynaecology
Royal Free Hospital
Liverpool Road
London N.1.
England

Kwashiorkor is "the most serious and widespread nutritional disorder known to medical and nutritional science."[1] However, kwashiorkor is only one of the clinical syndromes which follow severe protein deficiency in human infants. Protein deficiency, often combined with calorie deficiency and infection, may produce the syndrome of marasmus, which in many parts of the world is far more prevalent than kwashiorkor.[2] One might suppose that with so vast a number of protein malnourished children in the world it would be a fairly simple matter to study the effects of this protein deficiency, but this is not so, for reasons which are given below.

When J. B. Allison, in whose honor this book is produced, referred to a protein-depleted dog both he and everyone else knew what he meant. The beagles at the Bureau of Biological Research are reared on a synthetic diet which they eat with obvious relish. It is, therefore, possible to start with two normal littermates in metabolic cages, and to remove protein from the diet of one dog until any desired level of protein deficiency is reached, while the other dog serves for comparison and control. In clinical medicine the situation is very different. The pediatrician in tropical countries is confronted with large numbers of more or less severely malnourished children, and his first duty is to try to help them to recover as quickly and certainly as possible. He will, of course, try to find out how the child came to be malnourished but the histories of such cases are notoriously unreliable: the child is malnourished because his parents have been unable to care for him adequately, perhaps as a result of poverty or ignorance, and in the circumstances one cannot expect that an accurate record of the child's diet will have been kept. The situation is further complicated by the fact that the child is usually brought to a hospital because his clinical state has recently worsened; often this is a result of infection superimposed on the previous malnutrition.

These, then are the children which the pediatrician must study in order to understand more about human protein malnutrition. On clinical grounds they are obviously malnourished, but exactly how they are malnourished, and the extent to which nondietary factors contributed, we cannot accurately know. Nor have we any proper control to show what a given child would have been like had he not been malnourished. Finally, since severe infantile malnutrition carries a high mortality even with the best treatment, there is very little room for investigations of a purely academic nature.

The difficulties in studying the effects of protein depletion in children would not matter so much if it were possible to produce a closely analogous state in a laboratory animal. Unfortunately, there are obvious differences between the response to protein depletion of an infant, and, for example, a rat. There are not many occasions on which it is possible to conduct parallel investigations on the two species, but three specific examples of these differences can be quoted from our own experience.

The first concerns an effect on structure; in the rat a low protein diet has no significant effect on the protein content of the brain,[3] but in malnourished children the brain loses significant quantities of protein and potassium.[4,5,6] The second difference is in amino acid reutilization; if a rat on a low protein diet is given a tracer dose of radioactive selenomethionine it will excrete a far smaller proportion of the radioactive selenium in the urine than a similar rat on a stock diet. In the child, however, the level of dietary protein does not measurably affect the rate at which labelled selenium is excreted.[7] The final example is the long term effect of protein deficiency in the young animal. The rat which is undernourished in the few weeks after birth is permanently stunted even if it is subsequently well fed; [8] on the other hand, the infant who is malnourished for a comparable time after birth seems to be capable of remarkable recovery in favorable circumstances.[9]

It can be argued that in none of the above examples were the conditions exactly the same for the two species. This may be so, but the ease with which such widely divergent results can be obtained should serve as a warning against too confident extrapolation of the results obtained in animals to the problems encountered in man. In recent years, studies on malnourished children have lead to a much better understanding of the effects which protein depletion has on the structure and function of the human infant. This subject was excellently reviewed by Waterlow, Cravioto, and Stephen[10] in 1960; the following discussion will, therefore, concentrate mostly on work published since that date.

Body Size and Shape in Malnourished Children

A child who has been maintained on a diet deficient in protein is invariably small for his age. Body weight is a crude indicator of the degree of malnutrition, and for all its limitations (which are discussed below) probably the best practical basis for classifying degrees of malnutrition. The Mexican system is to classify as "first degree malnutrition" children whose body weight is 76–90% of the average weight of a normal Mexican child of the same age.[11] Children who are 61–75% of average, or less than 60% of average are called "second degree" or "third degree," respectively. It should be noted that these are *percentage* weights, not percentile weights. A typical child who required hospital treatment for malnutrition would weigh about 5 kg at the age of one year; this is less than 60% of the expected weight for that age and hence is third degree malnutrition. The extreme weight deficit compatible with recovery in this age range is probably about 30%. For example, a girl aged 9 months was admitted to the M.R.C. Tropical Metabolism Research Unit in Jamaica weighing only 2.47 kg. As she was born in a hospital, her birth date was accurately known. The 50th percentile value for girls in Boston aged 9 months is 8.71 kg, so by this standard her weight was 28.4% of "normal." During her first 9 months of life she had in fact lost 0.6 kg, but she recovered and when last seen at the age of 4 years weighed 10.3 kg, her height was 90 cm, and she was clinically well, although still small for her age.

Severe malnutrition affects body shape as well as size. In general, height is less affected than weight; in the extreme case quoted above the child had lost weight since birth but obviously could not lose height. In fact, her height on admission to the hospital was 53 cm; so despite her weight loss she had grown in length by about 3 cm since birth.

Graham [12] reports very interesting follow-up measurements on 53 malnourished Peruvian children, and shows that although at the beginning of treatment body weight was more affected than height, bone age, or head circumference, this situation usually changed during the recovery of the child.

It is possible, by comparing the extent to which a child is underweight and underheight, to estimate the length of time for which he has been malnourished. It is obvious that an acutely malnourished child will be of relatively normal height even though he may be severely underweight, and, on the other hand, a severe height deficit must signify a fairly prolonged period of undernutrition. This approach can be used to trace in retrospect the events which lead to the development of the clinical syndromes of kwashiorkor and marasmus. According to Dean,[13] among children in Uganda, kwashiorkor occurs after a long illness, but marasmus after a relatively short one. McLaren [2] reported that the same is true in Lebanon. In Jamaica, however, children with marasmus are more stunted in height than those of the same age with kwashiorkor, suggesting that in that island marasmus is the consequence of more chronic malnutrition. The average 1 year old child with marasmus has a height of 64 cm, the average for kwashiorkor is 68 cm, and a boy in the Boston 50th percentile at one year is 75 cm. Among a group of 71 Jamaican children with marasmus and 39 with kwashiorkor the difference in height for age was highly significant [14] ($p < 0.001$).

Body Composition in Malnourished Children

During the first year of extrauterine life the normal well-fed infant trebles his birth weight, and also continues the chemical maturation which is characteristic of intrauterine development. The principal changes brought about by chemical maturation are a reduction in the proportion of water in the body, especially extracellular water, a decrease in the concentration of sodium and chloride, and an increase in protein, intracellular electrolytes and fat.[15]

The changes brought about by protein malnutrition in the infant are the opposite of those involved in normal maturation [8,16,17] so the malnourished child has an increased proportion of water in the body, especially extracellular water, usually an increase in sodium and chloride, and a decrease in protein, intracellular electrolyte and fat.

If the changes due to malnutrition are to be expressed quantitatively it is necessary to compare the malnourished child with a "normal" one, and at this point difficulties arise. Consider a child such as "A.C.," upon whom a whole body analysis was performed.[4] The amount of water, fat, protein, and electrolytes in the whole body and in various tissues is known, but what is the correct normal with which they should be compared? This child was 1 year old, he weighed 6.3 kg and he was 70 cm tall. A normal child of the same age weighs about 10 kg and would be chemically more mature, so this would not be a valid comparison so far as biological age is concerned. On the other hand, a normal child of the same weight would be about 6 months old and probably chemically less mature. A more serious objection to using body weight as a basis on which to make comparisons is that it is greatly influenced by the body water content. A malnourished child is overhydrated, and as he recovers he often initially loses weight with the loss of excess water; it would be absurd to compare the composition of this recovering child with progressively younger and smaller "controls."

The best practical solution is to use height as a basis for comparison for reasons which have been discussed elsewhere.[4] A normal child whose height is 70 cm is about 9 months old and of comparable biological age and maturity to the child "A.C." The system based on height will be used in the following discussion.

The composition of a typical child with kwashiorkor and one with marasmus is illustrated in Figure 1. This diagram is based on data from whole body analyses [4,6,18] and from measurements of water and potassium *in vivo* by isotopic means.[19–22] The results of analyses of biopsy samples of liver [23] and muscle [24–26] generally fit in with the same picture.

Since the terms "kwashiorkor" and "marasmus" have now been used several times, the nonpediatric reader may well feel that he is entitled to a definition of what these terms mean. They have no exact meaning [14] but they are used by clinicians to describe two types of malnourished child; a protein malnourished child who is edematous and who has a fatty liver and depigmented skin and hair will be called kwashiorkor, whereas one who is extremely wasted, but without edema or depigmentation will be called marasmus. The majority of malnourished children do not exactly fit the classical description of either syndrome. Recently an admirable attempt has been made by McLaren and his colleagues [27] to introduce an objective scoring system which would place a given child exactly in the kwashiorkor-marasmus spectrum, but for present purposes the above very loose definition will serve.

The noticeable differences between the child with kwashiorkor and the one with marasmus, when they are compared with a normal child

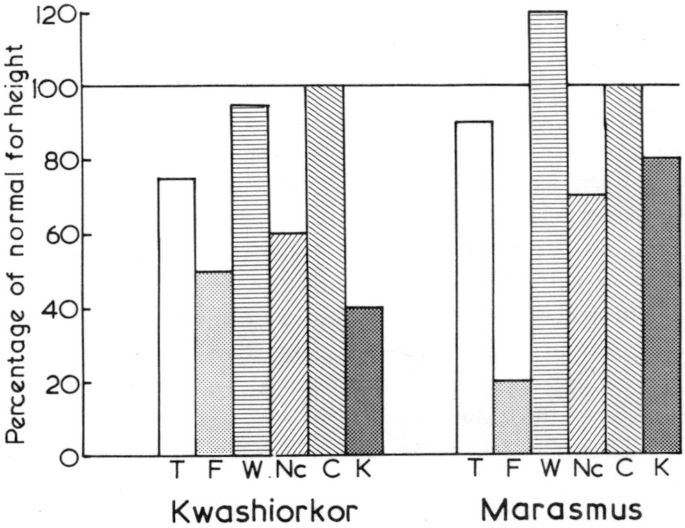

Figure 1. Representation of the body composition of a typical child with kwashiorkor, and of one with marasmus. The values for total body weight (*T*), fat (*F*), water (*W*), non-collagen protein (*Nc*), collagen (*C*) and potassium (*K*) are expressed as a percentage of the amount expected in a normal child of the same height. For source of data see text.

of the same height, are that the child with kwashiorkor has a larger deficit of body weight and of most body components except fat. The difference in fat content is real, but the other differences arise largely as a result of the height-based comparison. If both malnourished children were 1 year old the heights of the child with kwashiorkor and with marasmus would be about 68 cm and 64 cm, respectively,[14] so the weight of their respective height controls would be about 8 kg and 7 kg, respectively. Thus the kwashiorkor child, who is 75% of 8 kg, weighs 6 kg. The marasmic child, who is 90% of 7 kg, weighs 6.3 kg; the difference in absolute body weight is quite small.

On the other hand, the similarities between the two types of malnourished child in other respects are striking. In both children the total body weight is reduced relative to height. In both there is a great excess of water relative to body weight, and this water has been shown by dilution methods in living children [19,28–35] to be mainly extracellular. In both children the non-collagen part of total body protein is reduced to a greater extent than body weight, but the collagen content is that which would be expected in a normal child of the same height.[18]

Both children are potassium depleted, but the child with kwashiorkor more so than the marasmic one.[22] This does not necessarily mean that a sample of the tissues of the child with kwashiorkor would show a lower potassium content than that of the one with marasmus; indeed, when the comparison has been made, no significant difference has been found between the two syndromes.[36] This paradox may be explained by reference to Figure 2, which is based on a whole body analysis of a malnourished child "A.S." by Halliday.[6] In the figure the three columns represent the total amount of protein, non-collagen protein, and potassium in the child, and these columns are subdivided to show the proportion in which brain, viscera (that is, the contents of the thoracic and abdominal cavities), muscle, skin, and bone contribute to the total amount of these three materials in the body.

Figure 2 illustrates the point, made by Klose in 1914,[37] that the ratio of potassium to protein in different organs varies widely. Brain, which contains only a small proportion of the total body protein contributes a much larger fraction of the total body potassium. Conversely, bone

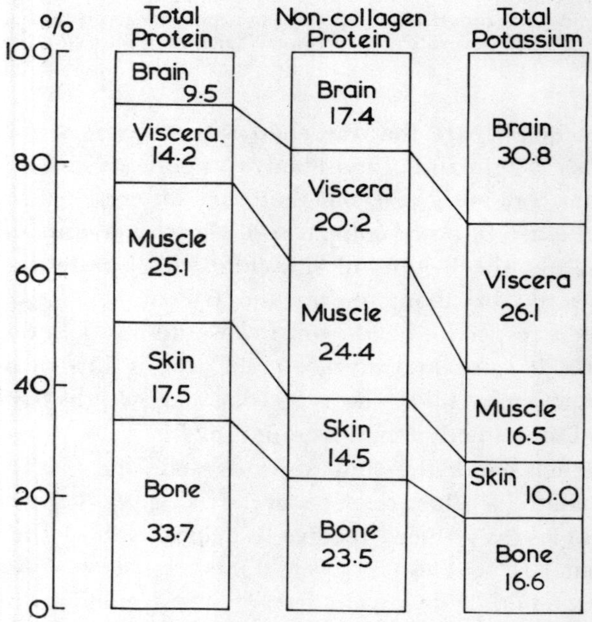

Figure 2. The distribution within the body of a malnourished child of protein, non-collagen protein and potassium. Calculated from whole body analysis data of Halliday (1967).

contributes a much smaller share of potassium than of protein to the body as a whole. The diagram also shows that those tissues which have a low collagen content, such as brain and viscera, are relatively potassium rich, while collagenous tissue like skin and bone are poor in potassium. This is to be expected since potassium is an intracellular ion and there is very little intracellular fluid in bone.

The relevance of this digression is that a child with kwashiorkor might have a low body potassium for two reasons: because his tissues had a low potassium concentration or because he had lost relatively more potassium-rich than potassium-poor tissue. We cannot measure in the living child the proportion of total body protein in brain and viscera, but in eleven whole body analyses of children who had died, there is a significant positive correlation ($r = 0.82$, $p < 0.01$) between the proportion of total body protein in brain and viscera and the whole body potassium/nitrogen ratio.[108] The analyses mentioned above of tissue from whole bodies or biopsies have usually, but not invariably,[27] shown a low potassium concentration. It appears, therefore, that for both reasons combined the potassium content of the child with kwashiorkor is particularly low.

There are three other changes in body composition in malnourished children which could not easily be indicated in the diagrams. Figure 1 shows that the child with kwashiorkor has much more fat than the one with marasmus, although not as much as would be expected on the basis of height or weight. It should be emphasized that the distribution of this fat may be quite abnormal; for example, one-third of the total body fat may be in the liver.[4] There is no clear relationship between the degree of fatty infiltration of the liver and the amount of subcutaneous fat. It has been suggested that a certain amount of subcutaneous fat is necessary for the development of clinically detectable edema[38] but results of whole body analysis do not support this view. However, there is an interesting association between fatty liver and edema which is statistically significant.[39,40]

The second point not indicated in Figure 1 is that the plasma volume in malnourished children is increased relative to body weight[29,41–43] unless they also have severe diarrhea. The expanded plasma volume should be borne in mind in the interpretation of plasma protein concentrations.

Thirdly, we should consider the evidence of magnesium depletion in malnourished children. Montgomery found a low concentration of magnesium in serum[44] and in muscle and urine[45] from malnourished children, and metabolic balance measurements [46–48] have shown that during recovery these children retain more magnesium than would be expected from the rate of nitrogen retention. This deficiency of magnesium causes remarkably little effect clinically [45] unless diarrhea is superimposed on the malnutrition.[49] Caddell [50] is an enthusiastic proponent of magnesium therapy for malnourished children, but her own results show no increase in the concentration of magnesium in muscle after treatment.[51] We may conclude that magnesium deficiency exists in malnourished children, but that the great reservoir of magnesium in bone mineral will usually prevent the effects of this deficiency from becoming evident unless diarrhea rapidly decreases the concentration in the extracellular fluid.

Before this discussion of body composition is concluded a final point should be made clear. Figure 2 is based on an analysis of a single child who died of malnutrition. For obvious reasons there is no comparable analysis of a normal child. We can guess, however, that in the corresponding diagram for a well-fed child the compartment representing muscle would occupy a much larger share of all the columns.

Disturbance of Function in Malnourished Children

The composition of the tissues of a child is greatly altered by protein malnutrition. There is also in almost every organ evidence of impaired function. The significant abnormalities are summarized below.

In children with kwashiorkor the heart is atrophic, and there are many reports of electrocardiographic changes,[52–55] some of which are attributed to magnesium deficiency.[50] Signs of poor circulation are common in malnourished children, and direct measurement of cardiac output [56,57] have shown very low values even when the plasma volume is perfectly adequate. It is evident, therefore, that the poor circulation is mainly a result of poor cardiac function.

Renal function is also seriously affected, and this must at least partly be a result of the small cardiac output. There is some doubt about the frequency of histological evidence of kidney damage; Stirling [58] and Ward [59] find a high incidence of renal lesions, but Bras *et al.*[60] report that these are no more common in malnourished children than among well-fed children from the same population. It is agreed, however, that renal blood flow and glomerular filtration rate are greatly reduced in children with malnutrition, and the more so if they also have gastroenteritis.[56,59,61,62] Tubular function is impaired in many respects. The ability to acidify the urine is reduced, probably as a result of potassium deficiency. Smith [63] showed a positive correlation ($p = 0.001$) between the minimum urine pH after a standard load of ammonium chloride and the potassium content of the child. Aminoaciduria is common among malnourished children [59,64,65] but the South African workers [66] consider that it is of the overflow, rather than the renal, type. Children with kwashiorkor produce urine hypotonic to plasma, but this is probably not entirely a result of tubular failure. In response to a dose of vasopressin they will produce hypertonic urine, but not as concentrated as that of a fully recovered child.[61,62]

The importance which has been attached to liver function in malnourished children has varied from time to time. The classical work on kwashiorkor in the West Indies by Waterlow [39] drew attention to the enormous fatty infiltration of the liver in what he then called "fatty liver disease." However, the Mexican workers [67] found no abnormality of liver function tests in malnourished children, and in later publications Waterlow showed that the degree of fatty infiltration was not significantly correlated with the severity of malnutrition.[40] Recently, however, interest in liver failure as a cause of death in these children has increased since McLean [68,69] showed that increased serum concentrations of certain liver enzymes and of bilirubin were associated with a very high mortality. Indeed, in a recent analysis,[70] serum bilirubin was the most significant of all prognostic factors.

Malnourished children often have diarrhea, even when no intestinal pathogens can be found. South African workers [71–73] have shown that one cause of this is lactose intolerance; there is a deficiency of intestinal lactase and of all pancreatic enzymes.[74] However, despite these enzyme deficiencies, and the atrophic state of the gut mucosa, the absorption of amino acids and electrolytes remarkably efficient.[46–48,75,76]

Numerous endocrine abnormalities have been reported in infantile malnutrition. Monckeberg *et al.*[77] in Chile found that malnourished children aged 5–11 months showed a greatly increased growth rate when they were given growth hormone. They concluded that there is, in malnutrition, a relative pituitary depression, perhaps as a protective mechanism. Histologically, however, the pituitary shows an increase in amphophilic cells.[78] Pimstone *et al.*[79] have shown an increased concentration of growth hormone in the plasma of children with kwashiorkor, and Hadden and Rutishauser [80] found that in malnourished children who were already recovering rapidly, the administration of growth hormone produced no further improvement in nitrogen balance.

Measurement of radioactive iodine excretion [81] or oxygen consumption [82,83] in malnourished children is difficult to interpret in terms of thyroid activity. The resting metabolic rate of a child is largely determined by the mass of metabolically active organs such as brain, liver, lungs, heart, and kidneys [84] and evidence has already been given that the weight and composition of these organs is very difficult to estimate during life.

The adrenal glands in children who have died from malnutrition are rather small [85] and urinary excretion of corticosteroids is usually low. However, in protein-depleted animals there is relative adrenal hypertrophy [86] and this state fits in better with the clinical features of infantile malnutrition.[87] The situation has been greatly clarified by the demonstration by Alleyne and Young [88] in Jamaican children that the plasma corticosteroid level is high (due to slow destruction rather than rapid synthesis) and that the low rate of excretion of the steroid in the urine is more a reflection of poor function of the kidney than of the adrenal cortex. Plasma insulin levels are low and there is impaired glucose tolerance in kwashiorkor [89] but hypoglycemia is a common and dangerous feature; [90] presumably, there is increased insulin sensitivity which is also seen in experimental protein-calorie deficiency.[91]

No mention has been made here of the influence of protein malnutrition on intellectual function: this very important subject is dealt with by another contributor to this book.

Discussion

In preceding sections some of the abnormalities of structure and function in the protein-depleted child have been described. The question now arises: Why do these changes occur? If we knew the answer to this question we would be better able to reverse the changes, or avoid them.

One answer to the question is that they occur because the child is malnourished; that this happens for complex reasons involving poverty, ignorance, and social disorganization, and when these things are set right and every infant receives at least his minimum requirement of attention and food,[92] the problem will disappear. This is a perfectly valid solution, but it is one which involves politicians, economists, and sociologists, and which applies to populations rather than individuals. As physiologists and clinicians we are entitled to ask of a particular malnourished child: Why does he have edema, or a fatty liver, or skin lesions? Why does he show the fluctuating states of irritability and apathy? Why does he die?

There are no firm answers to these questions, but an increasing amount of information is available which bears upon them. The edema of kwashiorkor obviously has a multiple etiology. Hypoproteinanemia, with reduced colloid osmotic pressure, is a contributary factor,[93] and so is potassium deficiency.[94] Probably the inability to excrete a water load[62] and the high plasma cortisol concentration[88] are also involved. The observation that edema is associated with fatty infiltration of the liver[40,70] is still unexplained. A step towards explaining the cause of the fatty infiltration was made by Fletcher[95] who showed that biopsies from fatty livers contained significantly less glucose-6-phosphatase than nonfatty livers. If this enzyme is limiting in children with kwashiorkor it may explain why dietary carbohydrate is deposited in the liver first as glycogen (which is *increased* in fatty livers[10]) and later as fat.

The skin changes of kwashiorkor were described by Cicely Williams in her original paper in 1933.[96] She was certain that the sores and areas of desquamation were "due to some dietetic deficiency," but that it was not pellagra, with which she was perfectly familiar. Stannus, who was an international authority on nutrition, made the mistake of contradicting her and saying that the lesions were "absolutely diagnostic" of pellagra,[97] without himself examining Williams's cases. Stannus proved to be wrong. The skin of recovering children heals without vitamin therapy, and conversely vitamins have no effect on the skin unless protein is also given. The reason may be that skin is a low-priority protein. In times of desperate protein depletion, the organism seems to concentrate its protein synthetic ability on certain metabolically active proteins in the viscera[3,98–100] while the proteins of slower turnover (like skin) receive even less than their proportionate share of the meager supply of amino acids. The appearance of the skin and hair of a child with kwashiorkor fits well with the idea that the renewal of skin is a luxury which this child can no longer afford.

A change in mental state is agreed to be a characteristic of kwashiorkor in all parts of the world.[101] It is also a helpful sign in assessing the progress of a patient: a child who can be persuaded to smile stands an excellent chance of recovery. The factors which may contribute to the depression of consciousness in severe cases include hypoglycemia,[90,102] liver failure,[69] or loss of brain protein or potassium.[5] No doubt progress will be made with this important and interesting problem when we have better methods for monitoring brain function, and composition, in living children.

The last question, and the one which most concerns the clinician, is why these children die. Any animal who is kept on a protein-free diet for long enough will die; an adult dog will not survive a loss of much more than a third of his total body protein.[99] Presumably the time comes when essential enzyme systems have been sacrificed to the limit compatible with life, and the next day's nitrogen loss must precipitate the animal into an irreversible state of protein depletion leading to death. If this is the sequence of events which usually leads to the death of malnourished children we would expect there to be some correlation between the mortality rate and an index of the degree of protein depletion, such as body weight or plasma protein concentration. No such correlation was found in a multiple regression analysis of a series of 248 severely malnourished Jamaican children.[70] The factors which were found to go with high mortality were a raised serum bilirubin concentration and a low serum sodium concentration; neither of these factors is obviously or directly linked with the degree of protein depletion. The damage done by severe protein depletion is not easily reversed by feeding protein; if it were, we would not see reports of kwashiorkor mortality rates approaching 50%.[103] As Metcoff says in a recent review,[104] "Future investigations must be directed at the metabolic sequences which preclude recovery. . . ." We need to know more about what is happening at cellular level in these children. This problem is being attacked along four different lines. Each way will be considered in turn, starting with the one which presents the least technical difficulties.

Whitehead[105] has developed somewhat empirical tests based on the ratio of free amino acids in the plasma, and on the ratio of hydroxyproline to creatinine in the urine of malnourished children. These tests have been represented as a means of making a more objective differential diagnosis between kwashiorkor and marasmus. It may be that these tests give different results when they are applied to groups of malnourished children in Uganda, to whom different clinical labels have already been applied, but this is a very minor by-product. The groups of Ugandan children,[105] in any case, differed significantly in such features as age and weight-for-age, and diet, and workers elsewhere[27,106] have suggested that the amino acid ratio reflects recent protein intake rather than true nutritional status, and that it fails to differentiate between kwashiorkor and marasmus in their cases. The current controversy on this point is not the main issue; the important fact is that the free amino acid pattern in plasma, and the rate of hydroxyproline excretion are observable phenomena which, if we could correctly interpret them, might yield information about the protein metabolism of the malnourished and recovering child.

In South Africa,[106] the amino acid pattern of 22 cases of kwashiorkor, 9 cases of marasmus, and 11 normal children was investigated by column chromatography. On admission to a hospital, both types of malnourished child showed the changes which have been described from many countries, namely a decrease in most free amino acids but particularly the essential group containing leucine, isoleucine, and valine, and an increase in the nonessential glycine. However, by the next day, when the malnourished children had received 50–100 ml of milk per kg, the picture changed dramatically towards the normal pattern, concentrations of all amino acids rose, except glycine which fell. These results are difficult to interpret quantitatively, but they show that the administration of a physiological quantity of a well-balanced protein can produce an immediate change in amino acid metabolism. It would be very interesting to know by what control mechanism this change is brought about. It is evident that there are several systems involved in the response of a child to a change in protein intake, some of which act almost immediately and some of which take several days to operation. This question will be discussed again with respect to the control of the albumin pool size.

The second line of attack on "the metabolic sequences which preclude recovery" is by chemical analysis of biopsy samples. Metcoff and his colleagues[107] analyzed muscle biopsies from 13 severely malnourished children, 8 of whom died. They showed that the most severe malnutrition was associated with an increase in intracellular water and sodium, and a decrease in potassium, phosphoenolpyruvate and oxaloacetate. In these analyses it is necessary to make the assumption that some marker substance, such as chloride, is totally extracellular, in order to calculate the distribution of the remaining ions between the intracellular and extracellular phase. This assumption is probably valid in normal tissue, but there is no way of testing whether it is true also of tissue from severely malnourished children. Certainly this type of calculation sometimes leads to erratic and implausible results, so caution is necessary.[108] If there is a great decrease in the intracellular concentration of potassium or magnesium, or if large quantities of sodium enter the cell, this would obviously very seriously impair the enzyme activity on which metabolism depends.

The third line of attack is technically still more difficult. Measurements of intracellular enzyme activity have been made on liver and muscle samples from malnourished children, and the results have been reviewed by Waterlow *et al.*[10] and by Metcoff.[104] There is some disparity between the results of different workers, and no very coherent picture emerges. However, phosphorylation is reduced in fatty livers, and ageing of the mitochondria by preincubation caused a much greater reduction of phosphorylation in mitochondria from malnourished than from normal livers.[109] The two great limitations to this approach are the skill required for the analysis, and the uncertainty as to whether the results *in vitro* give a true picture of the situation *in vivo.*

The fourth way of investigating protein metabolism in the human infant is with isotopic tracer techniques. This method has been successfully applied to plasma protein metabolism. The early studies of Gitlin *et al.*[110] and of Garrow and Waterlow [111] showed that the concentration of iodinated albumin in the plasma of malnourished children after the injection of a tracer dose fell at the same rate whether they were malnourished or recovered. The conclusion that the rate of turnover of albumin is unaffected by malnutrition was a wrong one; in fact, the malnourished child at the beginning of his stay in hospital breaks down the labeled albumin more slowly than he does after recovery, but since he is at the same time expanding his albumin pool and diluting the labeled albumin with new unlabeled albumin, the apparent catabolic rate is similar to that after recovery. The true situation is revealed if measurements of albumin catabolism are based on urinary excretion of radioiodine, and Picou and Waterlow [112] have shown by this method that the half-life of albumin in malnourished children is 24.6 days (S.E.M. ± 2.52), and after recovery 13.1 days (S.E.M. ± 0.67). Similar results were obtained by Cohen and Hansen.[113]

The calculation of albumin turnover rates from measurements of activity in plasma or urine is tedious and involves considerable inconvenience to the patients in the collection of samples. A much more convenient and sensitive way to follow changes in the amount of isotope in the body is to make serial measurements in a whole body counter. The curve of whole body activity against time together with some measurements on plasma, can be analyzed by computer to yield estimates of the rates of synthesis, catabolism, and transfer between intravascular and extravascular albumin pools.[7] Preliminary results using this technique show that when a child is changed from a high to a low protein diet the rate of catabolism of albumin falls by about 20%, but the synthetic rate falls still more, so the pool becomes smaller and turns over more slowly. On transfer from a low to a high protein diet the increase in synthetic rate is established more rapidly, and is of greater magnitude than the increase in catabolic rate.[7] These results, like the observations of Saunders *et al.*[106] on free amino acids in plasma, show that protein metabolism is capable of rapid adjustments. Perhaps we should think of "labile protein" as a consequence of a time lag between changes of rate constants, rather than a special store of protein as a protection in time of need.[7]

Iodinated proteins cannot provide information about the metabolism of intracellular protein, since they are confined to the extracellular space. In order to see into the cell it is necessary to use an isotopically labeled amino acid. In children the use of ^{14}C labeled material is to be avoided on account of the very long radioactive half-life of this isotope. An alternative is to use ^{35}S labeled methionine, since this isotope has a half-life of 87 days. Somewhat naive studies were done with this compound, and the results were interpreted as showing a shift in protein synthesis towards the visceral proteins in malnourished dogs[99] and children.[98] Subsequent work[3,100] has supported this view, but McFarlane[114] has pointed out that the results of experiments in which a labeled amino acid is given as a single dose are almost impossible to interpret quantitively. A much more promising technique is to give the labeled amino acid by a steady infusion[7] so that a known constant level of specific activity is reached in the precursor pool.

These are the four methods by which our understanding of the metabolism of the malnourished child has been advanced in the past few years, and they are also the ways which seem most likely to afford progress in the future. The problem of severe infantile malnutrition, which is most pressing in the underdeveloped parts of the world, will not be solved without the help of quite sophisticated biochemical techniques. We all hope that with better planning protein-calorie malnutrition will become a rare disease, but while the disease is still with us we have much to learn about human protein metabolism from the severely malnourished child.

Summary

The human infant who is given a diet grossly inadequate in protein does not react in the same way as an experimentally malnourished laboratory animal. It is, therefore, necessary to be cautious in extrapolating results between species.

In human malnutrition the deficiency of protein is usually complicated by other factors such as infection, vitamin deficiency, and parasitic infestation. So far as possible, the effects which have been described here are those in human infants which are attributable to protein deficiency (with a greater or lesser degree of calorie deficiency), rather than other complicating factors. The principal changes in body composition which occur in malnourished children are an increase in total body water, especially extracellular water, and a decrease in non-collagen protein, intracellular electrolytes, and fat. These changes are the opposite of those which characterize normal chemical maturation. However, the changes in malnutrition are more severe than those which could be explained simply by retarded chemical maturation; in particular, potassium depletion is an important feature of kwashiorkor.

Changes in function (other than intellectual function) are described and related to changes in structure. Mortality in severely malnourished children is very high even with the best treatment, and this illustrates our basic ignorance about the way in which human protein metabolism responds to protein deficiency, and the point at which this compensating mechanism breaks down.

Severe infantile malnutrition presents a challenge both to the clinician and the biochemist. More subtle techniques are being developed to investigate the metabolic lesions within the cell. As Claude Bernard observed, you cannot tell what happens inside a house by watching what goes in at the door and comes out by the chimney.

References

1. Brock, J. F. and Autret, M., W.H.O. Monograph, No. 8. Geneva, 1952.
2. McLaren, D. S., *Lancet,* **ii,** 465 (1966).
3. Waterlow, J. C., and Stephen, J. M. L., *Brit. J. Nutr.,* **20,** 461 (1966).
4. Garrow, J. S., Fletcher, K., and Halliday, D., *J. Clin. Invest.,* **44,** 417 (1965).
5. Garrow, J. S., *Lancet,* **ii,** 643 (1967).
6. Halliday, D., *Clin. Sci.,* **33,** 365 (1967).
7. Waterlow, J. C., Alleyne, G. A. O., Chan, H., Garrow, J. S., Hay, A., James, P., Picou, D., and Stephen, J. M. L., *Arch. Latinamer. Nutric.,* **16,** 175 (1966).
8. McCance, R. A., *Lancet,* **ii,** 621, 627 (1962).
9. Garrow, J. S., and Pike, M. C., *Lancet,* **i,** 1 (1967).
10. Waterlow, J. C., Cravioto, J., and Stephen, J. M. L., *Advances in Protein Chemistry,* **15,** 131 (1960).
11. Gomez, F., Ramos-Galvan, R., Cravioto, J., Frank, S., Chavez, R., and Vasquez, J. *J. Trop. Pediatr.,* **2,** 77 (1956).

12. Graham, G. G., *Fed. Proc.,* **26,** 139 (1967).
13. Dean, R. F. A., in "Recent Advances in Pediatrics," D. Gairdner, Ed., J. A. Churchill, London, 1965, pp. 234–265.
14. Garrow, J. S., *Arch. Latinamer. Nutric.,* **14,** 145 (1966).
15. Widdowson, E. M., and Spray, C. M., *Arch. Dis. Childh.,* **26,** 205 (1951).
16. Cravioto, J., *Am. J. Clin. Nutr.,* **11,** 484 (1962).
17. Friis-Hansen, B., "Human Body Composition," J. Brozek, Ed., Pergamon Press, Oxford, England, 1965, p. 191.
18. Picou, D., Halliday, D., and Garrow, J. S., *Clin. Sci.,* **30,** 345 (1966).
19. Smith, R., *Clin. Sci.,* **19,** 275 (1960).
20. Smith, R., and Waterlow, J. C., *Lancet,* **i,** 147 (1960).
21. Brinkman, G. L., Bowie, M. D., Friis-Hansen, B., and Hansen, J. D. L., *Pediatrics,* **36,** 94 (1965).
22. Garrow, J. S., *Lancet,* **ii,** 455 (1965).
23. Waterlow, J. C., and Weisz, T., *J. Clin. Invest.,* **35,** 346 (1956).
24. Frenk, S., Metcoff, J., Gomez, R., Ramos-Galvan, R., Cravioto, J., and Antonowicz, I., *Pediatrics,* **20,** 105 (1957).
25. Waterlow, J. C., and Mendes, C. B., *Nature,* **180,** 1361 (1957).
26. Vis, H., Dubois, R., Vanderborght, H., and de Maeyer, E., *Rev. Franc. Et. Clin. Biol.,* **10,** 729 (1965).
27. McLaren, D. S., Pellett, P. L., and Read, W. W. C., *Lancet,* **i,** 533 (1967).
28. Kerpel Fronius, E., and Kovack, S., *Pediatrics (Springfield),* **2,** 21 (1948).
29. Gollan, F., *J. Clin. Invest.,* **27,** 352 (1948).
30. Brinkman, G. L., *S. Afr. Med. J.,* **38,** 572 (1964).
31. Brinkman, G. L., and Hansen, J. D. L., *S. Afr. Med. J.,* **38,** 684 (1964).
32. Cokington, L., Hanna, F. M., and Jackson, R. L., *Ann. N. Y. Acad. Sci.,* **110,** 849 (1963).
33. Passaro, G., *Bol. Soc. Ital. Biol. Sper.,* **29,** 245 (1953).
34. Ramos-Galvan, R., and Cravioto, J., *Bol. Med. Hosp. Inf. (Mexico),* **15,** 763 (1958).
35. Hansen, J. D. L., Brinkman, G. L., and Bowie, M. D., *S. Afr. Med. J.,* **39,** 491 (1965).
36. Smith, R., M.D. Thesis, Cambridge, 1960.
37. Klose, E., *Jahrb. f. Kinderheilk.,* **80,** 154 (1914).
38. Metcoff, J., Frenk, S., Gordillo, G., Gomez, F., Ramos-Galvan, R., Cravioto, J., Janeway, C. A., and Gamble, J. L., *Pediatrics,* **20,** 317 (1957).
39. Waterlow, J. C., *M.R.C. Spec. Rep. Ser. 263,* H.M.S.O., London, 1948.
40. Waterlow, J. C., Bras, G., and De Pass, E., *J. Trop. Pediatr.,* **2,** 189 (1957),
41. Cohen, S., and Hansen, J. D. L., *Clin. Sci.,* **23,** 351 (1962).
42. Patel, B. D., Patel, J. C., Gaitonde, B. B., and Rao, G. S., *Ind. J. Child Health,* **9,** 407 (1960).
43. Alleyne, G. A. O., *Arch. Dis. Childh.,* **41,** 313 (1966).
44. Montgomery, R. D., *J. Pediat.,* **59,** 119 (1961).
45. Montgomery, R. D., *Lancet,* **ii,** 74 (1960).
46. Pretorius, P. J., Wehmeyer, A. S., and Theron, J. J., *Amer. J. Clin. Nutr.,* **13,** 331 (1963).
47. Pretorius, P. J., and Wehmeyer, A. S., *S. Afr. J. Lab. Clin. Med.,* **10,** 52 (1964).

48. Linder, G. C., Hansen, J. D. L., and Karabus, C. D., *Pediatrics,* **31,** 552 (1963).
49. Back, E. H., Montgomery, R. D., and Ward, E. E., *Arch. Dis. Childh.,* **37,** 106 (1962).
50. Caddell, J., *J. Pediatr.,* **66,** 392 (1965).
51. Caddell, J., and Goddard, D. R., *New Eng. J. Med.,* **276,** 533 (1967).
52. Schyns, Ch., and Demayer, E. M., *Acta Cardiologica,* **12,** 413 (1957).
53. Smythe, P. M., Swanepoel, A., and Campbell, J. A. H., *Brit. Med. J.,* **i,** 67 (1962).
54. Conje, R. E., and Pretorius, P. J., *S. Afr. J. Lab. Clin. Med.,* **9,** 11 (1963).
55. Swanepoel, A., Smythe, P. M., and Campbell, J. A. H., *Amer. Heart J.,* **67,** 1 (1964).
56. Kerpel Fronius, E., Vargas, F., Kun, K., and Vonoczky, J., *Acta Med. Acad. Sci. Hung.,* **5,** 27 (1954).
57. Alleyne, G. A. O., *Clin. Sci.,* **30,** 553 (1966).
58. Stirling, G. A., *Arch. Dis. Childh.,* **37,** 378 (1962).
59. Ward, E. E., M.D. Thesis, London, 1964.
60. Bras, G., Waterlow, J. C., and De Pass, E., *J. Trop. Ped.,* **2,** 147 (1956).
61. Gordillo, G., Soto, R. A., Metcoff, J., Lopez, E., and Antillon, L. G., *Pediatrics,* **20,** 303 (1957).
62. Alleyne, G. A. O., *Pediatrics,* **39,** 400 (1967).
63. Smith, R., *Lancet,* **i,** 764 (1959).
64. Edozien, J. C., Phillips, E. J., and Collis, W. R. F., *Lancet,* **i,** 615 (1960).
65. Edozien, J. C., and Obasi, M. E., *Clin. Sci.,* **29,** 1 (1965).
66. Saunders, S. J., Barbezat, G. O., Wittmann, W., and Hansen, J. D. L., *Amer. J. Clin. Nutr.,* **20,** 760 (1967).
67. Gomez, F., Ramos-Galvan, R., Bienvenu, B., and Cravioto Munoz, J., *Bol. Med. Hosp. Inf. (Mexico),* **7,** 485 (1950).
68. McLean, A. E. M., *Lancet,* **ii,** 1292 (1962).
69. McLean, A. E. M., *Clin. Sci.,* **30,** 129 (1965).
70. Garrow, J. S., and Pike, M. C., *Brit. J. Nutr.,* **21,** 155 (1967).
71. Bowie, M. D., Brinkman, G. L., and Hansen, J. D. L., *Lancet,* **ii,** 550 (1963).
72. Bowie, M. D., Brinkman, G. L., and Hansen, J. D. L., *J. Pediatr.,* **66,** 1083 (1965).
73. Bowie, M. D., Barbezat, G. O., and Hansen, J. D. L., *Amer. J. Clin. Nutr.,* **20,** 89 (1967).
74. Barbezat, G. O., *S. Afr. Med. J.,* **41,** 84 (1967).
75. Waterlow, J. C., and Wills, V. G., *Brit. J. Nutr.,* **14,** 183 (1960).
76. Waterlow, J. C., Wills, V. G., and Gyorgy, P., *Brit. J. Nutr.,* **14,** 199 (1960).
77. Monckeberg, F., Donoso, G., Oxman, S., Pak, N., and Meneghello, J., *Pediatrics,* **31,** 58 (1963).
78. Tejeda, C., and Russfield, A. B., *Arch. Dis. Childh.,* **32,** 343 (1957).
79. Pimstone, B. L., Wittman, W., Hansen, J. D. L., and Murphy, P., *Lancet,* **ii,** 779 (1966).
80. Hadden, D. R., and Rutishauser, I. H. E., *Arch. Dis. Childh.,* **42,** 29 (1967).
81. Montgomery, R. D., *Arch. Dis. Childh.,* **37,** 383 (1962).

82. Montgomery, R. D., *J. Clin. Invest.,* **41,** 1653 (1962).
83. Monckeberg, F., Beas, F., Horowitz, I., Dabancens, A., and Gonzales, M., *Pediatrics,* **33,** 554 (1964).
84. Holliday, M. A., Potter, D., Jarrah, A., and Bearg, S., *Pediatr. Res.,* **1,** 185 (1967).
85. Stirling, G. A., *J. Path. Bact.,* **77,** 555 (1959).
86. Platt, B. S., and Stewart, R. J. C., *J. Endocr.,* **38,** 121 (1967).
87. Gillman, J., Gilbert, C., and Savage, N., *Metabolism,* **11,** 800 (1962).
88. Alleyne, G. A. O., and Young, V. H., *Clin. Sci.,* **33,** 189 (1967).
89. Baig, H. A., and Edozien, J. C., *Lancet,* **ii,** 662 (1965).
90. Whitehead, R. G., and Harland, P. S. E. G., *Brit. J. Nutr.,* **20,** 825 (1966).
91. Heard, C. R. C., *Diabetes,* **15,** 78 (1966).
92. Chan, H., and Waterlow, J. C., *Brit. J. Nutr.,* **20,** 775 (1966).
93. Montgomery, R. D., *Arch. Dis. Childh.,* **38,** 343 (1963).
94. Hansen, J. D. L., and Brock, J. F., *Lancet,* **ii,** 477 (1954).
95. Fletcher, K., *Am. J. Clin. Nutr.,* **19,** 170 (1966).
96. Williams, C. D., *Arch. Dis. Childh.,* **8,** 423 (1933).
97. Stannus, H. S., *Arch. Dis. Childh.,* **9,** 115 (1934).
98. Garrow, J. S., "Amino Acid Malnutrition." XIII Annual Protein Conference, W. H. Cole, Ed., Rutgers University Press, New Brunswick, N. J., 1957, p. 14.
99. Garrow, J. S., *J. Clin. Invest.,* **38,** 1241 (1959).
100. Waterlow, J. C., *Nature,* **184,** 1875 (1959).
101. Waterlow, J. C., and Scrimshaw, N. S., *Bull. W.H.O.,* **16,** 458 (1957).
102. Gomez, F., Ramos-Galvan, R., Cravioto, J., and Frenk, S., *Ann. N. Y. Acad. Sci.,* **69,** 969 (1958).
103. Carr, W. R., and Gelfand, M., *Trans. Roy. Soc. Trop. Med. Hyg.,* **51,** 229 (1957).
104. Metcoff, J., *Ann. Rev. Med.,* **18,** 377 (1967).
105. Whitehead, R. G., *Arch. Dis. Childh.,* **42,** 479 (1967).
106. Saunders, S. J., Truswell, A. S., Barbezat, G. O., Wittman, W., and Hansen, J. D. L., *Lancet,* **ii,** 795 (1967).
107. Metcoff, J., Frenk, S., Torres Pinedo, R., Kaiser, E., and Hansen, J. D. L., *Medicine,* **45,** 365 (1966).
108. Garrow, J. S., Smith, R., and Ward, E. E., "Electrolyte Metabolism in Severe Infantile Malnutrition," Pergamon Press, Oxford, England, 1967.
109. Waterlow, J. C., *Proc. Roy. Soc. B.,* **155,** 96 (1961).
110. Gitlin, D., Cravioto, J., Frenk, S., Montano, E. L., Ramos-Galvan, R., Gomez, F., and Janeway, C. A., *J. Clin. Invest.,* **37,** 682 (1958).
111. Garrow, J. S., and Waterlow, J. C., *Clin. Sci.,* **18,** 35 (1959).
112. Picou, D., and Waterlow, J. C., *Clin. Sci.,* **22,** 459 (1962).
113. Cohen, S., and Hansen, J. D. L., *Clin. Sci.,* **23,** 351 (1962).
114. McFarlane, A. S., Garrow, J. S., and Waterlow, J. C., "Radioisotopes in Tropical Medicine," I.A.E.A., Vienna, 1962, p. 17.

II. Endocrinology

Hormones and Growth in Mammals *

JAMES H. LEATHEM
Bureau of Biological Research
Rutgers University
New Brunswick, New Jersey

Growth implies that hyperplasia and/or hypertrophy has occurred resulting in a net increase in protoplasm. However, one quickly recognizes that growth patterns vary. Thus, prenatal and postnatal growth rate and tissue composition differ from each other. Intrauterine growth is initially accomplished by cell division and fetal growth is noted as 90% water, 5–9% protein, 1–2% fat and 1–2% ash. After birth, the protein increases to 15–20% and total body growth is presented as a sigmoid curve.

Many factors control growth and these factors may be separated into genetic and nongenetic. The hormones, therefore, must be viewed as only one of many factors exerting an influence on growth. Furthermore, the hormones affect growth in many different ways but in order to have the semblance of a central theme, the hormonal role in protein metabolism is emphasized.

It is clear that hormones are not necessarily essential for a reaction but rather serve to stimulate or to inhibit a reaction. Consequently, hormones may be labeled as having an anabolic or catabolic action. These broad terms must be viewed with caution as in many cases the metabolic responses to endocrine gland secretions are organ or tissue specific. Certain hormones will inhibit protein metabolism in some organs while stimulating metabolism in others. Thus the stage of development and the manner of measurement, *i.e.,* body weight gain, nitrogen balance,

* These investigations were supported by a research grant AM 00462 from the Institute of Arthritis and Metabolic Disease of the National Institutes of Health, United States Public Health Service, and by a grant from E. R. Squibb and Sons.

organ protein, etc., must be considered in an examination of hormone action on growth.

Hormones and Fetal Growth

Obviously, considerable growth and development take place before the endocrine glands evolve in the embryo. However, in mammals, the maternal and placental hormones are available. Subsequently, fetal hormones are elaborated but the action of specific hormones on protein metabolism in the fetus remains to be elucidated.

Pituitary gland removal from the fetus by x-irradiation in the mouse or decapitation of the fetus in the rabbit has very little effect on growth. Nevertheless, hypophysectomy of the rat fetus [1] retarded fetal growth and growth hormone but not serum albumin injected into the fetus restored fetal size. Administration of growth hormone to pregnant rats causes a significant increase in the birth weight of the young. Alloxan-induced diabetes in rats on days 11–14 of pregnancy increased fetal weight as well as nitrogen and fat content. However, fetal growth was retarded by injecting rat fetuses with cortisone, by injecting the mother with insulin or by feeding the antithyroid drug thiouracil.[2]

The need for maternal hormones to sustain fetal growth varies with the species. Thus, hypophysectomy or ovariectomy after the first trimester of pregnancy in the primate has no serious effect in fetal growth. On the other hand, ovariectomy in the rat terminates the pregnancy. The placental lactogen which may have some growth promoting properties may be an important factor. The placental fraction mimics growth hormone in stimulating glucose uptake and conversion of fat and has physiological effects in man.[3]

Hormones and Prepubertal Growth

The rate of body weight gain after birth is linear and is responsive to muscle mass. For muscle, the classic concept has been that postnatal growth is due to an increase in muscle fibers but recent studies [4,5] indicate cell numbers also increase. During the first 8 weeks of postnatal life, muscle mass in the rat increases from 3.4 grams to 100 grams. Cell number increases 5-fold, cell size 8-fold.[6] Liver mitoses peak at 21 days and liver cell volume increases five times.

Pituitary dwarfism is well known clinically and pituitary ablation in the 6-day-old rat slows body growth rate. Hypophysectomy of an 80 gram rat holds body weight at 80 grams for 6 months while controls grow to 400 grams. Furthermore, hypophysectomy sharply reduces muscle cell size (P/DNA) relative to age. Body weight increase is also retarded by hypothyroidism which is well exemplified clinically by cretinism. Cretin rats can be induced experimentally by feeding thiouracil during pregnancy and lactation. Cretin rats would weigh 19 grams at weaning whereas control rats weighed 42 grams. Despite withdrawal of the drug the cretins gained only 6–11 grams during the following 20 days whereas the normal rat gained 70 grams.[7] Thyroid deficiency reduces muscle cell size and number.[6] Similar growth disabilities have been noted in diabetes.

The gonadal hormones express themselves in mammals by influencing body size and form. In general, girls remain shorter and thinner than boys and prepubertal castration in the male increases statural growth. On the other hand, growth in rodents is retarded by testis removal and accelerated by ovariectomy. However, examination of muscle RNA and DNA failed to reveal any effect of early castration.[6]

Hormonal excesses in the neonatal period can invoke long lasting growth effects. Thus, corticosterone reduces cholesterol/DNA brain ratio,[8] cortisone acetate given immediately after birth causes "runting" and retards body weight gain (Table I). Thyroxine administered on

Table I. Influence of Cortisone Acetate on Body Weight Gain of Neonatal Mice in 7 days

Treatment	Age Days	No. Mice	Body Weight, g Initial	Body Weight, g Final
None	1	8	1.8	6.5
1.25 mg	1	7	1.8	3.3
None	2	6	2.2	7.5
1.25 mg	2	3	2.2	4.9
None	3	10	2.7	6.1
1.25 mg	3	7	2.7	3.8
None	4	6	3.0	7.9
1.25 mg	4	6	3.0	4.2

days 1, 2, and 3 after birth invokes permanent stunting. Permanent sterility invoked by androgen on day 5 stimulates growth rate at anticipated puberty. Organ studies have not clarified the site of action.

Hypophysectomy and Growth

Before the turn of the century, clinicians had acquired evidence that pituitary dysfunction was associated with human dwarfism and gigantism. Pituitary ablation results in an abrupt suppression of growth and in adults may impose a body weight loss. Hypophysectomy is followed by a decrease in nitrogen retained; in muscle mass and the carcass reveals a greater loss in nitrogen but lesser loss of fat than a pair fed control. Liver and muscle exhibit a reduced capacity to incorporate amino acids into protein. Pituitary ablation is followed by a reduction in liver and muscle RNA and yield of ribosomes but not in ribosomal profile. Garren, *et al.*[9] found the rate of incorporation of phenylalanine into protein was lower with ribosomes from hypophysectomized rats, and added polyuridylic acid was less effective as a stimulant. Staehelin suggested that hypophysectomy causes a loss of m-RNA and a reduced translation but Korner[10] now suggests that hypophysectomy alters the ability of the ribosome to attach to synthetic m-RNA. Protein anabolism is restored by growth hormone.

At least part of the physiological changes ascribed to pituitary ablation do not occur if the protein stores of the body are depleted prior to the operation. Hypophysectomy does not prevent the refed animal from gaining weight or returning to a positive nitrogen balance[11] but may[12] or may not[13] influence amino acid incorporation into liver and muscle. Hypophysectomy does slow the rate of protein repletion and does induce protein shifts which are not due to a casein diet. Thus, kidney mass decreases despite a positive nitrogen retention[14] and synthesis of plasma albumin is reduced. However, liver protein concentration improves and a modest increase in total liver protein occurs.

The hypophysectomized rat fed *ad libitum* will respond to insulin with an increase in weight and growth of the skeleton. The insulin-treated rat exhibits a greater gain in fat and a lesser skeletal response than does a rat treated with growth hormone. The influence of insulin on the hypophysectomized rat previously protein deprived has had little consideration. If adult male rats are hypophysectomized at the end of 1 month of protein-free feeding, then refed an 18% casein diet for 20 days, a body weight gain of 30 grams was recorded. Administration of 0.5 units of protamine zinc insulin daily for 20 days resulted in a 44 gram body weight gain in contrast to the 103 gram gain exhibited by the normal pair fed control. Retention of nitrogen failed to distinguish the hypophysectomized treated or untreated animal. Furthermore, insulin in the dosage used, did not enhance protein content of liver or gastrocnemius muscle or prevent the decrease in kidney size associated with pituitary ablation (Table II). Insulin did increase liver glycogen.

Table II. Hypophysectomy, Insulin, and Protein Repletion

		0% Casein × 30; 18% Casein × 20		
	0% Casein × 60	Hypox Only	PZI (0.5 μ)	Normal Control
No. rats	14	10	8	8
Body wt. change, g	−62	+30	+44	+103
N retained, g/kg	0	8.0	7.6	9.0
Liver wt., g	9.3	8.4	8.9	9.8
Protein, g	1.3	1.9	1.9	2.0
Kidney wt., g	1.8	1.6	1.45	1.83
Protein, g	0.28	0.28	0.24	0.35
Muscle wt., g	0.90	1.0	0.82	0.95
Protein, g	0.21	0.27	0.17	0.21

Diabetes and Protein Repletion

Insulin is needed for normal growth and development of the body, for the action of growth hormone and for the control of nitrogen balance in the adult.[15] The deranged metabolism of the diabetic includes the loss of nitrogen and reduced protein synthesis. Transport of amino acids is altered and ribosomes lose their protein synthetic competence.[16,17] Insulin stimulates the synthesis of glycogen, lipids, proteins, and RNA but not of DNA. Insulin stimulates amino acid transport selectively despite a blocking of protein synthesis by puromycin.

The fact that hypophysectomy does not negate the anabolism of protein repletion does not rule out a hormonal contribution as insulin is still present. To investigate this aspect it was necessary to study the adult male rat made diabetic one month after feeding a protein free diet. Alloxan (150 mg/kg) was administered subcutaneously and diabetes was indicated by a blood sugar of 400 mg% 6 days later. The rats were refed 18% casein for 20 days and compared with pair fed controls. Body weight gain by the diabetic rats was 45 grams whereas the control animal gained 122 grams. However, nitrogen retention was not adversely influenced by diabetes. These data were very similar to those obtained with hypophysectomized rats. However, an improvement in serum albumin and in A/G ratio occurred in the diabetic rat in a manner comparable to the control rat and in contrast to the effects of hypophysectomy (Table III). Again the nutritional state of the animal preceding an endocrine gland removal can be shown to modify anticipated results.

Table III. Protein Repletion, Hypophysectomy, and Diabetes

	0% Casein × 30	0% Casein × 30; 18% Casein × 20			
		Hypox	Control	Diabetic	Control
No. rats	8	8	8	6	6
Body wt. change, g	−62	+50	+102	+45	+122
N balance, g/kg	0	8.0	8.0	8.4	8.2
Serum protein					
Total g/100 ml	4.92	6.21	5.75	5.24	5.44
Albumin	1.95	2.07	2.55	2.31	2.32
A/G	0.67	0.50	0.81	0.83	0.75

Thyroidectomy and Protein Repletion

It is well known that thyroid hormone is essential for the maintenance of a normal basal metabolic rate thus implicating the hormone with growth. There is essential agreement that thyroid hormones are needed for normal growth and differentiation.

A subnormal body growth and poor muscle development is associated with hypothyroidism and is due, in part, to a voluntary reduction in food intake motivated by the decrease in metabolic rate and a secondary lack of growth hormone. Force feeding a thyroidectomized rat improved body weight gain but *decreased* carcass nitrogen from 21% to 12.5% while *increasing* carcass fat from 12% to 38%.[18]

A much less severe effect is imposed by hypothyroidism in adulthood. Normal male rats continue to grow indefinitely and thyroidectomy inhibited but did not block this growth over a 1 year period.[7] Examination of nitrogen balance revealed the influence of food intake and a suggestion that the refilling of protein stores was abnormal after thyroidectomy. Hypothyroidism is associated with a reduced catabolism of serum globulin and a reduction in kidney and liver size and protein content. It has been shown that rate of release of amino acids from peripheral tissues and release of nitrogen and amino nitrogen from the liver and the isolated diaphragm are reduced by hypothyroidism. Thyroidectomy decreases synthesis of protein and RNA.[19] Incorporation of amino acids into protein is reduced as is microsomal protein and RNA synthesis as well as RNA polymerase activity in the liver.[9]

If adult male rats are thyroidectomized at the end of one month of protein-free feeding, then refed an 18% casein diet, body weight gain and nitrogen retention simulates the pair fed control (Table IV).

Table IV. Protein Repletion and Thyroidectomy

	0% Casein × 30	0% Casein × 30; 18% Casein × 20	
		Thyroidx	Control
No. rats	14	5	5
Body wt. change, g	−62	+32	+43
N retained, g/kg	0	7.6	7.3
Liver wt., g	9.3	10.5	12.1
Protein, g	1.3	1.9	2.0
Protein, %	36.2	60.7	60.4
Kidney wt., g	1.80	1.86	2.20
Protein, g	0.28	0.31	0.38
Protein, %	62.5	72.5	72.6
Muscle wt., g	0.90	0.82	0.83
Protein, g	0.21	0.26	0.19
Protein, %	85.7	83.7	78.2

Nitrogen retention does not assure uniform distribution of the nitrogen retained.[18] Furthermore, the prior protein depletion reduces total protein content of muscle, kidney, and liver and decreases liver protein concentration. Thus, the repletion process involves a considerable physiological correction. Nevertheless, definite tissue protein correction occurs with refeeding despite thyroidectomy. Thus, liver protein concentration was corrected and total protein was increased significantly. Muscle protein of thyroidectomized rats exceeded that of euthyroid rats whereas kidney weight and total protein were modestly retarded although protein concentration returned to control levels.

Adrenalectomy and Protein Repletion

If the adrenal glands are removed, growth is largely suppressed if for no other reason than that life itself is generally suppressed. Young animals fail to grow after adrenalectomy but this lack of body weight change is simulated by pair fed controls. When adult male rats were fed a protein free diet for 40 days, then adrenalectomized and fed 18% casein the average survival was 11 days. During this time the rats lost 38 grams body weight as compared with a 2 gram loss exhibited by the pair fed control. Nevertheless a slightly positive nitrogen balance was observed in both groups.

Adrenal cortical hormones are not vital for growth as saline alone will correct many of the deficiencies of adrenalectomy as indicated by growth rate, urea nitrogen formation, serum protein formation and distribution of N^{15} from glycine and S^{35} from methionine in carcass and viscera.[20] Adrenal removal acutely enhances the incorporation of adenine into the RNA of the diaphragm and increased amino acid incorporation in protein of diaphragm and heart.[2]

The quality of protein fed to an adult animal influences the nitrogen balance. The feeding of 20% lactalbumin to normal and adrenalectomized rats given saline resulted in a retention of 6.0 g/kg of nitrogen in 20 days in both groups. However, when 20% gelatin was fed in place of lactalbumin the adrenalectomized rat lost 1.8 grams of nitrogen while the pair fed control retained 0.4 grams. Thus, a protein of lower biological value appears to be less well utilized when saline is provided as replacement for cortical hormones.

Although saline can maintain the adrenalectomized rat in a reasonably normal physiological condition, the imposition of stress signals the lack of corticoids. Protein deprivation can be considered a stress with corticosterone levels increased in rats,[21] but the absence of corticoids in a subsequent refeeding has not been examined. Thus, adult male rats were adrenalectomized after one month of protein-free feeding, then refed casein for 20 days while being maintained with 1% NaCl. All animals survived the postoperative period and gained 74 grams body weight compared with a 99 gram body weight gain exhibited by pair fed controls. Nitrogen balance studies revealed a 7.8 g/kg and 7.7 g/kg retention by adrenalectomized and control rats respectively. Liver and gastrocnemius muscle weights and total protein content were comparable in refed rats with and without adrenal glands. Retardation or loss of kidney size and protein exhibited by the protein refed hypophysectomized or thyroidectomized rat was not observed in saline maintained adrenalectomized rats.

Summary

The refeeding of protein, following a period of protein deprivation, reveals an enhanced tendency for tissue protein anabolism. However, the contribution of the endocrine secretions to the protein repletion process is unclear. These data emphasize that considerable nitrogen retention can be attained during protein repletion despite hypophysectomy. However, internal shifts of protein occur which can be ascribed to hormonal influences. Protein synthesis occurred in protein repleted rats despite induction of diabetes, thyroidectomy, or adrenalectomy. The adrenalectomized rats maintained on saline exhibited a body weight gain, nitrogen balance and organ proteins comparable to normal rats. Thyroidectomy retarded protein repletion primarily by reducing food intake although kidney mass was not improved. Alloxan diabetes retarded body weight gain but not nitrogen retention in a manner simulating hypophysectomy but serum protein synthesis revealed the endocrine deficiencies as separate entities. These data emphasize the importance of body protein reserves in the evaluation of an endocrine problem.

References

1. Wells, L. J., in "Comparative Endocrinology," A. Gorbman, Ed., Wiley, New York, 1959, p. 444.
2. Leathem, J. H., in "Mammalian Protein Metabolism," H. Munro and J. B. Allison, Eds., Academic Press, New York, 1963, p. 343.
3. Samaan, H., Yen, C. C., Gonzalez, D., and Pearson, O. H., *J. Clin. Endoc. Met.,* **28,** 485 (1968).
4. Enesco, M., and Puddy, D., *Am. J. Anat.,* **114,** 235 (1964).
5. MacConnachie, H. F., Enesco, M., and Leblond, C. P., *Am. J. Anat.,* **114,** 245 (1964).
6. Cheek, D. B., Powell, G. K., and Scott, R. E., *Bull. Johns Hopkins Hospt.,* **117,** 306 (1965).
7. Leathem, J. H., in "Protein Metabolism, Hormones and Growth," W. H. Cole, Ed., Rutgers U. Press, New Brunswick, N. J., 1953, p. 17.
8. Howard, E., *J. Neurochemistry,* **12,** 181 (1965).
9. Garren, L. D., Richardson, A. P., and Crocco, R. M., *J. Biol. Chem.,* **242,** 650 (1967).
10. Korner, A., *Mem. Soc. Endocrinology,* **15,** 9 (1967).
11. Leathem, J. H., *Rec. Prog. Hormone Res.,* **14,** 141 (1958).

12. Wagle, S. R., *Arch. Biochem. Biophys.*, **102,** 373 (1963).
13. Wool, I. G., *Am. J. Physiol.*, **198,** 357 (1960).
14. Leathem, J. H., in "Protein Metabolism," F. Gross, Ed., Springer-Verlag, Berlin, 1962, p. 202.
15. Randle, P. G., Garland, P. B., Halas, C. N., Newsholme, E. A., Denton, R. M., and Pogson, C. I., *Rec. Prog. Hormone Res.*, **22,** 1 (1966).
16. Scharff, R., and Wool, I. G., *Biochem. J.*, **99,** 173 (1966).
17. Rampersad, O. R., and Wool, I. G., *Science,* **149,** 1102 (1965).
18. Scow, R. O., *Am. J. Physiol.*, **169,** 859 (1959).
19. Tata, J. R., Proceedings of the Second International Congress of Endocrinology. *Excerpta Medica International Congress Series,* **85,** 46 (1964).
20. Fritz, I., *Endocrinology,* **58,** 484 (1956).
21. Riggs, L. J., and Leathem, J. H., *Proc. Penna. Acad. Sc.*, **40,** 14 (1967).

Endocrine and Nutritional Interrelationships

LEONARD J. LERNER
Squibb Institute for Medical Research
New Brunswick, New Jersey

Growth, repair and function of the body and its various components require adequate nutrition. Optimal growth and physiological activity require not only adequate quantities of the essential nutrients but also that the nutrients be qualitatively balanced. In times of stress or illness the qualitative and quantitative requirements for nutrients may be altered. Indeed, such bodily needs may change with the normal stages of growth and development, age, reproduction and under the influence of external factors such as drugs.

The endocrine system plays an active or permissive role in most physiological functions. Therefore, it is of interest to examine the influence of dietary protein and calorie intake on the endocrine system and on the effects of exogenous hormones on endocrine-responsive tissues. Interpretation of data obtained from studies that combine the effects of hormones and nutritional states, however, should be made with care; for example, alterations in dietary regimens may modify the growth and secretory potential of one tissue but not another. If the change in diet involves the reduction or elimination of an essential element, the body may conserve the supply of that element for one physiological function to the detriment of another function. Nelson and Evans [1] demonstrated that dietary protein in the form of casein at levels of 6–12% were required for full maintenance of pregnancy in the rat. Yet maintenance of maternal body weight and normal birth weight of the pups required a level of 18–24% casein in the diet. Another example of two systems competing for available nutrients can be shown by comparing the growth of the uterus under estrogen-stimulation in the normal animal and in an animal bearing a rapidly growing tumor. In our laboratory [2] ovariectomized mice were implanted with Sarcoma 180 and on days 9, 10, and 11 after such transplantation these animals were injected subcutaneously with either estradiol benzoate, estrone, or estriol at various dosages. On the twelfth day the animals were sacrificed and the weights of uteri were compared with those of tumor-free animals treated similarly. The results (Table I) demonstrate that the uteri were unable to respond maximally to estradiol benzoate treatment in the presence of a growing tumor; similar results were obtained with the use of the other two estrogens. This antagonism of uterine growth by the tumor was greatest when the higher doses of estrogens were employed and at a time when the nutritional demand for growth of this hormonally responsive tissue was greatest. The effect of the tumor on uterine response could not be attributed to decreased food consumption, because the presence of the neoplasm partially prevented the estrogen-induced uterine response even with equalization of the intakes of food (Table II).

An effect similar to that obtained in the above tumor system resulted when animals were deprived of dietary protein.[3] In this study immature mice were ovariectomized and then fed protein-free diets for 7 days partially to deplete body stores of protein. For the next 3 days half of

Table I. Effect of Sarcoma 180 Tumor Growth on the Uterotrophic Response to Estradiol Benzoate in Ovariectomized Mice

Total Dose * of Estrogen (μg)	Uterine Weight of Tumor Free Mice (mg)		Uterine Weight of Mice with 12-day Tumor Growth (mg)		*P*
–	10.6 ± 0.7 †	(10) ‡	11.0 ± 1.1	(29)	
0.01	12.2 ± 1.3	(8)	9.9 ± 1.5	(10)	
0.03	19.1 ± 1.5	(9)	16.8 ± 1.9	(9)	
0.09	45.8 ± 1.4	(9)	36.3 ± 2.8	(10)	0.02
0.27	69.3 ± 5.3	(10)	48.9 ± 3.8	(10)	0.005
0.81	64.2 ± 3.7	(10)	43.7 ± 3.2	(15)	0.001
2.40	53.6 ± 6.1	(10)	39.3 ± 3.9	(10)	0.05

* Total dose administered subcutaneously over 3-day (days 9, 10, and 11) period.

† Standard error of the mean.

‡ Number of animals shown in parentheses.

Table II. Effect of 12-day Sarcoma 180 Tumor on the Uterotrophic Response to Estradiol Benzoate in Pair-Fed Mice

Experimental Group		Estradiol Benzoate Total Dose (μg)	Δ Body Weight (g)	Tumor Weight (g)	Uterus Weight (mg)
Tumor	(8) *	–	1.4	1.99 ± 0.23 †	9.6 ± 0.4
Tumor	(15)	0.81 ‡	1.2	1.86 ± 0.15	40.6 ± 1.9 §
No tumor	(11)	0.81	1.0	–	60.5 ± 3.5

* Number of animals.
† Standard error of the mean.
‡ 0.27 mcg administered subcutaneously on days 9, 10, and 11.
§ $P < 0.001$ comparing uteri from tumor-bearing mice with those of tumor-free mice.

the groups of animals were placed on a 20% casein diet and the other groups were maintained on an isocaloric but protein-free regimen. The casein-fed groups were pair-fed to their respective control or estrogen-treated but protein-deprived groups. The low dose of estradiol benzoate increased uterine weight to a similar extent in both the protein-fed and protein-deprived animals (Table III). Increasing the dose of the estrogen, however, resulted in uterine responses that were significantly less in those animals that received no dietary protein. This effect of dietary protein was evident even if the uterine weights were calculated relative to body weight. There was also a greater quantity of estrogen-stimulated uterine luminal fluid in the mice consuming the casein diets. Again, reduction in uterine response with dietary protein deprivation was greatest when the nutritional requirements for uterine growth were increased through the administration of the larger doses of the estrogen.

Inadequate dietary regimens could alter hormonal requirements for physiological functions by affecting either the efficiency of utilization of nutrients or the metabolic rate. An adequate diet can reduce the metabolic efficiency of the liver, thereby lowering the rate of degradation of the hormone and prolonging its half-life. Prolonged undernutrition, however, leads to atrophy of the endocrine glands. It is difficult to determine why the reduction of dietary nutrients leads to the decline of a specific endocrine function, since adequate nutrients are required for all metabolic reactions. The synthesis of hormones, the integrity and

Table III. Effect of Dietary Protein on the Estrogen-Induced Uterotrophic Response in Ovariectomized Pair-Fed Mice Partially Protein Depleted Prior to Start of Treatment

Total Dose * Estradiol Benzoate (μg)	Uterine Weight of 20% Casein-Fed Mice (mg)	Uterine Weight of Protein Free-Fed Mice (mg)	Uterine Wt.-mg / Body Wt.-g of 20% Casein-Diet Mice	Uterine Wt.-mg / Body Wt.-g of Protein Free-Diet Mice
0.0	8.2 ± 0.5 †	7.9 ± 0.5	0.7 ± 0.04	0.8 ± 0.1
0.03	16.9 ± 1.0	15.6 ± 1.2	1.6 ± 0.1	1.7 ± 0.1
0.1	40.1 ± 2.1	31.4 ± 2.2 ‡	3.7 ± 0.1	3.1 ± 0.2 ‡
0.3	42.8 ± 1.6	27.0 ± 1.2 §	3.9 ± 0.1	3.0 ± 0.1 §

* Total dose administered subcutaneously over 3-day (days 8, 9, and 10) period.
† Standard error of the mean.
‡ $P < .05$.
§ $P < .01$.

responsiveness of the hormone-target tissue or cell, and the system for catabolism and elimination of the hormone are all dependent upon external sources of nutriments.

The rate of oxygen utilization is altered by the thyroid state in mice. Hyperthyroid animals placed in a closed vessel die of anoxia sooner than euthyroid animals.[4] Thyroidectomized mice resist anoxia and live longer.[5] The effects of the elimination of dietary protein and of thyroid hormones on anoxia in mice were studied by Lerner and Leathem.[6] A closed vessel technic[7] was employed. Prior to hormonal treatment, the animals were fed either a protein-free diet or a 20% casein diet for 14 days and the dietary regimen was continued after treatment up to the time of the anoxia test. The mice deprived of dietary protein survived 45–63% longer than those fed protein diets (Table IV). Administration of 10 μg L-triiodothyronine or 50 μg L-thyroxine to the animals on the protein-free regimen restored oxygen utilization, as measured by a decreased survival time at 24, 48, 72, and 96 hours post-injection, which approximated that of mice consuming 20% casein diets. These results seem to indicate that, in the absence of dietary protein, the thyroid gland secretes less hormone and a state paralleling that of thyroidectomy is obtained. Reisfield and Leathem,[7] using the goitrogenic compound propylthiouracil, were unable to alter the survival time in a closed vessel of mice on normal diets. This anti-thyroid drug, however, has a number of activities and may, therefore, cause effects that are dissimilar to those produced by removal of the thyroid gland.

Table IV. Survival Time of Young Adult Mice * in a Closed Vessel †

Diet (14 days prior to and during test period)	Subcutaneous Injection at Start of Test Period	Survival time in minutes at			
		24 hours	48 hours	72 hours	96 hours
20% casein	Water	151 ± 3.8 ‡ (16) §	152 ± 5.4 (12)	145 ± 5.7 (12)	153 ± 6.1 (11)
Protein free	Water	246 ± 13.8 (14)	221 ± 9.4 (10)	222 ± 6.8 (10)	223 ± 15.2 (10)
Protein free	L-triiodothyronine 10 μg	160 ± 6.4 (16)	154 ± 9.6 (16)	144 ± 6.2 (15)	182 ± 10.3 (16)
Protein free	L-thyroxine 50 μg	188 ± 15.8 (15)	162 ± 10.0 (15)	134 ± 5.5 (15)	134 ± 6.7 (15)

* Approximately 40–45 days old at time of anoxia test period.
† Volume of 940–950 ml.
‡ Standard error of the mean.
§ Number of animals.

The effect of the anti-thyroid drug thiouracil is similar to thyroidectomy in one important aspect only, namely, a reduced production of thyroid hormone. Removal of the thyroid gland causes a cessation of thyroid hormone influence on body physiology. Thiouracil decreases or eliminates thyroid hormone synthesis depending on the dose administered. One study in rats indicated that 0.3 μg of L-triiodothyronine daily would be required to counteract the goitrogenic effect of 0.1% thiouracil in the diet.[10]

Reduction of food intake by rats resulted in a lessening of the hypertrophy and hyperplasia of the thyroid gland induced by thiouracil treatment.[11] Underfeeding, however, did not reduce the response of the thyroid to thiouracil when this was expressed relative to the body weight of the rat.[12] Mighorst reported that the goitrogen methylthiouracil had a definite effect upon the thyroid gland of cockerels only if the protein content of the feed was sufficiently high.[13] Histologically, the thyroid gland activity of these birds fed the methylthiouracil was markedly increased when the diet contained 10% casein. The thyroid activity was further increased by elevating the protein content of the feed to 50% in spite of the body growth-inhibiting effect of so large an intake of casein.[14,15] In contrast to the results on the cockerel, Mighorst[15] found that high protein intake did not increase the response of the mouse thyroid to methylthiouracil.

Animals not subjected to goitrogen treatment also demonstrated a thyroid sensitivity to lowered nutrition. Restriction of food intake reduced thyroid weight and induced morphological evidence of inactivity.[16,17,18] Uptake of I^{131} by the thyroid gland was also reduced by underfeeding.[19] D'Angelo reported that acute starvation of the rat and mouse caused significant reduction in hypophyseal thyrotrophic (TSH) hormone content.[20] Srebnik and Nelson, however, noted only a slight decrease of the trophic hormone in the pituitaries of male rats deprived of dietary protein.[21]

A series of studies was conducted at the Bureau of Biological Research of Rutgers University to determine the influence of food intake, dietary protein, dietary protein content, and quality of dietary protein on the thyroid gland and on the pituitary-thyroid relationship during goitrogen administration.[8,9]

Immature male rats weighing 50–60 g, in treatment groups of 8–15 were fed a purified diet containing 0, 5, 10, or 20% casein with or without 0.1% thiouracil. Whenever part of the casein was eliminated from the diet, an equal weight of dextrose was added in order to form isocaloric feeding regimens. *Ad libitum* and paired feeding regimens were employed during the 14 day experimental periods. The results of these studies indicated that reduction of dietary protein, elimination of dietary protein, or limiting the intake of calories did not prevent the thyroid from responding to thiouracil administration (Table V). The thyroid weight increased in direct relationship to the levels of protein in the diet but the percent increases (200–300%) of these glands over the glands of rats fed similar levels of protein without added thiouracil were similar for all the groups. Moreover, since dietary protein or calorie restriction adversely affected body weight gain, analysis of the thyroid weight relative to body weight demonstrated that these elements of the feeding regimens had little specific action on the pituitary-thyroid axis. These data suggested that the pituitary was able to synthesize TSH and the thyroid gland was able to respond to stimulation under conditions of restricted caloric and protein intake or even in the absence of dietary protein.

Lerner and Leathem[8] and Lerner[9] demonstrated that hypertrophy and hyperplasia of the thyroid gland induced by thiouracil could be influenced by dietary protein and food consumption. It was of interest therefore, to study the effect of dietary protein deprivation upon the antigoitrogenic action of L-triiodothyronine.[10] Inhibition of thiouracil-induced thyroid hypertrophy was employed for the assay of L-triiodothyronine in immature male rats fed protein-free diets or diets con-

Table V. Effect of Thiouracil and Diet on Immature Male Rat Thyroid Weight

Treatment in Diet		Thyroid Weight	
Percent Casein	0.1% Thiouracil	mg	mg/100 g Body Weight
0	−	3.0 ± 0.2 *	7.6 ± 0.6
0	+	9.4 ± 0.4 §	22.4 ± 1.5 §
5	−	4.2 ± 0.1 §	7.6 ± 0.5
5	+	10.8 ± 0.1 §	21.8 ± 1.2 §
10	−	6.4 ± 0.3 §	9.4 ± 0.5 ‖
10	+	13.4 ± 1.6 §	20.3 ± 2.0 §
20	−	7.6 ± 0.5 §	7.6 ± 0.3
20	+	18.2 ± 1.3 §	18.2 ± 1.4 §
5	− †	3.6 ± 0.1	7.5 ± 0.3
5	+ ‡	10.7 ± 1.2 §	20.5 ± 2.0 §
10	− †	3.4 ± 0.2	6.5 ± 0.3
10	+ ‡	12.1 ± 1.0 §	20.7 ± 1.2 §
20	− †	3.9 ± 0.2 ‖	6.1 ± 0.4
20	+ ‡	12.4 ± 1.2 §	18.2 ± 1.5 §

* Standard error of the mean.
† Pair-fed with protein free diet group.
‡ Pair-fed with protein free + thiouracil group.
§ $P = <0.01$.
‖ $P = <0.05$.

taining 20% casein. The results of this study confirmed those of the earlier investigations [8,9] in that the actual thyroid weights were smaller in the animals deprived of dietary protein but, relative to body weights, these glands were similar in size to those of animals fed protein. This study also showed once again that, even in the absence of dietary protein for 14 days, the hypophysis could respond to the goitrogen by increasing synthesis and release of thyrotrophic hormone.

The endocrine system is more important to the animal's homeostasis than the skeletal system or other less vital tissues. Therefore, it is possible that in the absence of dietary protein there is a shift of body protein stores to uses of higher priority. The increased requirement for synthesis of TSH under conditions of anti-thyroid treatment may have induced such a call on body protein stores. In order to test this hypothesis, immature male rats were subjected to a 10-day period of feeding with diets devoid of protein prior to the treatment period. During the next

14 days the animals were fed diets containing 20% casein or no protein at all and half of these groups also received 0.1% thiouracil in their diets. The animals that were continued on the protein-free regimens lost weight; however, the animals repleted with dietary protein rapidly regained their weight (Table VI). These latter groups of animals

Table VI. Effect of Prior Protein Depletion on the Response of the Rat Thyroid to Protein Repletion and Thiouracil Treatment

14-Day Treatment in Diet			Thyroid Weight	
Percent Casein	0.1% Thiouracil	Δ Body Weight * (g)	mg	mg/100 g Body Weight
0	−	−20.2	3.2 ± 0.2 †	8.4 ± 0.4
0	+	−19.8	5.2 ± 0.3 ‡	13.3 ± 0.9 ‡
20	−	+58.8	7.9 ± 0.3 ‡	6.7 ± 0.2 ‡
20	+	+52.5	24.2 ± 1.3 ‡	21.9 ± 1.4 ‡

* Change in body weight over the 24-day study period.
† Standard error of the mean.
‡ $P = <0.01$.

responded to the goitrogen to a much greater extent than did the depleted animals. This 2-fold greater increase of thyroid weight under the thiouracil treatment when protein was supplied in the diet, was evident even when the data were calculated relative to the animals' body weight. It can be concluded from this study that dietary protein is important for maximal response of the hypophyseal-thyroid axis and that a period of protein depletion increases the sensitivity of the thyroid response during repletion with dietary protein.

Proteins differ in their nutritional value and in their utilization by different tissues and systems of the body. In studies of body growth or repletion of nitrogen stores of the body in rats, lactalbumin is of greater value than casein and casein is better than wheat glutin. In studies utilizing the experimental design outlined on the previous pages, protein depleted rats were repleted for 14 days with diets containing 5 or 20% of lactalbumin, casein, or wheat glutin with or without thiouracil incorporated into the diet. The diets containing the 5% level of protein did not allow the animals to grow enough to overcome the loss in body weight sustained during the initial 10 days when the rats were fed protein-free regimens. The animals fed the higher level of protein did gain enough during their repletion period to overcome the loss in body weight induced by protein deprivation. Thyroid response to thiouracil treatment in these latter groups of animals was greatest in the rats fed casein, lower in those fed lactalbumin, and lowest in those fed wheat glutin. It is interesting that the influence of the quality of protein was altered when only 5% of the diet was protein. In this case the responses of the thyroid to the goitrogen were greatest when the rats were fed lactalbumin, least when they were fed casein with wheat glutin diets occupying an intermediate position of influence on this response (Table VII). These differences in responses of the thyroid could relate to the differences in the amino acid composition of these proteins. Lactalbumin, casein and wheat glutin contain 6.6, 3.6, and 4.1% of sulfur amino acids respectively.[22] The relative response of the thyroid to thiouracil treatment (Table VII) appeared to be related to the sulfur

Table VII. Effect of Protein and Methionine on the Thyroid Weight Response to Thiouracil in Protein Depleted Immature Male Rats

14-Day Treatment in Diet				Thyroid Weight	
Protein *	0.1% Thiouracil	1% Methionine	Body Weight † (g)	mg	mg/100 g Body Weight
5% C	−	−	−10.2	3.5 ± 0.1 ‡	7.8 ± 0.3
5% C	+	−	−8.9	8.1 ± 0.6 §	18.1 ± 1.0 §
5% C	−	+	−1.8	3.7 ± 0.2	7.2 ± 0.3
5% C	+	+	−4.5	12.1 ± 0.8 §	24.9 ± 1.8 §
5% W	−	−	−14.7	3.1 ± 0.1	7.8 ± 0.2
5% W	+	−	−12.6	9.4 ± 0.5 §	22.0 ± 1.0 ‖
5% W	−	+	−11.7	3.7 ± 0.2	8.5 ± 0.6
5% W	+	+	−8.7	11.0 ± 0.7 §	25.6 ± 1.7 §
5% L	−	−	−6.8	4.2 ± 0.2	8.3 ± 0.5
5% L	+	−	−10.5	11.7 ± 0.4 §	25.7 ± 1.0 §
5% L	−	+	−1.9	4.7 ± 0.2 ‖	8.9 ± 0.4
5% L	+	+	−6.2	12.3 ± 0.5 §	25.5 ± 0.9 §

* C = casein; W = wheat glutin; L = lactalbumin.
† Change in body weight over the 24-day study period.
‡ Standard error of the mean.
§ $P = <0.01$.
‖ $P = <0.05$.

amino acid content of the dietary protein. This was interesting to explore further since it had been shown that small amounts of methionine added to diets containing low levels of casein increased the TSH content of rat pituitaries.[23] Addition of 1% methionine to a 5% casein diet fed rats during the protein repletion period improved the response of the thyroid to thiouracil by 50% (Table VII). Supplementation of 5% wheat glutin diets with methionine elicited only borderline enhancement of the response to the goitrogen and addition of the sulfur amino acid to the lactalbumin diets failed to alter the thyroid response. This order of response inversely follows that of the sulfur amino acid content of the several proteins. This points to the need for these amino acids as well as for a relative balance of amino acids in order to enable the hypophysis to form TSH, and the thyroid to respond optimally to thiouracil. The supplementation of the more adequate 20% protein diets with 1% methionine failed to influence the goitrogenic response to thiouracil. Lack of correlation between the results of methionine supplementation at the 5% and 20% dietary protein levels may be due to the ample supply of sulfur amino acids at the higher protein level. It could be also due to toxicity in consequence of a large amount of methionine inducing an imbalance of amino acids.

Studies of the type described above emphasize the importance of the relationship of nutrition to the functioning of the endocrine system. They also demonstrate the importance of the nutritional state and bodily stores in relation to the ability of the animal to respond to hormonal stimuli. This ability becomes critical during periods when maximal responses are required such as in times of stress. Rarely in the study of endocrine drugs or in the *in vivo* study of other pharmacological agents is consideration given to the nutrients required by the test animal in order to assure full reaction. This variable should always be considered since drugs or hormones are many times employed in clinical situations where the patient is either in a debilitated state or where the disease has decreased the metabolic efficiency of the body. Certainly, nutrition must be considered an important environmental factor that can influence the physiology of the body and alter the conditions of biological experimentation. The endocrine system and hormonal state also influence the nutritional requirements of the animal. It is therefore paramount in the evaluation of any nutritional study to consider the status of the endocrine systems. Hormones can greatly alter metabolism, induce protein loss or nitrogen retention, and produce nutritional imbalances. Biological research should never become so specialized that the investigator loses sight of all of the factors that can influence his study. The endocrinologist should recognize the influence of nutrition and nutritional state on his studies and the nutritionist should appreciate the effects of hormonal alteration on his investigations.

References

1. Nelson, M. M., and Evans, H. M., *J. Nutr.,* **51,** 71 (1953).
2. Hilf, R., Lerner, L. J., Lang, E., and Borman, A., *Cancer Res.,* **23,** 304 (1963).
3. Lerner, L. J., and Turkheimer, A. R., *Endocrinol.,* **76,** 539 (1965).
4. Smith, A. V., Emmens, C. W., and Parkes, A. S., *J. Endocrinol.,* **5,** 186 (1947).
5. Leblond, C. P., *Proc. Soc. Exper. Biol. and Med.,* **55,** 114 (1944).
6. Lerner, L. J., and Leathem, J. H., Unpublished data.
7. Reisfield, D. R., and Leathem, J. H., *Endocrinol.,* **46,** 122 (1950).
8. Lerner, L. J., and Leathem, J. H., *Fed. Proc.,* **13,** 86 (1954).
9. Lerner, L. J., Ph.D. Thesis, Rutgers University, New Brunswick, N. J., 1954.
10. Lerner, L. J., and Leathem, J. H., *Endocrinol.,* **73,** 652 (1963).
11. Gomez-Mont, F., Paschkis, K. E., and Cantarow, A., *Endocrinol.,* **40,** 225 (1947).
12. Meites, J., and Agrawala, I. P., *Endocrinol.,* **45,** 148 (1949).
13. Mighorst, J. C. A., *Acta Endocrinol.,* **8,** 97 (1951).
14. Mighorst, J. C. A., and van der Veen, H. E., *Acta Endocrinol.,* **9,** 395 (1952).
15. Mighorst, J. C. A., *Verslag, Landbouwk. Onderzoek.,* **59,** 1 (1953).
16. Meites, J., and Reed, J. O., *Proc. Soc. Exp. Biol. and Med.,* **70,** 513 (1949).
17. Mulinos, M. G., and Pomerantz, L., *J. Nutr.,* **19,** 493 (1940).
18. Stephens, E., *Endocrinol.,* **26,** 485 (1940).
19. Meites, J., and Wolterink, L. F., *Science,* **111,** 175 (1950).
20. D'Angelo, S. A., *Endocrinol.,* **48,** 341 (1951).
21. Srebnik, H. H., and Nelson, M. M., *Endocrinol.,* **70,** 723 (1962).
22. Allison, J. B., *Borden's Rev. Nutritional Research,* **14,** 61 (1953).
23. Aschkenasy, A., Benhamou, J. P., and Rolland, J. P., *Compt. Rend. Soc. Biol.,* **146,** 44 (1952).

III. Immunology

Immunoelectrophoretic Analysis of Cytoplasmic Protein Synthesis in the Liver *

CURTIS A. WILLIAMS
The Rockefeller University

Abstract

An important problem in the study of protein biosynthesis is the identification of newly finished products. With immunoelectrophoretic analysis (IEA), followed by autoradiography of the patterns, positive identification of many of the labeled products is possible, and in some cases, rough estimates of relative amounts produced are obtained. This investigation has centered on the synthesis of proteins by the liver of mice and rats. The objectives to date were: (1) to compare qualitatively the biosynthetic activity of tissue *in vitro* and cell-free incorporating systems with *in vivo* patterns of biosynthesis; and (2) to determine mechanisms of regulation by the study of induced abnormal states. The high sensitivity and resolving power of IEA, as employed in this study, have led to several observations which we consider important. Most significant of these are: (1) the activity of cell-free systems reflect transient *in vivo* states more faithfully than do tissue cultures; (2) proteins produced for export, *i.e.*, the plasma proteins, are produced only on membrane bound ribosomes and are subsequently bound or enclosed by the microsomes; (3) many soluble proteins produced for use by the cells are made only on free ribosomes.

* Modified from a manuscript read at the Analytical Chemistry Award Symposium, American Chemical Society, San Francisco, April 1, 1968.

Supported by grants from the National Science Foundation (GB 1383 and GB 5290) and U. S. Public Health Service (AI-07346).

Background

The main purpose of this paper is to demonstrate the effectiveness of an immunochemical approach to a challenging and exciting biological problem—the mechanisms of protein biosynthesis. It will be difficult, however, to avoid discussing some details of the problem and some of the biological implications of our findings so far. Nevertheless, I will emphasize rather the significance of the investigation to methodology, citing only direct influences and collaborations, and innovations of technique. Nearly all of the material has been published or is in press, and there we have focused more on the biological aspects of the work.

In the study of protein biosynthesis only in recent years have most investigators been concerned with the specific identity of newly finished products. In complex systems, convenient and defining chemical isolations are applicable to very few proteins. Faced with this difficulty, a few biochemists turned to immunochemical methods—precipitation by specific antibody the most favored.[1,2] This is still the best for quantitative separation of proteins from mixture if specific antibody is available and if great care is taken to avoid nonspecific coprecipitation.

Immunoelectrophoretic analysis (IEA), while it does not provide reliable quantitative data as customarily applied, it does give critical qualitative information. In IEA, an antigen mixture is placed in a well in a layer of agar and is resolved by zone electrophoresis in a manner similar to that described by Gordon and co-workers in 1950.[3] After separation, an antiserum is placed in a trough, lateral to the migration route, from which antibodies diffuse into the gel, encountering the corresponding antigen diffusing from its characteristic electrophoretic position. The intersecting diffusion fronts produce an arc of precipitate for each antigen-antibody reaction independently of the others. The method which Grabar and I developed in 1953,[4] made use of 13 × 18 cm gels, but the more economical method adapted by Scheidegger[5] to microscope slides was used for the studies described here.

In 1961, with Courtney Wemyss, we observed very marked qualitative and quantitative changes in mouse plasma protein patterns 4 days following staphylococcal infection.[6] It was expected from many references in the literature (for reviews, see references 7 and 8) that we should find elevations notably in the alpha and beta and eventually in the gamma region. Densitometric scans of the simple gel electrophoresis patterns shown in Figure 1 demonstrate the large quantitative changes observed. "Normal" mice in this context means young (6–8 weeks of

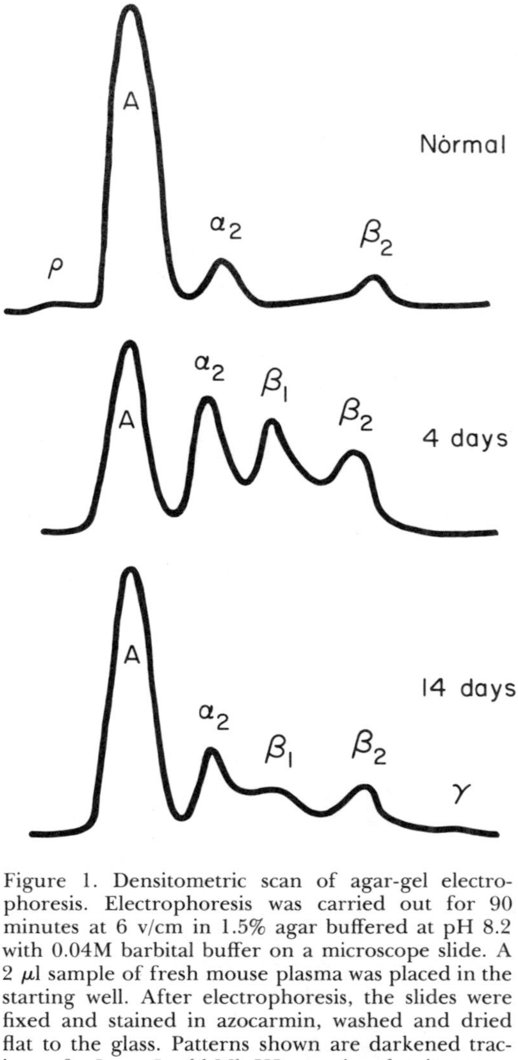

Figure 1. Densitometric scan of agar-gel electrophoresis. Electrophoresis was carried out for 90 minutes at 6 v/cm in 1.5% agar buffered at pH 8.2 with 0.04M barbital buffer on a microscope slide. A 2 μl sample of fresh mouse plasma was placed in the starting well. After electrophoresis, the slides were fixed and stained in azocarmin, washed and dried flat to the glass. Patterns shown are darkened tracings of a Joyce-Loebl Mk III scanning densitometer. Times refer to duration of an experimental staphylococcal infection.

age) and inexperienced with the usual infectious stresses most animals encounter early in life. Actually the pattern after recovery from infection at 14 days has a more familiar "normal" appearance.

About this time, Hochwald, Thorbecke, and Asofsky[9,10] were incubating various human and monkey tissues in a medium containing radioactive C^{14} amino acids in order to determine the sites of synthesis of the various plasma proteins. They had employed immunoelectrophoresis to analyze the protein secreted into the culture medium. After the slides were washed and dried, they obtained autoradiograms of the patterns. This technique, first described by Rejnek and Bednarik,"[11] seemed ideal for the study of the source of the various proteins Wemyss and I had seen so profoundly affected by the staphylococcal infections. Accordingly, with Asofsky and Thorbecke, we set about to see what mouse liver slices, spleen minces, and lymph nodes from infected mice produced in culture.[12] The interesting question, of course, was whether these bits of tissues, removed from the animal, would continue to reflect altered biosynthetic activity.

Results obtained with liver slices indicated greatly increased radioactivity in the alpha and beta globulins after three days of infection. This level of incorporation subsided markedly after 11 days when the animals were near recovery. Similarly, cultured spleen minces labeled immunoglobulins—the four-day infected spleens in a manner indistinguishable from the control spleens. Spleens from ten-day infected mice, however, incorporated large amounts of C^{14} amino acids into immunoglobulin G and β_1-C globulin (a subcomponent of complement). The tissues in culture, therefore, appear to be making plasma proteins of defined characteristics in the relative amounts which would account for the corresponding density scans shown in Figure 1.

At this point it was decided to follow two directions at once: 1) to obtain autoradiographic records corresponding to quantitative data directly from the mouse; and 2) to reduce the phenomenon one step further to a cell-free incorporation system.

Amino Acid Incorporation Studies in Mice

We reasoned that at some time after the start of the infectious stress, we should be able to detect a sharp decrease in the relative amount of labeled amino acids incorporated into albumin and an equal or greater increase of incorporation into globulins. Figure 2 demonstrates this to occur after about 24 hours with the rates of incorporation approaching the nadir in the case of the albumin and the peak in the case of the globulins after about 4 days.[13] Initially, however, there is a sharp increase in the rate of albumin synthesis.

It should be mentioned at this point that, while the selection of parameters and the handling of data are not rigorous, we believe these rela-

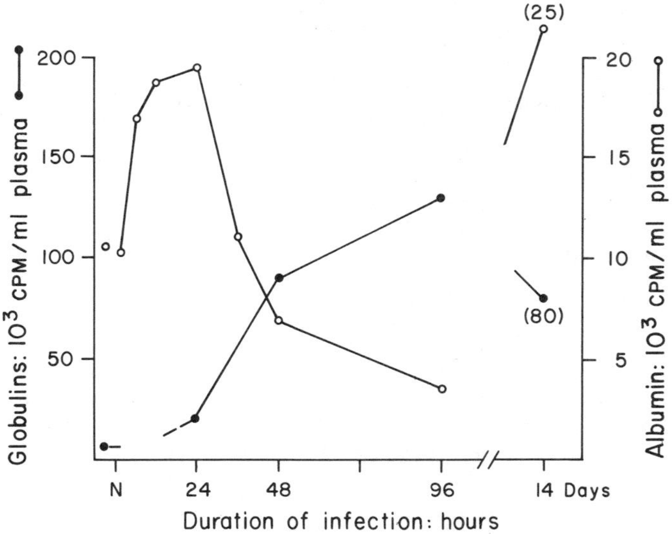

Figure 2. Incorporation of amino acids into plasma proteins during one hour. Mice were infected intravenously with 2.7×10^7 viable units of *Staphylococcus aureus* (Giogio). After 2, 6, 12, 24, 36, 48, 96 hours, and 14 days, 2 microcuries of a C^{14} amino acid mixture was injected into the tail veins of 3 mice. One hour later, the mice were exsanguinated. Total incorporation into plasma protein was determined on 25μl samples. Albumin (o——o, right-hand scale) was extracted from cold TCA precipitates with absolute ethanol. The TCA-ethanol precipitate residues were considered as globulins (●——●, left-hand scale). Radioactivity was determined by liquid scintillation at 55% efficiency. *N* stands for noninfected control.

tionships reflect real fluctuations in rates of biosynthesis of different plasma proteins. Accessible amino acid pools in the plasma and liver fluctuate, surely, but probably no more than in prolonged fasting (see references 14, 15). The acutely infected animals do not eat much and they lose weight at about the same rate as starved mice.

The effect of small fluctuations in specific amino acid pools on apparent rates of protein synthesis, should be largely overcome by the use of mixture of radioactive amino acids for incorporation studies. We have routinely employed a mixture of 13 purified C^{14}-labeled precursors ("Reconstituted protein hydrolysate"—Schwartz Bioresearch, Inc.).

Table I compares results of *in vivo* incorporation in several abnormal states. The data are expressed in terms of percent of injected label incorporated. Note particularly that the "live" infection is required for the massive globulin response, but that killed staphylococci will depress

Table I. Percent of C^{14} Amino Acid Pulse Incorporated into Mouse Plasma Proteins in One Hour

Treatment of Mice	Albumin	Globulins
None	.43	.33
Staphylococcal infection (4 days)	.15	5.42
Killed staphylococcus (4 days)	.18	.33
Fasted (4 days)	.53	.47
Protein-free diet (6 days)	.10	.15

Incorporation of C^{14} amino acids administered intravenously in a single pulse. Mice were bled from the infraocular venous plexus after 60 minutes. Incorporation is expressed as percent of administered dose. Albumin was extracted from ice-cold TCA precipitates by absolute ethanol. Infected mice received 2.7×10^7 viable units of bacterial culture. 10^8 killed units were given to group 3.

albumin synthesis. Fasting has no effect, but a protein-free diet markedly curtails apparent plasma protein synthesis.

The autoradiographic patterns of *in vivo* labeled proteins in the plasma are not very helpful but they confirm earlier studies and identify components into which isotope is most readily incorporated. However, when we prepare the liver microsomes after the administration of C^{14} amino acids, we can see what has just come off the assembly line, so to speak. The patterns of protein released by sonication of the microsomes are shown in Figure 3. The IEA pattern is from a mouse infected for 48 hours. The autoradiogram of the normal pattern shows strong radioactivity in the albumin, fibrinogen, and transferrin and weaker activity in the alpha-1 and alpha-2 region. After 24 hours and 48 hours of infection the patterns indicate that incorporation into albumin had been greatly reduced, transferrin remains essentially the same, fibrinogen synthesis appears progressively increased, and a beta-1 component we identify as hemopexin (a hemin binding protein) appears with a structurally related low mobility component. The structural relationship is indicated by the conjoining of the two arcs of precipitate.

Cell Free Systems of Mouse Liver

Campbell and his colleagues had shown that cell-free liver systems would incorporate C^{14} amino acids into serum albumin,[16] and von der Decken used autoradiography of IEA patterns for positive identifica-

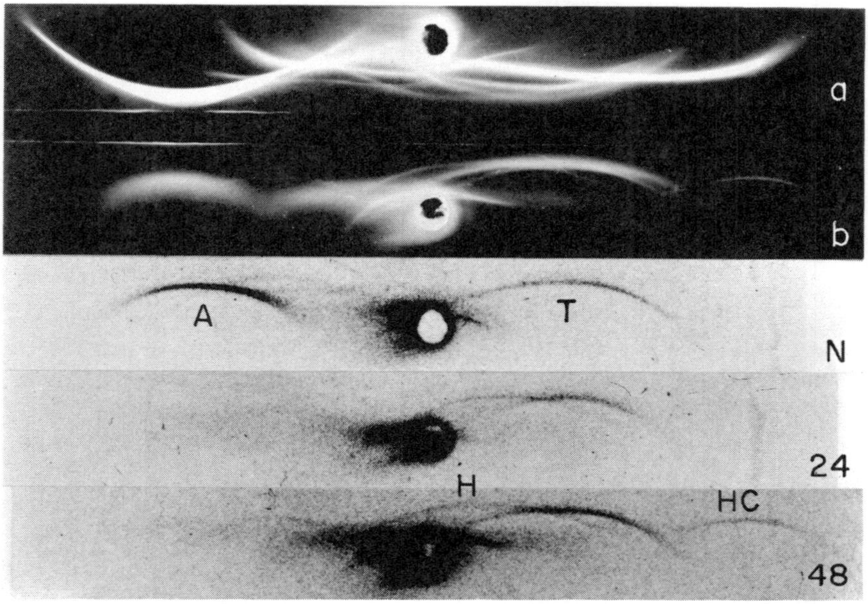

Figure 3. C^{14}-labeled plasma proteins released from mouse liver microsomes by sonication. Mice were infected intravenously with staphylococci and after 24 hours and 48 hours were injected with C^{14}-labeled amino acid (see Figure 2). One hour later, mice were exsanguinated and the livers were homogenized. The microsomal fractions were prepared by ultracentrifugation,[18] resuspended in saline and disrupted by sonication. The suspension was cleared by ultracentrifugation and the released supernatant proteins were concentrated by precipitation with ammonium sulfate at 75% saturation. The immunoelectrophoretic pattern of such a preparation from 48-hour infected mice, *b*, is compared with the whole plasma pattern, *a*, both developed by rabbit antiserum against whole mouse plasma proteins. The lower three patterns are autoradiograms of slides obtained with proteins from normal, *N*, 24-hour infected, *24*, and 48-hours infected mice, *48*. The bottom panel is the autoradiogram of pattern *b* in the second panel. *A*, albumin; *T*, transferrin; *H*, hemopexin; *HC*, hemopexin crossreacting component.

tion.[17] With Ganoza and Lipmann we demonstrated that many plasma proteins were labeled in a rat liver cell-free system.[18] We decided to study the staphylococcal infection phenomena in mouse liver incorporation systems.[19] Liver microsomes and soluble (105,000 × *g*) extracts from infected and control animals were prepared. These were recombined and cross-combined in the incorporation mixture. After incubation at 37° for 50 minutes, microsomes were collected and disrupted by sonication. Released protein was dialyzed, lyophilized, and reconstituted in a minimal volume for analysis. It can be seen in Figure 4a that the normal microsomal system produces only albumin and transferrin in easily detectable amounts. When microsomes from infected mice are combined with soluble extract of normal liver, the labeling pattern shown in Figure 4b is obtained. There is striking similarity of these autoradiograms to those of proteins obtained from liver slice culture or from liver microsomes after *in vivo* incorporation, even down to details such as marked reduction of albumin synthesis and the production of the slow component related to hemopexin.

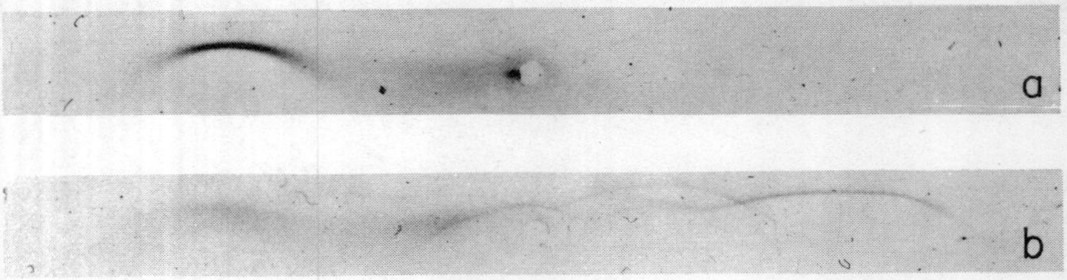

Figure 4. Autoradiograms of IEA patterns of mouse plasma proteins made by mouse liver cell-free systems. Microsomal incorporation systems were prepared as described in reference 19 from livers of normal and 4-day staph-infected mice. Proteins recovered from the microsomes following the incorporation period were dialyzed and concentrated by lyophilization. IEA patterns were developed by rabbit antiserum against whole mouse plasma. *a*, autoradiogram of normal pattern. *b*, autoradiogram of 4-day infection pattern. Compare *a* with *N* and *b* with *48* on Figure 3.

Interestingly, supernatant protein preparations revealed only faint traces of labeled albumin and transferrin, probably resulting from slight disruption of microsomes during incubation and centrifugation.

We conclude from these experiments that transitory states of plasma protein synthesis in the intact animal can be investigated with considerable confidence in organ culture and in cell-free systems. There was more perfect correspondence between the patterns of incorporation into plasma proteins by cell-free systems and the intact mouse than there was between either of these and the tissue culture results. This difference was mainly seen in the amount of label incorporated into serum albumin, probably because this protein is the only one whose production radically curtailed early in infection. Infected liver slice culture patterns show strong labeling of albumin, contrary to results obtained by other methods. This may be because the intact liver cells of the slice culture once removed from the infected animal, partially "recover" from albuminism "suppression" effect during the prolonged incubation period.

Rat Liver Systems

We concluded also from the experiments above that serum proteins are made on the microsomes, where they are held until secreted. This makes sense for proteins made by the parenchymal cells for export.

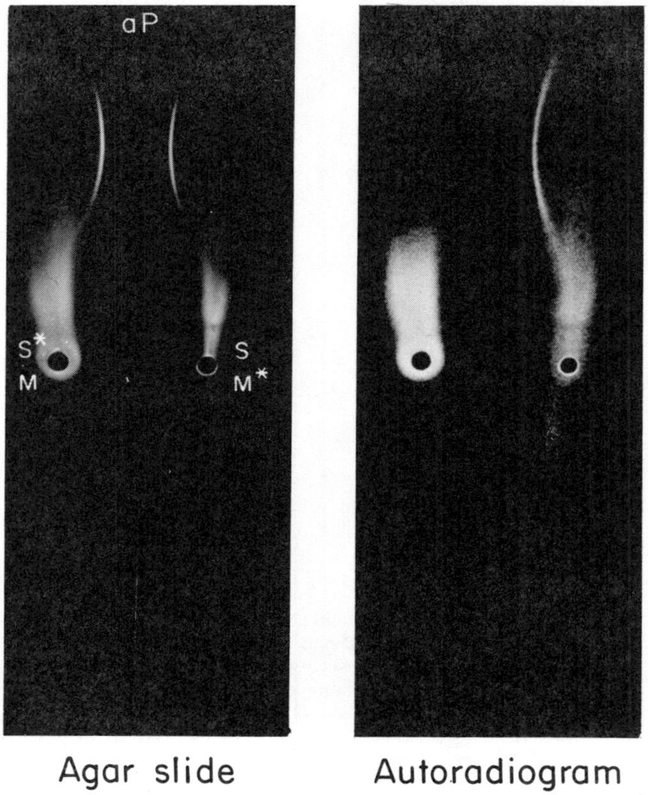

Figure 5. IEA patterns (left) and autoradiograms of mixtures of the proteins of the supernate, *S*, and the proteins obtained from sonicated microsomes, *M*, after cell-free incorporation. *S** and *M** are fractions from cell-free systems made with C^{14} amino acids; *S* and *M* are corresponding nonradioactive fractions. Patterns are developed with anti-mouse plasma, *aP*. Experiment shows that newly made plasma proteins are not released to the supernate during cell-free biosynthesis. The slight trace of label associated with the albumin arc of the *S*M* pattern could be due to some microsomal disruption during fractionation, or it could be labeled material from *S** bound nonspecifically to cold albumin from the *M* fraction. The latter explanation is thought unlikely since no binding could be demonstrated to a cold human albumin precipitate band.

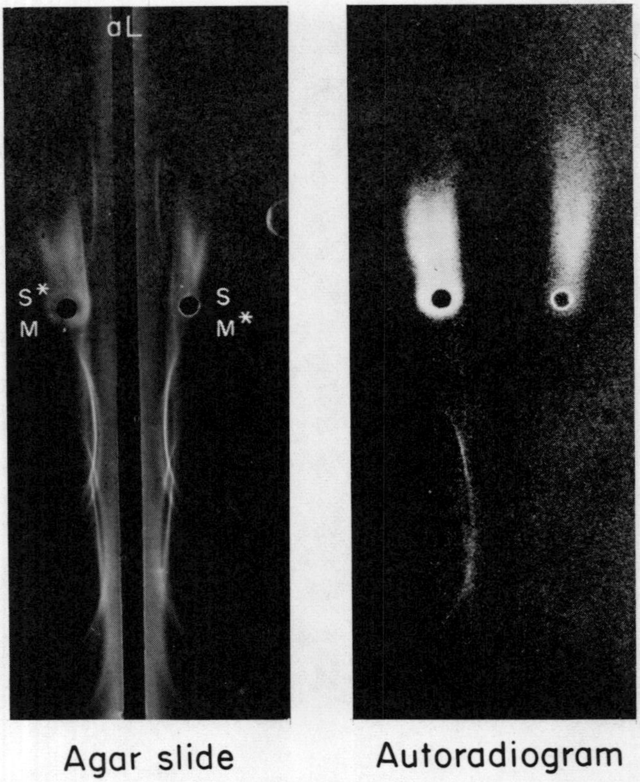

Figure 6. IEA patterns and autoradiograms of mixtures described under Figure 5. Patterns were developed with rabbit antisera against soluble mouse liver proteins, *aL*. This experiment shows that no identifiable liver proteins are bound by microsomes after their biosynthesis or that they are not released by sonication of the microsomes as are the plasma proteins. There are labeled liver proteins in the *S** fraction, however, apparently released spontaneously to the fluid phase.

But what of the proteins that work at home in the liver? To answer this question, we prepared antisera in rabbits to soluble extracts of rat liver, some of which detected as many as 20 distinct antigenic components.

Cell-free incorporation and concentration of products was performed as with mouse liver. Both radioactive and nonradioactive systems were prepared. For immunoelectrophoresis labeled supernatant proteins were mixed with unlabeled microsomal proteins and, conversely, labeled microsomal proteins were mixed with supernatant proteins from the incubation with cold amino acids. Appropriate carriers, rat plasma or liver extract, were added for development of patterns with anti-rat plasma or anti-rat liver.

Figure 5 shows that labeled plasma proteins are obtained in significant amounts only from sonicated microsomes (*M**). The supernate (*S**) contributes practically nothing which can be identified as a plasma protein. It is important to note that these experiments serve as their own controls for nonspecific coprecipitation of labeled material.

Figure 6 is the pattern developed by the antiserum against liver antigens. Here it is seen that the labeled liver proteins are detected only in the $105{,}000 \times g$ supernate of the cell-free system.

From this experiment we can conclude that the soluble protein made for use in the liver cell is made in some compartment other than the microsomes, or that they are spontaneously released rather than bound to microsomes.[20] Since we favored the former explanation, our recent experiments with Ganoza have been directed to determining the role of the "free" ribosomes. At this point, we can say conclusively, that at least some of the soluble liver proteins are made on the free, unbound ribosomes in the liver. Conversely, membrane bound ribosomes do not make the soluble liver proteins we detect with our multispecific antisera.[20]

The quantitative analyses of materials used for these experiments support all of the conclusions drawn above. The purpose of this presentation, however, was not merely to summarize results of recent research from this laboratory, but rather to demonstrate the power of a qualitative immunochemical tool to provide the significant framework. The ability to discriminate between several native proteins and to distinguish groups of proteins with different origins and destinations is of critical importance in studies of protein biosynthesis.

References

1. Cohn, M., and Torriani, A. M., *Compt. Rend.,* **232,** 115 (1951).
2. Campbell, P. N., and Stone, N. E., *Biochem. J.,* **66,** 19 (1957).
3. Gordon, A. H., Keil, B., Sebesta, K., Kuessl, O., and Sorin, F., *Coll. Czech. Chem. Comm.,* **XV,** 1 (1950).
4. Grabar, P., and Williams, C. A., *Biochim. Biophys. Acta,* **10,** 193 (1953).
5. Scheidegger, J. J., *Intern. Arch. Allergy,* **7,** 103 (1955).
6. Williams, C. A., and Wemyss, C. T., *J. Exp. Med.,* **114,** 311 (1961).
7. Peterman, M. L., in "The Plasma Proteins," F. W. Putnam, Ed., Academic Press, New York, 1960, Vol. 2, p. 309.
8. Owen, J. A., *Adv. Clin. Chem.,* **9,** 1 (1967).
9. Hochwald, G. M., Thorbecke, G. J., and Asofsky, R., *J. Exptl. Med.,* **114,** 459 (1961).

10. Asofsky, R., and Thorbecke, G. J., *J. Exptl. Med.,* **114,** 471 (1961).
11. Rejnek, J., and Bednarik, T., *Clin. Chim. Acta,* **5,** 250 (1960).
12. Williams, C. A., Thorbecke, G. J., and Asofsky, R., *J. Exptl. Med.,* **118,** 315 (1963).
13. Williams, C. A., *Fed. Proc.,* **24 (2),** 506 (1965).
14. Henderson, L. M., Schurr, P. E., and Elvehjem, C. A., *J. Biol. Chem.,* **177,** 815 (1949).
15. Steele, B. F., Reynolds, M. S., and Baumann, C. A., *Arch. Biochem.,* **25,** 124 (1950).
16. Campbell, P. N., Greengard, O., and Kernot, B. A., *Biochem. J.,* **74,** 107 (1960).
17. Decken, von der, A., *Biochem. J.,* **88,** 385 (1963).
18. Ganoza, M. C., Williams, C. A., and Lipmann, F., *Proc. Natl. Acad. Sci.* (*U. S.*), **53,** 619 (1965).
19. Williams, C. A., Ganoza, M. C., and Lipmann, F., *Proc. Natl. Acad. Sci.* (*U. S.*), **53,** 622 (1965).
20. Ganoza, M. C., and Williams, C. A., *Proc. Natl. Acad. Sci.* (*U. S.*)

A Review of the Nature and Significance of Hemagglutinins of Selected Invertebrates *

ELIAS COHEN
Roswell Park Memorial Institute
Buffalo, New York

Invertebrate natural antibody-like agglutinins to cellular components of other species have been reviewed extensively by Bisset,[1] Huff,[2] Cushing and Campbell.[3] Serologic response among invertebrates, other than insects, has been reviewed by Bang[4] and by Cushing.[5] The immune response in the invertebrates has not been characterized by specific antibodies, as in the vertebrates according to observations of a number of workers.[2,6–9] The hemocyanin of invertebrate hemolymph tested by Tyler and Metz[10] did not demonstrate agglutination activity for sheep erythrocytes and spermatozoa of other maring species. It was also shown by Cohen *et al.*[11] and Marchalonis and Edelman[12] that hemocyanin is not the hemagglutinating component of *Limulus* hemolymph.

No attempt will be made in this paper to comprehensively review hemagglutinins described for all invertebrate species. Selected observations in my laboratory on hemagglutinins of *Limulus polyphemus* (Horseshoe Crab) and *Birgus latro* (Coconut Crab) will be reported. These observations will be discussed in relation to the work of other investigators, with *Limulus* and other invertebrate hemagglutinins.

Limulus polyphemus, a marine arthropod, sometimes classified as a

* Supported in part by Grant HE-07728, National Heart Institute; and, in part, by the Atomic Energy Commission, Eniwetok Marine Biological Laboratory, Eniwetok Atoll, Marshall Islands and the University of Hawaii.

member of the Arachnida is often called a "living" fossil. Evidence of this animal's existence dates back to the Triassic period of 200 million years ago. It is indigenous to the Atlantic seacoast, as well as East coast of Africa and Asia, the Malayan Archipelago and Indian Ocean.

Birgus latro is a large hermit crab indigenous to many islands of the Pacific Ocean. It has adapted to living on land without employing gastropod shells as shelter, except when in its juvenile stages. *Limulus* and *Birgus* were selected because of large size and concomitantly large volumes of hemolymph.

Noguchi [12] gave the earliest description of hemagglutinins of the serum of *Limulus*. Erythrocytes of amphibians and reptiles were differentiated, without hemolysis. Cohen *et al.*[11] and Marchalonis [13] independently corroborated Noguchi's major observations and reported agglutination of erythrocytes of a variety of vertebrate species. Cohen [14] reported that the order of most avid hemagglutination was of mouse, human, guinea pig, rat, rabbit, and sheep erythrocytes by *Limulus*. The hemagglutination pattern for *Birgus* species was similar for mouse, human, and sheep erythrocytes.

Cohen [14] observed that *Limulus* serum could differentiate between human red cells and those of rhesus and stump-tail monkeys. In contrast, alligator erythrocytes were not agglutinated by every lot of *Limulus* serum. Marchalonis and Edelman [15] found that horse erythrocytes gave higher titers than rabbit, chicken, or sheep erythrocytes when agglutinated by *Limulus* serum. It is of interest that McDade and Tripp [16] observed that *Crassotrea* (oyster) hemolymph gave the highest titers with horse erythrocytes, with mouse and sheep next, followed by rabbit and human group A, B or O cells and least with monkey erythrocytes. Li and Fleming [17] found that rainbow trout red cells gave higher titers than rabbit or human erythrocytes.

Common receptor sites for the attachment of *Limulus* agglutinin are evident by the variety of species whose cells were tested. However, alligator red cells can absorb agglutinins, but may not be agglutinated. Cohen [14] demonstrated that protease treatment of alligator erythrocytes that were not clumped, would make them agglutinable to the same titer and score as human cells. Yet, the enzyme-treated cells did not absorb any more agglutinin from *Limulus* serum than did untreated cells. The interpretation was made that receptor sites on alligator erythrocytes, could be reached by agglutinin molecules, whether the cells are protease-treated, or not. The enzyme altered the erythrocyte surface so that cell-agglutinin-cell lattice formation is possible with subsequent agglutination of cells. This mechanism of action was also proposed for agglutination of bacterial lipopolysaccharide coated alligator red cells by homologous antibodies for the lipopolysaccharide that was used, by Neter, *et al.*[18]

Some relationship of red cell receptor sites to saccharides of the ABO blood group substances was suggested by the inhibition of human red cells agglutination by n-acetyl-d-glucosamine (NADG), with *Limulus* agglutinins and with *Crassotrea* agglutinins. Cohen [14] reported that neuraminidase-treated human erythrocytes gave minimal to no agglutination, with *Limulus* serum. Furthermore, supernatants of neuraminidase-treated cells inhibited agglutination. Our laboratory and Miss Marjory Stroup of Ortho Foundation found more specificity for A and B human blood group antigens than for O(H) antigens. Cations of lithium, potassium, and sodium decreased *Limulus* and human (Anti-A and Anti-B) titers, but calcium and magnesium ions increased avidity of human and *Limulus* agglutinins. Yokoyama and Finlayson [19] also observed that hemagglutination was influenced by the qualitative and quantitative nature of cations. Marchalonis and Edelman [15] observed that an addition of $0.1M$ calcium chloride to the hemagglutination system increased, but that additional $0.1M$ magnesium chloride decreased the titer of *Limulus* serum for agglutination of horse erythrocytes. Cohen [14] did not work with horse erythrocytes, but used cation addition of a broad range of molarities. *Crassotrea* (oyster) hemagglutinin was reported by McDade and Tripp [16] to require calcium to agglutinate human red cells, but calcium was not required to agglutinate cells of nonhuman species. Also, in absence of calcium, *Crassotrea* (oyster) hemagglutinin is heat-labile and can be degraded to protein sub-units without agglutinin activity. It may be expected that species differences of the test erythrocytes, as well as invertebrate agglutinins may influence the nature of the agglutination. Li and Fleming [17] fractionated oyster hemagglutinin into seven protein fractions, but only the salt-soluble and a minor water-soluble fraction possessed hemagglutination activity.

Shuster [31] who studied the systematic serology of *Limulidae,* reported at least four arcs in the agar-gel diffusion precipitin reaction between rabbit anti-*Limulus* serum and homologous antigen. Some noncoalescing bands were found in double-diffusion comparisons of serum of *Limulus* with sera of two Indo-Pacific species. Cohen *et al.*[11] reported that the *Limulus* serum contained 4 components, by ultracentrifugal analysis with sedimentation rates of 14, 51, 70, and 81, respectively. Eriksson-Quensel and Svedberg [20] gave the earliest ultracentrifugal pattern of

Limulus, with four components with sedimentation rates of 6, 16, 34, and 57.

Limulus agglutinins were eluted by 56°C saline washes of erythrocyte agglutinates by Cohen [14] and were analyzed by immunoelectrophoresis with rabbit anti-*Limulus* serum. The eluate protein gave one arc with the mobility properties of an alpha or beta human globulin. Marchalonis and Edelman [15] isolated the natural hemagglutinin of *Limulus* by ultracentrifugation and gel filtration in order to compare it structurally with vertebrate immunoglobulins. They describe it as a protein of approximately 400,000 molecular weight, with sedimentation characteristics of S°20, W = 13.5S. Electrophoretically it migrates as a beta globulin. The molecule is composed of subunits, each of 23,000 molecular weight, with linkage by monovalent interactions. Starch gel electrophoresis, amino acid analysis, 2-dimensional high voltage electrophoresis of tryptic peptides, and immunologic analyses are made of the purified hemagglutinin. Its structure was found to be different from that of mammalian immunoglobulins. However, until comparisons are completed of amino acid sequences, the relationship between *Limulus* hemagglutinin and vertebrate immunoglobulin can not be completely excluded. On the basis of their work, Marchalonis and Edelman [15] conclude that *Limulus* hemagglutinin and mammalian immunoglobulin are not related evolutionary developments. Ultracentrifugal analyses of 2-mercaptoethanol treated *Limulus* serum showed an increase in the concentration of components slower than 575, accompanied by a decrease in hemagglutinin activity.[14] However, treatment of isolated *Limulus* hemagglutinin by Marchalonis and Edelman,[15] with reducing agents 2-mercaptoethanol or dithiothreitol at pH 8, did not reduce hemagglutination titer or sedimentation coefficient of the preparation.

Tripp considered the hemagglutinin in oyster blood as similar to that of the agglutinin of *Limulus,* as reported by Cohen *et al.*[11] because both agglutinate erythrocytes of several vertebrate species, due to common antigens; both are stable at 56°C, but are thermolabile at 65°C; both have electrophoretic mobility comparable to mammalian vertebrate alpha or beta globulins. However, the oyster hemolymph contains no hemocyanin respiratory pigments as does *Limulus.* Marchalonis and Edelman,[15] as well as McDade and Tripp,[16] demonstrated inhibition of invertebrate hemagglutinin of activity by appropriate rabbit antisera. Normal rabbit serum or saline did not inhibit hemagglutination. Cohen [14] did not find any cross reaction in agar gel diffusion tests between *Birgus* and *Limulus* components. Simultaneous cellulose acetate electrophoresis of *Limulus, Birgus,* and human serum revealed a faster mobile component than human albumin in *Birgus* serum, whereas *Limulus* serum contained components of slower mobility than human albumin.

The first report on *Birgus* hemagglutinins [14] suggests a marked similarity to those of *Limulus.* However, the opportunity to collect young and older specimens of *Birgus,* at Eniwetok Atoll revealed no hemagglutination activity in sera of individuals 400 g or less in weight. Of 18 individuals collected in 1966, agglutinin activity was directly proportional to weight of animals. Of 15 individuals collected in 1967, a direct relationship could be shown between hemagglutination score and carapace width. No hemagglutinin could be obtained with the sera of seven of the eight smallest individuals collected. Since *Limulus* has been selectively collected to favor large adults, no data is available on agglutinin titers of juvenile specimens.[21]

Cohen [14] speculated that the agglutinins of *Limulus* or *Birgus* may give some selective advantage to the animals during their evolution. However, it was also considered possible that the hemagglutinating proteins are part of a saccharide transport or storage mechanism, part of the shell formation of those invertebrates. Data on variations in serum protein and hemagglutinin levels during moulting of *Limulus* and *Birgus* are not available. Nevertheless, it may be significant to note that Zuckerkandl [22] reported that hemocyanin virtually disappeared in *Maia squinado* (spider crab) during the moulting cycle. The biochemistry, paralleled by immunochemical study, of the protein constituents of the moult cycle is a prime target for future study, possibly with young stages of *Limulus* and *Birgus.*

McDade and Tripp [16] also speculate on the significance of oyster hemagglutinin as a mobilizer of saccharides for carbohydrate metabolism, since the oyster is herbivorous. The importance of calcium for the stability of the hemagglutinin molecule, also, may suggest its role in transportation of calcium ions for shell formation. Tripp [8] found that oyster serum could opsonize foreign material, suggesting a selective advantage, as part of a defense mechanism.

Collaborative work with Dr. A. Watne,[23] J. A. Migailo, and V. E. Lyon [24] has indicated that *Limulus* agglutinins could be used to harvest leukocytes and/or tumor cells from blood of patients with cancer, without lysing red cells. Unfortunately, the use of whole *Limulus* serum additive is not as effective as desired. The purified hemagglutinin of Marchalonis and Edelman [15] may be worthy of application to that problem.

At this time, one can only speculate as to how *Limulus* or *Birgus,* or other invertebrate hemagglutinins, are involved in the phylogeny of the immune response. Overall, the hemagglutinins of *Limulus, Birgus,* and *Crassostrea* are serologically similar to popular phytohemagglutinins, *i.e.,* avid agglutination, without lysis. Biomedical applications of natural cellular agglutinins may well be part of our "harvest from the sea." Certainly, the invertebrate hemagglutinins should be investigated, as comprehensively as phytohemagglutinins.

Hemagglutinins of expressed fluids of invertebrates have been studied by other workers. Examples are the agglutinin of human group "weak" A erythrocytes obtained from a clam (*Asxidemis gigantus*) by Johnson.[25] Boyd *et al.,*[26] also Chattaraj *et al.,*[27] reported strong agglutinins in fluid expressed from the snail *Otala lactes,* that could distinguish human group A_1 and A_2 from A_3 and A_x erythrocytes. Recently, Broun *et al.*[28] found five examples of anti-H specific agglutinin and one possible source of anti-M reagent in marine invertebrates. Cushing, Calaprice, and Trump[29] reported a blocking action of a substance in the serum of *Octupus bimaculatus* that could inhibit the action of human anti-A serum. Taxonomic application of *Limulus* serum was reported by Friar,[30] who distinguished cells of individual specimens of *Testudo hermanni* by *Limulus* hemagglutinins.

The possible utilization of *Limulus, Birgus,* and other invertebrate sera, or other body fluids, as immunohematologic reagents for blood grouping, tissue culture, and cytogenetic study are in progress in my laboratory.

Summary

The hemolymph of selected marine invertebrates as *Limulus polyphemus* (Horseshoe Crab) and *Birgus latro* (Coconut Crab) contain alpha or beta globulin-like proteins, that avidly agglutinate erythrocytes of a variety of vertebrate species. Some hemagglutinins require calcium for maximal serologic activity and for stability. Saccharides as N-acetyl-D glucosamine that are moieties of blood group substances can inhibit hemagglutination. It is speculated that the "natural" hemagglutinins of marine invertebrates, as crustaceans and mollusca, may be involved in saccharide transport and storage, possibly involved in shell formation. In addition, there may be some selective advantage to hemagglutinins by their ability to trap and opsonize bacteria or other foreign particles.

Marine invertebrate hemagglutinins of biomedical value should become an integral part of our "harvest from the sea."

References

1. Bisset, K. A., *J. Hyg.*, **45,** 128 (1947).
2. Huff, C. G., *Physiol. Rev.*, **20,** 68 (1940).
3. Cushing, J. E., and Campbell, D. H., "Principles of Immunology," McGraw-Hill Book Co., N. Y., N. Y., 1958.
4. Bang, F., *Fed. Proc.*, **26,** 1680 (1967).
5. Cushing, J. D., *Fed. Proc.*, **26,** 1666 (1967).
6. Briggs, J. D., *J. Exptl. Zool.*, **138,** 155 (1958).
7. Stephens, J. M., and Marshall, J. H., *Can. J. Microbiol.*, **8,** 719 (1962).
8. Tripp, M. R., *Ann. N. Y. Acad. Sci.*, **113,** 467 (1963).
9. Good, R. A., and Papermaster, B. W., *Adv. Immunol.*, **4,** 1 (1964).
10. Tyler, A., and Metz, C. B., *J. Exptl. Zool.*, **100,** 387 (1945).
11. Cohen, E., Rowe, A. W., and Wissler, F. C., *Life Sci.*, **4,** 2009 (1965).
12. Noguchi, H., *Zentr. Bakt. Abt. I. Orig.*, **34,** 286 (1903).
13. Marchalonis, J. J., *Fed. Proc.*, **23,** 1468 (1964).
14. Cohen, E., *Trans. N. Y. Acad. Sci., Ser. II,* **30,** 427 (1968).
15. Marchalonis, J. J., and Edelman, G. M., *J. Mol. Biol.*, **32,** 453 (1968).
16. McDade, J. E., and Tripp, M. R., *J. Invert. Path.*, **9,** 523 (1967).
17. Li, M. F., and Fleming, C., *Canad. J. Zool.*, **45,** 1225 (1967).
18. Neter, E., Cohen, E., Westphal, O., and Liideritz, O., *J. Immunol.*, **82,** 85 (1959).
19. Yokoyama, M., and Finlayson, J. S., *Transfusion,* **1,** 175 (1961).
20. Eriksson-Quensel, J. B., and Svedberg, T., *Biol. Bull.*, **70,** 498 (1936).
21. Boyden, M. G., *Serol. Mus. Bull.*, **37,** 9 (1967).
22. Zuckerkandl, E., *C.R. Soc. Biol. Paris,* **151,** 460 (1957).
23. Watne, A. L., and Cohen, E., *Proc. Am. Assoc. Can. Res.*, **5,** 261 (1964).
24. Watne, A. L., Cohen, E., Migaiolo, J. A., and Lyon, V. E., *Acta Cytologica,* **10,** 255 (1966).
25. Johnson, H., *Science,* **146,** 548 (1964).
26. Boyd, W. C., Broun, R., and Boyd, L. G., *J. Immunol.*, **96,** 301 (1966).
27. Chattoraj, A., Gilbert, R., Jr., and Josephson, A. M., *Transfusion,* **8,** 37 (1968).
28. Broun, R., Almodovar, L. R., Bhatia, H. M., and Boyd, W. C., *J. Immunol.*, **100,** 214 (1968).
29. Cushing, J. E., Calaprice, N. L., and Trump, G., *Biol. Bull.*, **125,** 69 (1963).
30. Friar, W., *Science,* **140,** 1412 (1963).
31. Shuster, C. N., Jr., *Zoologica,* **47,** 1 (1962).

IV. Physiology

Some Effects of Physiological State on Various Metabolic Systems

HAROLD L. ROSENTHAL
Dept. of Physiological Chemistry
Washington University
School of Dentistry

Many years ago, J. B. Allison and I began to use the term "physiological state" in order to define the changing metabolism of an animal during its lifetime, recognizing that the metabolic systems of an organism change in response to its internal and external environment—the amount and quality of its nutriment, activity, recovery from disease, growth, etc. Of the sum total of events that affect the physiological state of an animal, chronic semi-starvation and total starvation appeared to be the most amenable for study. The early studies in Allison's laboratory, primarily devoted to clarifying the relationship between protein nutrition and physiological state, were expanded by my colleagues and myself to include some aspects of vitamin B_{12} metabolism and, more recently, studies of enzyme induction and regulation. Perhaps this essay will demonstrate the influence of Allison as a teacher and scientist on at least one of his students.

Numerous studies have demonstrated that retention of nitrogen in the animal is, in part, a function of the caloric intake. This function has been emphasized by Allison *et al.*,[1,2] who demonstrated in the dog that urinary nitrogen excretion increased as the caloric intake decreased, but the nitrogen balance index of the dietary protein was not altered if the calories amounted to 50% or more of the normal requirement. Bosshardt *et al.*[3] pointed out that these results suggest two responses to a systematic reduction in calories. The first response involves catabolism of nitrogenous constituents without an alteration in the index, while the second response, associated with a severe caloric restriction, is correlated with a marked decrease in the nitrogen balance index. This second response has been reported in the mouse,[3,4] in the rat,[3,5–7] in the dog,[2] and in man.[8,9] The first response may be the result of increased catabolism of a portion of the nitrogenous constituents of the metabolic pool. The second response could be associated with a shift in catabolism following partial depletion of the pool. Some animals however, may resist the depleting effects of a restricted caloric intake much more than others.[10]

The marked loss of nitrogen in animals fed a low caloric intake is interpreted to indicate changes in physiological state and adjustment of the caloric and protein economy of the animal. In the early years (some 20 years ago) it was difficult to envisage how to approach the complex problem of physiological state from the viewpoint of intermediary metabolism. Nonetheless, the need to explore the problem was evident. We decided to explore the metabolism of vitamin B_{12} because little information was available concerning the physiological state of the animal and its effect on tissue vitamin concentrations.[11] Surprisingly, we found that total starvation for 10 days resulted in increases of vitamin B_{12} concentration in rabbit kidneys as much as 6-fold but other tissues remained unchanged (Table I).

Table I. Effect of Starvation on Tissue Vitamin B_{12} in Rabbits

Tissue	Control	Starved *	Refed *
Kidney	0.38 ± 0.08	2.71 ± 0.42	0.94 ± 0.18
Liver	0.37 ± 0.04	0.78 ± 0.07	0.61 ± 0.07
Heart	0.16 ± 0.03	0.33 ± 0.03	–
Spleen	0.18 ± 0.02	0.16 ± 0.02	–
Plasma	0.017 ± 0.003	0.025 ± 0.003	–

Average values ± S.E. for 8 animals in terms of μg B_{12}/g tissue.
* Animals were starved for 10 days and refed for 9 days.

During the starvation period, the rabbits lost 25% of their body weight (Table II). The excretion of vitamin B_{12} in urine more than doubled following starvation but the excretion of alkaline-stable material (deoxyribosides) decreased significantly. On refeeding, the body weight was regained and the excretion of the vitamin and alkaline-stable material returned almost to pre-starvation levels. It is interesting to note that rabbits maintained on a diet practically devoid of vitamin

B_{12} excrete 375 ng/kg of body weight/day in urine as compared with 25 ng/kg/day for dogs [12] and 1 ng/kg/day for man.[13,14] Although the creatinine excretion decreased to some extent during the experiment, the decrease is of doubtful significance and the evidence is good that urine collections were essentially complete.

Table II. Effect of Starvation on Urinary Excretion of Vitamin B_{12} by Rabbits

Treatment	Body Wt. (kg)	Urine Creatinine (mg/day)	Urine Vitamin B_{12} (μg/day)	Alkaline-Stable Activity (mμg/day)
Fed	2.81 ± 0.12	153 ± 8	1.05 ± 0.12	42 ± 4
Starved	2.12 ± 0.14	122 ± 12	2.48 ± 0.26	33 ± 4
Refed	2.71 ± 0.18	114 ± 10	1.63 ± 0.34	48 ± 5

Average values ± S.E. obtained from 7 or 8 animals.

Injected Co^{57} labeled cyanocobalamin was accumulated in rabbit tissue in a manner similar to that for native vitamin B_{12}. The accumulation was greatest in rabbit kidney mitochondria although microsomes and cell sap also increased in activity (Table III). For comparison, the lower accumulation in liver fractions are also shown.[15] The studies demonstrated that parenteral administration of radioactive cyanocobalamin to rabbits starved for 10 days resulted in markedly increased incorporation of radioactivity in kidney but not in liver.

Table III. *In vivo* Incorporation of Co^{57} Cyanocobalamin into Fractions of Rabbit Tissues

	Kidney		Liver	
	Fed	Starved	Fed	Starved
Whole organ	0.15 ± 0.09	1.92 ± 0.63	0.08 ± 0.03	0.16 ± 0.02
Nuclear fraction	0.15 ± 0.07	2.47 ± 1.14	0.13 ± 0.06	0.21 ± 0.03
Mitochondria	0.35 ± 0.23	3.99 ± 1.20	0.17 ± 0.04	0.50 ± 0.10
Microsomes	0.07 ± 0.04	0.38 ± 0.05	0.08 ± 0.07	0.07 ± 0.01
Cell sap	0.09 ± 0.05	0.50 ± 0.19	0.02 ± 0.01	0.11 ± 0.01

The values were determined 3 days after intramuscular injection of 0.04 μg cyanocobalamin/kg. Average values ± S.E. presented as ng cyanocobalamin/100 mgs protein.

In order to elucidate this phenomenon more fully, a series of experiments were performed to study the *in vitro* incorporation of $Co^{60}B_{12}$ into kidney and liver slices from fed and starved rabbits.[16] The uptake of $Co^{60}B_{12}$ by kidney slices increased rapidly with time, approaching maximum values after 1–2 hours incubation and remained at the maximum for at least 7 hours (Figure 1). Kidney slices from normal and

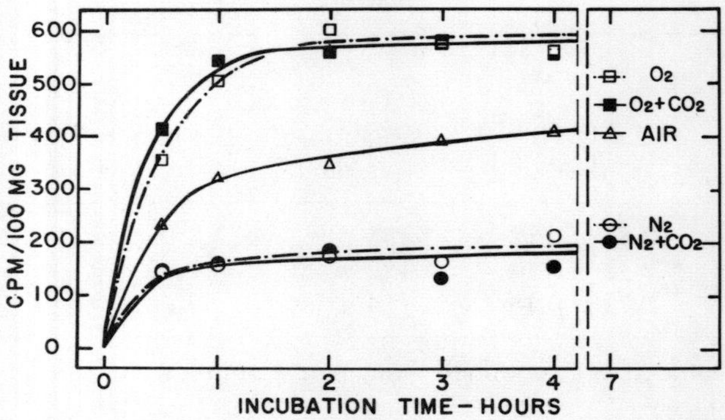

Figure 1. Uptake of $Co^{60}B_{12}$ versus time by kidney slices from normal rabbits with various gas phases. Each 250-ml Erlenmeyer flask contained 12 slices in 18 ml solution. Conditions: Krebs-Ringer phosphate, pH 7.4; 5.56 mM glucose; 4 mμc. $Co^{60}B_{12}$/ml: Temp. 37°C.

starved rabbits accumulate the greatest amount of $Co^{60}B_{12}$ when oxygen is the gas phase, while the uptake in nitrogen is approximately 30% of that obtained in oxygen. In contrast to the data obtained for kidney slices, the uptake of $Co^{60}B_{12}$ by liver slices from normal and starved rabbits is not altered following incubation in various gas mixtures (Figure 2).

In kidney slices (Table IV), $Co^{60}B_{12}$ uptake was inhibited 24 to 39% for tissues from normal and starved animals, respectively, in the presence of azide (2×10^{-4} M). Cyanide (1×10^{-4} M) did not inhibit uptake in kidney slices from normal rabbits but did inhibit (29%) the uptake in slices from starved animals. At higher concentrations of azide and cyanide, inhibition is increased in kidney slices from both normal and starved animals to approximately the same degree. Dinitrophenol was markedly inhibitory to kidney slices from normal and starved animals at all concentrations studied. Iodoacetate (1×10^{-3} M) inhibited the uptake by 42% for kidney slices of normally fed animals, but no inhibition was apparent for tissues from starved animals. At higher concentrations of iodoacetate, inhibition increased to approximately the same extent in tissue from both normal and starved animals.

In liver slices, inhibition by azide and cyanide for tissues from normal and starved animals was less than that obtained for kidney slices. Inhibition with dinitrophenol (3×10^{-5} M) was barely detectable and was less apparent or absent at higher concentrations. Iodoacetate does not in-

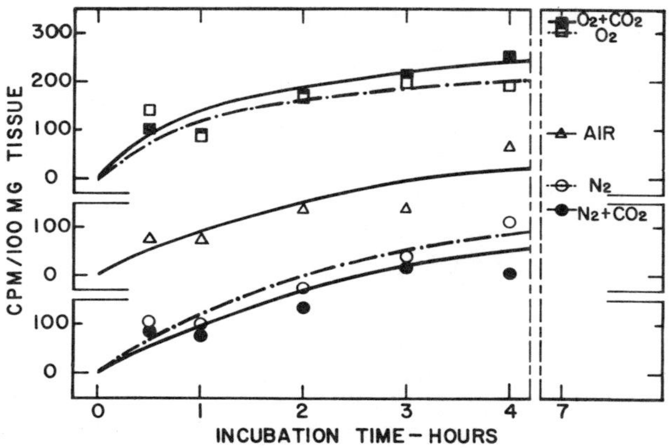

Figure 2. Uptake of $Co^{60}B_{12}$ versus time by liver slices from normal rabbits with various gas phases. Each 250-ml Erlenmeyer flask contained 12 slices in 18 ml solution. Note separation of ordinate for liver slices to separate curves. Conditions: Krebs-Ringer phosphate, pH 7.4; 5.56 mM glucose; 4 mμc. $Co^{60}B_{12}$/ml: Temp. 37°C.

hibit the uptake of $Co^{60}B_{12}$ by liver slices from normal or starved rabbits.

It is apparent from the results of this study that the mechanism for *in vitro* incorporation of $Co^{60}B_{12}$ by rabbit kidney and liver slices is markedly different. On the other hand, the rate of uptake of $Co^{60}B_{12}$ by rabbit kidney and liver slices is similar to the findings of Miller and Hunter,[17] Herbert,[18] and Latner and Raine [19] for rat liver slices.

The incorporation of $Co^{60}B_{12}$ by rabbit kidney and liver slices may be associated with different metabolic pathways. Because the inhibitors used in the study are generally known to interfere with oxygen-dependent metabolic systems—azide and dinitrophenol uncoupling oxidative phosphorylation [20,21] and azide and cyanide blocking electron transfer via the cytochrome system [22,23]—$Co^{60}B_{12}$ incorporation by kidney slices is apparently dependent on the integrity of oxidative metabolism. The uptake of $Co^{60}B_{12}$ by liver slices appears to be independent of oxidative metabolism. This is further substantiated by the stimulation of

Table IV. The Inhibition of Co^{60} Cyanocobalamin Uptake by Tissues from Normal Rabbits and Rabbits Starved for 10 Days

Substance	Conc. M	Kidney		Liver	
		Normal	Starved	Normal	Starved
Azide	2×10^{-4}	24	39	7	20
	2×10^{-3}	58	61	30	21
	2×10^{-2}	63	75	27	7
Cyanide	1×10^{-4}	0	29	0	26
	1×10^{-3}	46	67	17	33
	1×10^{-2}	53	51	15	21
Dinitrophenol	3×10^{-5}	55	40	13	17
	3×10^{-4}	65	57	0	10
	3×10^{-3}	77	45	0	5
Iodoacetate	1×10^{-3}	42	0	0	1
	1×10^{-2}	28	15	0	0
	1×10^{-1}	62	51	0	0

Each 50-ml Erlenmeyer flask contained two kidney and two liver slices from the same animal in 6 ml solution. The data are average values of duplicate experiments. The values are expressed as the percent inhibition compared to slices in the absence of inhibitor.

Conditions: Krebs-Ringer phosphate, pH 7.4; 5.56 mM glucose; 5 muc Co^{60} cyanocobalamin/ml; temp., 37°C; incubation time, 4 h; gas phase, O_2.

$Co^{60}B_{12}$ uptake by oxygen in kidney slices but not in liver slices. The lack of inhibition of $Co^{60}B_{12}$ uptake by dinitrophenol in rabbit liver is in agreement with Herbert's findings with rat liver. In kidney slices, the inhibitory effect of iodoacetate indicates that sulfhydryl groups [24] are associated with $Co^{60}B_{12}$ uptake. On the other hand, sulfhydryl groups do not appear to be involved in the uptake of $Co^{60}B_{12}$ by liver slices. The physiological significance of these observations remains obscure.

Some aspects of possible relationships between oxidative metabolism and dietary regimens have been studied by O'Dell *et al.*[25] who found that the cytochrome oxidase activity of liver in growing rats is not altered by food restriction. However, Wainio *et al.*[26] found that caloric and protein restriction lowers the succinic dehydrogenase activity in liver homogenates from mature rats. Cambria [27] also found significant decreases for succinic dehydrogenase activity for heart muscle homogenates of starving pigeons.

In view of the possibility that changes in physiological state of the animal may alter the extractability or other properties of enzymic activity, a series of experiments was performed to characterize the cytochrome *c* reductase system for kidney and liver tissue from fed and starved rabbits [28] and other animals.[29,30] Early experiments showed that incubation of kidney homogenates at 4°C for 48 hours resulted in liberation of maximum amounts of enzyme activity (Figure 3). The enzyme activity

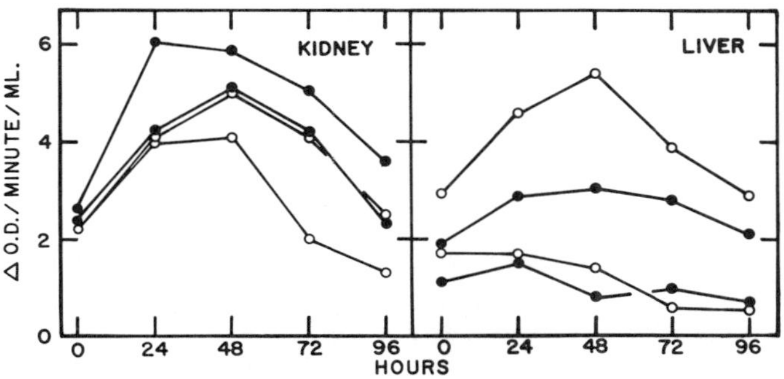

Figure 3. Effect of aging on enzyme activity. The data represent individual values obtained in two experiments for tissues from starved (●) and fed (○) rabbits. System: 240 μmoles phosphate, 3 μmoles KCN, 20 μmoles succinate, and 8 μmoles cytochrome *c* in 3.0 ml water at pH 7.5 and 27°C. Reactions started by addition of 50 μl of homogenate.

does not remain stable but decreases when aging is extended to 96 hours. It may be noted that the activity after 96 hours of aging is approximately the same as the original activity. The aging effect for kidney and liver preparations from starved animals is essentially similar to that obtained in fed animals. The succinic cytochrome *c* reductase activity of freshly prepared homogenates of kidney and liver homogenates of animals starved for 5 days is essentially the same as the values obtained in tissues of fed animals (Table V). However, starvation for 10 days resulted in significant decreases of enzyme activity in both kidney and liver homogenates.

It is apparent from these data that comparisons of enzyme activity of animals of different physiological states depends, to a certain extent, on the ability to extract comparable quantities of enzyme for quantitative analysis. Because aging of the homogenates tends to eliminate much of the difference for enzyme activity between kidney tissue of fed and starved animals suggests the need for caution in interpreting

Table V. Succinic Cytochrome *c* Reductase Activity of Freshly Prepared Kidney and Liver Homogenates from Fed and Starved Rabbits

	Enzyme Activity ± S.E.*	
	Kidney	Liver
Exp. 1 (5 days fed or starved)		
Fed	.400 ± .078 (4)	.240 ± .088 (4)
Starved	.409 ± .074 (4)	.302 ± .079 (4)
Exp. 2 (10 days fed or starved)		
Fed	.337 ± .031 (33)	.235 ± .031 (33)
Starved	.205 ± .034 (18)	.115 ± .023 (17)
"P"	<.01	<.02

* Enzyme activity is expressed as ΔOD/min/mg protein for number of animals in parentheses.

data based on enzyme content of tissue. These findings suggested the need to study enzyme systems which could serve as an indicator of specific metabolic activities such as the inducible enzymes of liver.

It is well known that liver tryptophan pyrrolase (LTP) activity is induced by the substrate and is also mediated by hormonal control under the influence of adrenal stimulation or injections of glucocorticoids. Preliminary experiments [31] showed that tryptophan pyrrolase could be induced in rabbits undergoing total starvation in as little as 18 hours. Chiancone,[32] however, found no alteration of LTP activity in rats subjected to fasting for 8–13 days. This difference in response between rabbits and rats may be due to the severe degree of starvation of the rats in Chiancone's experiments. In our experience 50% of totally starved rats will die in 8 days, while rabbits may easily withstand total starvation for 15 days or more.

As shown in Figure 4, the induction of LTP activity in female rabbits is a function of age of the animal and length of time of starvation.[33] Young 30-day-old female rabbits respond maximally to starvation in a regular fashion with significant increases being apparent within 24 hours from the beginning of starvation. In older animals, ranging between 4–10 months of age, starvation for as long as 11 days is associated with minimal LTP induction. The bimodal response of 2-month-old

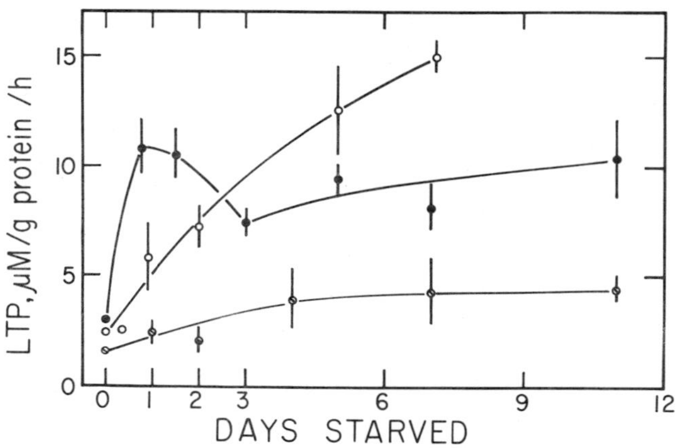

Figure 4. Liver tryptophan pyrrolase activity versus days of starvation for female rabbits. 30-day-old —○—, 2 months —●—, 4–10 months —⦶—. Each point represents average values ±1 S.E. (vertical bars) obtained with 4 to 10 animals.

female rabbits is not apparent in either 30-day-old or 4–10-month-old animals.

The LTP activity in rabbits starved for 24 hours (Figures 5, 6) is maximal in young animals between 30 and 75 days of age. Although male

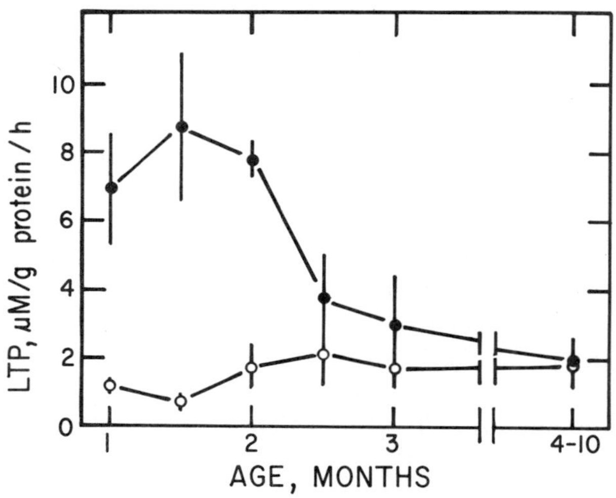

Figure 5. Liver tryptophan pyrrolase activity versus age of male rabbits fed *ad libitum* (—○—) or starved for 24 hours (—●—). Each point represents average values ±1 S.E. (vertical bars) obtained with 5–6 animals.

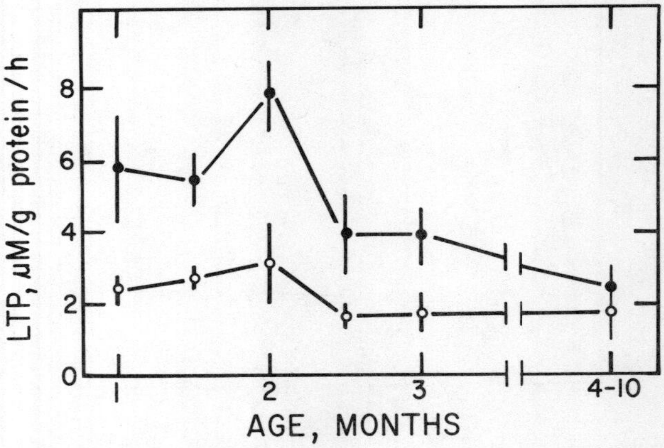

Figure 6. Liver tryptophan pyrrolase activity versus age of female rabbits fed *ad libitum* (—○—) or starved for 24 hours (—●—). Each point represents average values ±1 S.E. (vertical bars) obtained with 6–10 animals.

rabbits appear to respond somewhat sooner than female rabbits, this difference is of doubtful significance.

The greater LTP activity of starved rabbits suggests an increased catabolism of tryptophan for energy purposes but it is difficult to reconcile the observed LTP changes with the pathway for tryptophan metabolism. This phenomenon is in contrast to the activity of other enzyme systems, notably cytochrome *c* reductase [28] and succinic dehydrogenase [26,27] that are reduced during starvation thus reflecting an overall depression of catabolism with a concomitant conservation of energy. The dynamic aspects of adaptive control of metabolic systems by intact animals has recently been presented in detail by Knox [34] who has indicated the complexity of control mechanisms.

In male rats fed *ad libitum,* the LTP activity averaged 1.04 μM kynurenine formed/g/h which increased to 1.86 μM/g/h after 8 days' starvation, but this difference is of doubtful significance (Figure 7). In *ad libitum* fed female rats, however, the LTP activity (2.70 μM/g/h) was about three times higher than in *ad libitum* fed males. The value decreased to a minimum (1.98 μM/g/h) after 24 hours starvation before again increasing to 5.75 μM/g/h after 8 days' starvation. For male and female rats, body and liver weight decreased in a regular fashion during the starvation period.

The variability and the high values of *ad libitum* fed female rats suggested the possibility that a bimodal response may occur very early and,

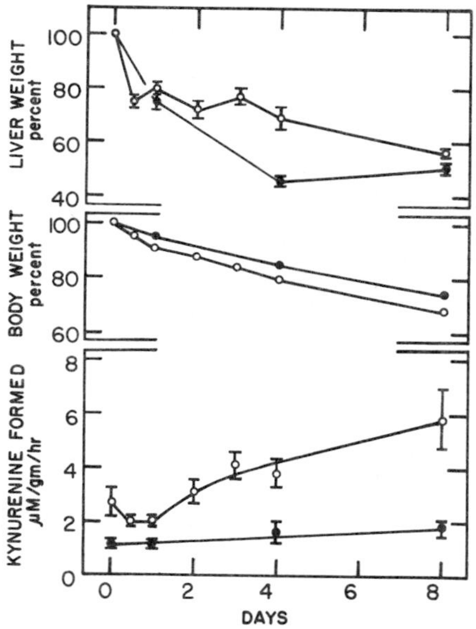

Figure 7. Kynurenine formation, body weight and liver weight of male (—●—) and female (—○—) rats versus days of starvation. Each point represents average data obtained for 6 to 9 determinations ±1 standard error.

because rats eat regularly between periods of 2–4 hours, the *ad libitum* fed female rats may have been voluntarily starved for a short period of time. In order to test this hypothesis, and to stabilize the animals in a more constant physiological state, refeeding techniques were initiated.[35] Rats were initially starved for 24 hours and then refed *ad libitum* for 48 hours, thus giving assurance that the animals were well fed, as shown by a rapid regain of body weight after refeeding (Table VI—Group C). The effect of short periods of starvation (4 and 24 hours) on the LTP activity, liver weight, and body weight of such *refed* rats, is presented in the table.

In male rats, the LTP activity was not altered significantly by this procedure, but in female rats (Group C) the enzyme activity decreased to 1.06 µM/g/h. It may be noted that the LTP activity of both male and female rats (Group C) was similar following the starvation-refeeding regimen. The LTP activity in *refed* male rats subjected to 4 hours' starvation is unaffected, but in similarly treated female rats, the LTP activity increased from 1.06 to 2.26 µM/g/h (Group D), a value not significantly different from that found in *ad libitum* fed untreated ani-

Table VI. Liver Tryptophan Pyrrolase Activity of Male and Female Rats

Group	Treatment	No.	Body Weight (g)	Liver Weight (g)	LTP Activity (μM/g/h ± S.E.)
Female					
A	Fed *ad libitum*	12	266	9.9	2.70 ± 0.63
B	Starved 24 h	11	242	7.8	1.98 ± 0.19
C	B + refed 48 h	6	269	10.5	1.06 ± 0.11
D	C + restarved 4 h	7	269	8.5	2.26 ± 0.34
E	C + restarved 24 h	7	253	7.2	1.06 ± 0.18
Male					
A	Fed *ad libitum*	7	452	15.9	1.04 ± 0.12
B	Starved 24 h	7	425	12.0	1.06 ± 0.12
C	B + refed 48 h	4	454	11.0	1.24 ± 0.40
D	C + restarved 4 h	4	453	10.8	1.36 ± 0.17

Significance. Females: C *vs.* A, $P < 0.05 > 0.01$; C *vs.* B, $P < 0.01$; C *vs.* D. $P < 0.01$. Males: Groups A, B, and D not significantly different from C.

mals (Group A). In refed female rats starved for 24 hours (Group E), the LTP activity again decreased to the refed values of Group C.

The present data confirm the report of Chiancone[32] that LTP activity of male rats is not significantly altered during starvation. However, the enzyme activity in female rats is markedly affected by the stress of starvation in accord with the responses found in female rabbits.[31]

Female rats and rabbits subjected to the stress of fasting show a bimodal response of LTP activity which varies with the duration of starvation and the time interval for each response is dependent on the species. Although the starvation response for LTP occurs only in female animals, Schor and Frieden[36] obtained a bimodal response of LTP activity in male and female rats injected with alloxan. The induction of LTP activity following parenteral injections of L-tryptophan[37,38] or cortisone[39–41] also occurs in both male and female animals. Oelkers and Nolten[42] have recently demonstrated *in vitro* depression of tryptophan pyrrolase by free estrogens and estrone sulfate. It is conceivable that starvation of female animals may alter estrogenic activity. Our data suggest that the difference in LTP response between male and female animals during starvation differs from other inducing systems for LTP by mechanisms that are as yet ill-defined and unknown.

The studies described in this essay demonstrate that alterations of physiological state modify many diverse metabolic systems. Allison and I considered the concept of "physiological state" to encompass, in the broadest context, all factors that result in the adaptation of metabolic systems to a changing physiological milieu. More recently, circadian rhythm has been shown to influence many metabolic systems.[43] Such daily or seasonal cyclical changes and their effect on metabolism are, in fact, part of the physiological state as defined by Allison and myself. Obviously, this area of interest needs considerable study and clarification.

References

1. Allison, J. B., and Anderson, J. A., *J. Nutrition,* **29,** 413 (1945).
2. Allison, J. B., Anderson, J. A., and Seeley, R. D., *Ann. N. Y. Acad. Sci.,* **47,** 245 (1946).
3. Bosshardt, D. K., Paul, W., O'Doherty, K., and Barnes, R. H., *J. Nutrition,* **32,** 641 (1946).
4. Bosshardt, D. K., *J. Nutrition,* **36,** 773 (1948).
5. Stevenson, G., Swanson, P. P., Willman, W., and Brush, M., *Fed. Proc.,* **5,** 240 (1946).
6. Rosenthal, H. L., and Allison, J. B., *Agri. and Food Chem.,* **4,** 792 (1956).
7. Benditt, E. P., Humphreys, E. M., Wissler, R. W., Steffee, C. H., Jr., Frazier, L. E., and Cannon, P. R., *J. Lab. Clin. Med.,* **33,** 257 (1948).
8. Schwimmer, D., and McGavack, T. H., *N. Y. J. Med.,* **48,** 1797 (1948).
9. Johnson, R. M., Deuel, H. J., Jr., Morehouse, M. G., and Mehl, J. W., *J. Nutrition,* **33,** 371 (1947).
10. Rosenthal, H. L., and Allison, J. B., *J. Nutrition,* **44,** 423 (1951).
11. Rosenthal, H. L., and Cravitz, L., *J. Nutrition,* **64,** 281 (1958).
12. Rosenthal, H. L., and Hampton, J. K., Jr., *J. Nutrition,* **56,** 67 (1955).
13. Register, U. D., and Sarett, H. P., *Proc. Soc. Exp. Biol. Med.,* **77,** 837 (1951).
14. Unglaub, W. G., Rosenthal, H. L., and Goldsmith, G. A., *J. Lab and Clin. Med.,* **43,** 143 (1954).
15. Rosenthal, H. L., Unpublished data.
16. Jud, L. A., and Rosenthal, H. L., *Arch. Biochem. Biophys.,* **95,** 526 (1961).
17. Miller, O. N., and Hunter, F. M., *Proc. Soc. Exp. Biol. Med.,* **96,** 39 (1957).
18. Herbert, V., *J. Clin. Invest.,* **37,** 646 (1958).
19. Latner, A. L., and Raine, L. C. D. P., *Biochem. J.,* **71,** 344 (1959).
20. Loomis, W. F., and Lipmann, F., *J. Biol. Chem.,* **173,** 807 (1948).
21. Spiegelman, S., Kamen, M. D., and Sussman, M., *Arch. Biochem.,* **18,** 409 (1948).
22. Stotz, E., Altschul, A. M., and Hogness, T. R., *J. Biol. Chem.,* **124,** 745 (1938).
23. Keilin, D., *Proc. Roy. Soc. (London),* **B121,** 165 (1936).
24. Colowick, S. P., and Kaplan, N. O., *Ann. Rev. Biochem.,* **20,** 528 (1951).
25. O'Dell, B. L., Gordon, J. S., Breummer, J. H., and Hogan, A. G., *J. Biol. Chem.,* **217,** 625 (1955).

26. Wainio, W. W., Eichel, B., Eichel, H. J., Person, P., Estes, F., and Allison, J. B., *J. Nutrition,* **49,** 465 (1953).
27. Cambria, A., *Boll. Soc. Ital. Biol. Sper.,* **35,** 2222 (1960).
28. Allen, J. R., and Rosenthal, H. L., *Proc. Soc. Exp. Biol. Med.,* **111,** 556 (1962).
29. Takeuchi, K., and Rosenthal, H. L., *Aichigakuin Daigaku,* **15,** 209 (1967).
30. Takeuchi, K., and Rosenthal, H. L., *Aichigakuin Daigaku,* **15,** 226 (1967).
31. Rosenthal, H. L., Barack, B. M., and Haessler, I., *Proc. Soc. Exp. Biol. Med.,* **117,** 222 (1964).
32. Chiancone, F. M., *Boll. Soc. Ital. Biol. Sper.,* **31,** 1310 (1955).
33. Wu, S. Y., and Rosenthal, H. L., *Proc. Soc. Exp. Biol. Med.,* **122,** 414 (1966).
34. Knox, W. E., *Trans. N. Y. Acad. Sci. Ser. II,* **25,** 503 (1963).
35. Pinto, P. V. C., and Rosenthal, H. L., *Experientia,* **21,** 511 (1965).
36. Schor, I. M., and Frieden, E., *J. Biol. Chem.,* **233,** 612 (1958).
37. Feigelson, P., Dashman, T., and Margolis, F., *Arch. Biochem. Biophys.,* **85,** 478 (1959).
38. Seschwind, I. I., and Li, C. H., *Nature,* **172,** 732 (1953).
39. Feigelson, P., Feigelson, M., and Greengard, O., *Recent Progr. Hormone Research,* **18,** 491 (1962).
40. Knox, W. E., and Auerbach, V. H., *J. Biol. Chem.,* **214,** 307 (1955).
41. Rosen, F., and Nichols, C. A., *Vitamins and Hormones,* **21,** 135 (1963).
42. Oelkers, W., and Nolten, W., *Hoppe-Seyler's Z. Physiol. Chem.,* **338,** 105 (1964).
43. Potter, V. R., Gebert, R. A., and Pitot, H. C., *Advances in Enzyme Regulation,* **4,** 247 (1966).

Role of Calcium in the Regulation of Muscle Contraction

C. PAUL BIANCHI
Department of Pharmacology
University of Pennsylvania
School of Medicine
Philadelphia, Pennsylvania

The regulation of muscle contraction and relaxation involves an interrelationship between three major structural elements of the muscle fiber, the surface membrane (S.M.), the sarcoplasmic reticulum (S.R.), and the contractile proteins (C.P.). The spatial relationships between these three structural elements are diagramed in Figure 1. The surface membrane, when excited, gives rise to a conducted action potential (MAP) which in turn, causes contraction to take place as it depolarizes the region of the surface membrane that invaginates into the fiber interior to form the transverse tubular element (TT) of the sarcoplasmic reticulum. A depolarization of the transverse tubular element causes calcium to be released from the inner surface of the transverse tubular membrane and from the terminal sacs (TS) of the sarcoplasmic reticulum which lie on either side of the transverse tubular element. The sarcoplasmic reticulum consists of longitudinal tubules that are 500–800Å in diameter and run from one end of the sarcomere to the other and fuse at the level of the Z-band to form the terminal sac. The terminal sac abuts against the flattened surface of the transverse tubule; it is across this junction that the action potential becomes coupled to the release of internally stored calcium of the terminal sacs. The sarcoplasmic reticulum represents about 6% of the skeletal muscle fiber volume.

The arrangement of the two terminal sacs abutting onto a transverse tubule is termed a triad.[1] In amphibian muscles the triad is located over

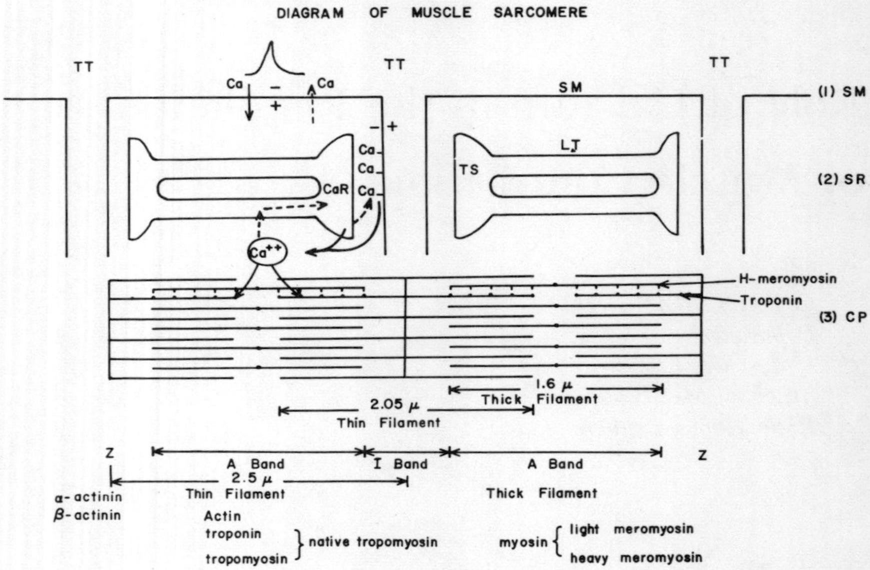

Figure 1. Diagrammatic sketch of skeletal muscle sarcomere showing the relationships between surface membrane (1), sarcoplasmic reticulum (2), and the contractile proteins (3) of the myofibril.

the Z-line of the sarcomere, while in most mammalian skeletal muscles the triad is located over the A-I band junction. Fast skeletal muscle fibers with long sarcomere lengths contain two triads per sarcomere (one at each A-I band junction) which decreases the diffusion path length along which calcium released from terminal sacs must diffuse to reach contractile proteins for contraction, *i.e.,* the region of overlap in the A-band (Figure 1).

In addition to the release of calcium from the terminal sacs, the sarcoplasmic reticulum can sequester calcium from the myoplasm by means of an ATP-dependent calcium transport system. The site of the calcium ATP-dependent transport is believed to reside in the longitudinal tubular network of the sarcoplasmic reticulum, and another proposed site of calcium transport is across the surface membrane. Repolarization of the transverse tubular membrane, following the passage of the muscle action potential, restores calcium to binding sites on the inner aspect of the membrane, thus resetting the system for another response to a second muscle action potential. The binding of calcium to these sites is believed to be dependent upon tubular membrane repolarization and external calcium in the lumen of the transverse tubular membrane.[3]

In Figure 1, the solid arrows indicate the pathway for calcium release which is under control of the membrane action potential, and the broken arrows indicate calcium resequestration or the transport of calcium from the cell across the surface membrane. The main source of supply of calcium for release from the inner aspect of the transverse tubular element is believed to be from the terminal sacs; during depolarization of the transverse tubular element, calcium is released from the inner aspect of the transverse tubule as well as from the terminal sacs, the released calcium then diffuses to the contractile protein binding sites and to resequestration sites of the longitudinal reticulum. Calcium also enters the myoplasm from the surface membrane in conjunction with the membrane action potential.

The contractile proteins of the myofibrils comprise the third structural element involved in muscle contraction. The myofibrils are longitudinally oriented, have a diameter of 1 μ, and consist of repeating units called sarcomeres (Figure 1). The sarcomere consists of an I-band region and an A-band region and is the unit delineated by the Z-lines that are present in the center of the I-bands. The A-bands are almost isotropic.[4,5] The myofibrils consist of thick filaments 100–200Å in diameter and 1.6 μ in length,[6,7,8] and thin filaments, 50–70Å in diameter and 2.05 μ in length.[9,10] The thick filaments are located in the A-band and lie 450Å apart in a hexagonal array. The thin filaments extend from the Z line across the I-band, and into A-band where they interdigitate with the thick filaments. The central region of the A-band contains only thick filaments and is called the H-zone. Crossbridges extend at right angles from the thick filaments and have a periodicity of 374Å. The thin filaments contain "native tropomyosin" (troponin + tropomyosin) in which the troponin has a periodicity of 376Å.[11] In vertebrate muscle the thin filaments are found in a square pattern in the I-band, while in the A-band each thick filament is surrounded by a hexagonal array of thin filaments.

The thick filaments contain myosin, which upon cleavage by mild tryptic digestion forms two fragments: light meromyosin and heavy (H) meromyosin. The backbone of the thick filament is formed by light meromyosin and the crossbridges consist of heavy meromyosin.[12]

The thin filaments consist of two helical chains of actin monomers interlaced with a second protein, consisting of tropomyosin and troponin. The thin filaments have a periodicity of 400Å which is attributed to the troponin bound to tropomyosin and actin.[13]

Ebashi[14] has isolated a protein that he calls "native tropomyosin," which actually consists of tropomyosin and a globular protein, troponin. In the absence of tropomyosin or troponin, actin and myosin can interact in the presence of Mg^{++} and ATP and shortening will occur; when "native tropomyosin" is added to the system, the actomyosin filaments relax, and contract only when calcium at a concentration of 10^{-7} M or greater is added to the system.[15] Thus, the interaction between the thick and thin filaments of the myofibrils requires the presence of ATP, Mg^{++}, and calcium bound to the "native tropomyosin." Removal of calcium from the tropomyosin binding site leads to relaxation of the muscle fiber. The functions of the minor structural proteins α actinin and β actinin are believed to be the regulation of the formation of actin polymers. α Actinin promotes the lateral association of actin while β actinin promotes the polymerization of actin.[16,17]

Contraction and relaxation of the muscle fiber is considered to be the result of the sliding of thick and thin filaments past one another.[18,19] Tension develops as a result of the interaction of crossbridges of thick filaments with sites on thin filaments. A relative force between adjacent filaments is generated at the crossbridges within the region of overlap of thick and thin filaments. Under physiological conditions, interaction between thick and thin filaments takes place when the calcium binding sites of native tropomyosin bind calcium as the myoplasmic free calcium rises above 10^{-7} M, binding of calcium to the native tropomyosin and actomyosin interaction becomes maximal as the calcium concentration increases to 10^{-6} M.[20] The crossbridges undergo a conformational change as the result of the interaction with thin filaments, ATP, Mg^{++}, and Ca^{++}; the energy from the breakdown of ATP is utilized for the conformational change that results in a relative movement between thick and thin filaments.[22,23] The rate of movement appears to be a function of myosin ATP'ase activity located in the heavy meromyosin fraction.[24,25]

The addition of a pulse of calcium solution directly to the myoplasm of a fast contracting muscle fiber results in a phasic contraction which lasts 5 sec and occurs at a rate of 6–12 μ/sec/sarcomere. In slow contracting muscles of the tonic type, similar treatment results in a contraction which only partially relaxes 3 seconds after application of calcium. The rate of shortening is approximately one-tenth the rate of the fast muscle fiber (0.5–1.5 μ/sec/sarcomere). Thus the rate of shortening is primarily determined by the intrinsic myosin ATP'ase activity and not by the appearance of calcium. If the myosin ATP'ase activity is high then the rate at which calcium diffuses from membrane release sites to the crossbridges may be the major factor determining the rate of tension development.

The electrolyte composition of various types of muscle fibers is shown in Table I. The magnesium and calcium contents are quite consistent, whereas, sodium and potassium values show a greater degree of variation. Magnesium is the major intracellular divalent cation of muscle and exists in a free and bound form. The amount of free magnesium in muscle has been calculated to about 3μmole/g.[26] The regulation of

Table I. Electrolyte Composition of Various Types of Muscle

(μmole/g wet weight)

Muscle	Species	K^+	Na^+	Mg^{++}	Ca^{++}
(1) Sartorius	frog	92	22	8.6	1.64
(2) Rectus	frog	78	31	6.8	1.70
(3) Ventricle	frog	70	28	9.0	1.20
(4) Stomach	frog	69	67	—	1.54
(5) Ventricle	rat	76	42	8.7	0.75
(6) Superior rectus	cat	72	37	6.1	0.74
(7) Leg muscle	cat	118	25	8.9	0.84

myoplasmic calcium levels is believed to be under the control of three active calcium transport sites, two of which are shown in Figure 1. The surface membrane transport site moving calcium out of the muscle fiber, and the calcium pump of the longitudinal reticulum which resequesters calcium released following a contraction. The third calcium transport system is located in the mitochondria of the muscle fiber. The mitochondrial calcium transport system is relatively unimportant for the control of myoplasmic free calcium in skeletal muscle, since the number of mitochondria is small, but it may play an important role in cardiac muscle which has a high mitochondrial content.[3]

The calcium transport system that has been most studied resides in the sarcoplasmic reticulum and is the transport system most associated with calcium accumulation from the myoplasm.[27,28,29] The accumulation of calcium by the vesicles of the disrupted sarcoplasmic reticulum accounts for their relaxing effect on the isolated actomyosin system.[21] Podolsky and Costantin[30] demonstrated that calcium applied internally to the myoplasm of the muscle fiber becomes sequestered in the region of the triad of the sarcoplasmic reticulum (Figure 1). Disruption of the transverse tubular element of the triad results in a loss of coupling between the muscle action potential and contraction.[31] The action potential still persists, although the negative after potential typical of the normal muscle action potential is no longer present.[32]

Amphibian skeletal muscle is composed of two types of muscle fibers, fast and slow. The fast muscle fiber gives a twitchlike response to a muscle action potential. The slow muscle fibers have multiple end plates that are depolarized as a function of the frequency of nerve stimulation; the contractile response is tonic, *i.e.*, sustained as long as the surface membrane is depolarized. In contrast sustained depolarization of the fast muscle fiber leads to a phasic response. The fast muscle fiber relaxes in spite of maintained depolarization, whereas the tonic fiber relaxes only upon repolarization of surface membrane.[3] In fast twitchlike fibers the action potential is associated with a calcium influx and efflux of apparently equal magnitude.[31] However, the magnitude of calcium influx during the action potential is too small to account for the ensuing contraction.

In Table II, the muscle parameters related to excitation-contraction coupling are listed. The sartorius muscle of the frog is the fastest contracting muscle of the types listed. The surface membrane area (exclusive of the transverse tubular system) is the least of all the muscle fiber types listed. The frog ventricle, which lacks a sarcoplasmic reticulum [36] has the largest surface membrane area. The type of coupling system found in the sartorius muscle is of the indirect type, *i.e.*, a depolarization across the transverse tubular membrane is coupled to the release of calcium from a second membrane structure, the terminal cisternae. In the frog ventricle the surface membrane releases calcium as a function of depolarization and modulates myoplasmic calcium by an influx of calcium, resequesters calcium during repolarization, and pumps calcium out of the cell in order to maintain cell calcium in a steady state. Thus, in the absence of the sarcoplasmic reticulum, the surface membrane appears to take over the major functions of the sarcoplasmic reticulum, *i.e.*, calcium release and resequestration. Smooth muscle also appears to be a system where depolarization across the surface membrane is directly coupled to the release of calcium.

If the frog sartorius muscle contains the same actomyosin content as mammalian skeletal muscle (70 mg/g wet weight),[36] then the amount of calcium bound to actomyosin under relaxed conditions is 0.9 μmole/g actomyosin. As the myoplasmic free calcium rises from 10^{-7} M to 10^{-6} M, the myofibrils reach a maximum state of contraction and the amount of calcium bound to actomyosin (in the presence of native tropomyosin)

Table II. Muscle Cell Parameters Related to Excitation-Contraction Coupling

Muscle	Species	Fiber Diameter (μ)	E_m mV	Actomyosin Content (mg/g *)	Intra Fiber Water (ml/g)	Ca^{++} Required for Coupling ($m\mu$-mole/g)	Ca^{++} Required for $[CaH]_i = 10^{-6}$ M ($m\mu$mole/g)	Surface Fiber Area (cm^2/g)	Fiber Volume (cm^3/g)	Calcium Influx per A.P. ($m\mu$-mole/g)	Coupling System
Sartorius (31)	frog	100	(92)	70	0.65	91	0.65	300	0.81	0.1	I
Ventricle (32)	frog	2–5	(85)	38	0.65	50	0.65	11,400	0.81	2	D
Auricle (33)	guinea pig	25	(85)	38	0.55	50	0.55	1,000	0.70	1	D&I
Taenia coli (34)	guinea pig	5	(50)	10	0.42	13	0.42	3,300	0.50	4	D

* Wet weight.

increases to 2.2 μmole/g of actomyosin. The minimal amount of calcium bound to contractile proteins that is required for maximal activation is 1.3 μmole/g actomyosin. The amount of calcium required for coupling of excitation of skeletal, cardiac and smooth muscle for maximal contraction is shown in Table II. Skeletal muscle fibers require approximately 91 mμmole/g wet weight of muscle, heart muscle 50 mμmole/g, and smooth muscle 13 mμmole/g.

The amount of calcium that flows into muscle during the action potential is insufficient to supply calcium for coupling of excitation and contraction, therefore, the major source of calcium for contraction must be released from intracellular sites during excitation. In the *taneia coli* calcium influx during the action potential is of sufficient quantity to cause a twitch tension equal to 30% of the maximum tension. The calcium influx in cardiac and smooth muscle is of sufficient magnitude to modulate the myoplasmic free calcium and thus enhance or maintain a given degree of contractility. Thus, once the contractile proteins bind calcium to the interaction sites, calcium influx can modulate muscle contraction by balancing calcium removed from the myoplasm by intracellular calcium sinks, *i.e.*, the sarcoplasmic reticulum, the externally directed calcium pump in the surface membrane and the mitochondrial sequestration of calcium. Thus, the degree of contraction is determined by the existing concentration of myoplasmic free calcium.

In cardiac and smooth muscle the factors that regulate calcium influx may markedly modify muscle contraction whereas in skeletal muscle, modulation is primarily brought about by factors that govern the internal cycle of calcium binding and release. The cardiac and smooth muscle are primarily open systems with regard to calcium metabolism since both markedly depend on calcium influx and efflux from the muscle cell to regulate muscle contraction. Skeletal muscle in contrast is a closed system in which calcium release and resequestration is primarily an intracellular cycle, with only sufficient calcium pumped out to balance the relatively small calcium influx that occurs during the action potential.

Since the frog ventricle lacks a sarcoplasmic reticulum, the intracellular calcium must be contained either in mitochondria or bound to the inner aspect of the excitable surface membrane. Calcium stored in the mitochondria would not be as readily available for release as the calcium stored in the inner surface of the surface membrane during excitation. Calcium released from the surface membrane, as well as calcium influx can be directly regulated by the cardiac action potential.[32] In mammalian heart muscle a sarcoplasmic reticulum is present,[37,38,39] consequently, excitation-contraction coupling can be thought of as being a combination of the direct (single membrane undergoes electrical change and releases bound calcium) and indirect type (surface and/or transverse tubular membrane undergoes electrical change and which in turn regulates calcium release from a second internal membrane).

Smooth muscle is more comparable to the frog ventricle in that there is no well-developed sarcoplasmic reticulum, but there is a relatively large surface membrane area. Goodford[34] proposes that calcium released from the surface sites (inner aspect) combined with calcium influx during the action potential supplies sufficient calcium to maintain the contractile proteins in a contracted state. Thus, excitation-contraction coupling appears to be of the direct type in smooth muscle.

Myoplasmic free calcium when increased from 10^{-7} to 10^{-6} M also enhances the phosphorylase B kinase activity.[40] Thus, the release of calcium during excitation of muscle, causes a parallel increase in muscle contraction and glycogenolysis.

In summary, calcium plays a major role in the regulation of muscle function. Interaction of calcium with the surface membrane stabilizes the membrane with regard to ion permeability and plays a role in the events that lead to the muscle action potential. Calcium release from intracellular binding or sequestering sites in combination with calcium influx results in parallel increase in muscle contractility and breakdown of glycogen to supply the energy that is utilized for contraction.

References

1. Porter, K. R., and Palade, G. E., *J. Biophys. Biochem. Cytol.,* **3,** 269 (1957).
2. Winegrad, S., *J. Gen. Physiol.,* **51,** 65 (1968).
3. Bianchi, C. P., "Cell Calcium," Butterworth, London, 1968.
4. Dobie, W. M., *Ann. Mag. Nat. Hist.* (*Ser. 2*), **3,** 109 (1849).
5. Krause, W., "Die Motorischen Endplatten Der querestriften Muskelfasern," Hannover, Hahn, 1869.
6. Huxley, H. E., *J. Molec. Biol.,* **7,** 281 (1963).
7. Page, S., and Huxley, H. E., *J. Cell Biol.,* **19,** 369 (1963).
8. Huxley, A. F., and Peachey, L. D., *J. Physiol.,* **156,** 150 (1961).
9. Huxley, H. E., *J. Biophys. Biochem. Cytol.,* **3,** 631 (1957).
10. Page, S., and Huxley, H. E., *J. Cell Biol.,* **19,** 369 (1963).
11. Endo, M., Nonomura, Y., Mosaki, T., Ohtsuki, I., and Ebashi, S., *J. Biochem.,* **60,** 605 (1966).
12. Peper, F. A., *J. Cell Biol.,* **28,** 505 (1966).
13. Cohen, C., and Longeley, W., *Science,* **152,** 794 (1966).
14. Ebashi, S., *J. Biochem.,* **58,** 107 (1965).

15. Ebashi, S., and Kodama, A., *J. Biochem.,* **60,** 733 (1966).
16. Ebashi, S., and Ebashi, F., *J. Biochem.,* **58,** 7 (1965).
17. Ebashi, S., in "Symosion über progressive muskeldystrophie," E. Kuhn, Ed., Springer-Verlag, Berlin, 1966.
18. Huxley, H. E., and Hanson, J., *Nature, London,* **173,** 973 (1954).
19. Huxley, A. F., and Niedergerke, R., *Nature, London,* **173,** 971 (1954).
20. Weber, A., and Winicur, S., *J. Biol. Chem.,* **236,** 3198 (1961).
21. Weber, A., and Herz, R., *J. Biol. Chem.,* **238,** 599 (1963).
22. Huxley, A. F., *Prog. Biophys.,* **7,** 255 (1957).
23. Davies, R. E., *Nature, London,* **199,** 1068 (1963).
24. Barany, M., *J. Gen. Physiol.,* **50,** 197 (1967).
25. Constantin, T., Franzini-Armstrong, C., and Podolsky, R. J., *Science,* **147,** 158 (1965).
26. Nanninga, J. B., *Biochim. Biophys. Acta,* **54,** 338 (1960).
27. Ebashi, S., *Prog. Theoret. Phys., Suppl.,* **17,** 35 (1961).
28. Nagai, T., Makinose, M., and Hasselbach, W., *Biochem. Biophys. Acta,* **43,** 223 (1960).
29. Ebashi, S., and Lipmann, F., *J. Cell Biol.,* **14,** 389 (1962).
30. Podolsky, R. J., and Costantin, T. T., *Fed. Proc.,* **23,** 933 (1964).
31. Bianchi, C. P., and Shanes, A. M., *J. Gen. Physiol.,* **42,** 803 (1959).
32. Niedergerke, R., *J. Physiol.,* **167,** 551 (1963).
33. Winegrad, S., and Shanes, A. M., *J. Gen. Physiol.,* **45,** 371 (1962).
34. Goodford, P. J., *J. Physiol.,* **192,** 145 (1967).
35. Staley, N. A., and Benson, E. S., *J. Cell Biol.,* **38,** 99 (1968).
36. Needham, D. M., and Williams, J. M., *Biochem. J.,* **89,** 534 (1963).
37. Simpson, F. O., and Oertelis, S. J., *Nature, London,* **189,** 758 (1961).
38. Simpson, F. O., and Oertelis, S. J., *J. Cell Biol.,* **12,** 91 (1963).
39. Nelson, D. A., and Benson, E. S., *J. Cell Biol.,* **16,** 297 (1963).
40. Ozawa, E., Keizo, H., and Ebashi, S., *J. Biochem.,* **61,** 531 (1967).

Relationship Between Protein Metabolism in Muscle Tissue and the Concept of "Protein Reserves" *

R. W. WANNEMACHER, JR., and
W. K. COOPER
Bureau of Biological Research
Rutgers—The State University
New Brunswick, New Jersey

Introduction

The concept of protein reserves was first suggested by Voit[1] in 1866. His data indicated that the initial excretion of nitrogen was very high during the first day of fasting and, after 5 to 6 days, the nitrogen excreted was markedly reduced. The nitrogen excreted during the early days of fasting was related to the protein content of the diet prior to fasting. Thus, Voit concluded that the nitrogen lost during the first 5 or 6 days of a fast represented a depletion of labile body proteins (protein reserves), which do not exceed 5% of the total body nitrogen. Many investigators agree with Voit that only 5 to 6% of the total body nitrogen is associated with the protein reserves and that these reserves are not needed for normal maintenance of an individual.[2,3,4] Other workers question the existence of protein reserves,[5] while still others have suggested that this represents the loss of non-protein nitrogen.[6,7]

This concept was of major scientific interest to Dr. James B. Allison,

* This work was supported by U. S. Public Health Service Research Grant AM-04341 from the Institute of Arthritis and Metabolic Diseases, by a grant from the Gaines Division of General Foods and by a U. S. Public Health Service Predoctoral Fellowship 5-FI-GM-33, 284-02 to W. K. Cooper.

who during the latter years of his life devoted a major portion of his research efforts to the understanding of this problem. He demonstrated: (a) that dogs with greater reserves had a rapid initial loss of nitrogen when fed a protein-free diet, but were more resistant to severe depletion [8] and toxicity of certain compounds such as N-(3-oxapentamethylane)-N′,N″-diethylene,[9] or 2-aminofluorene; [10] (b) that certain body proteins, such as serum albumin [11] or certain tissue enzymes,[12] could be rapidly depleted by feeding a low protein diet while other proteins, such as serum globulin [13] or brain proteins [14,15] could not be decreased even in severe depletion. Allison defined protein reserves "as those tissue proteins that can be reversibly depleted and repleted, thereby contributing to the free amino acid pools of each cell for the synthesis of certain essential proteins that may be needed for maintenance of cellular integrity during periods of malnutrition or stress." [16] These reserves could amount to as much as 40–50% of the total body protein [17] and would include the initial high excretion of nitrogen that is noted when an animal is fasted or fed a low protein diet. Recent data [17,18] support the concept that changes in nitrogen excretion of an animal placed on a low protein diet represent altered kinetics of protein metabolism in different tissues of the animal.

As a new-born animal grows, the cells from various tissues and organs increase in both number and size.[19,20] From the 17th to 95th day of life of male rats, the total DNA content of various muscles increased by a factor of 1.8 for the biceps brachii, 3.7 for the gastrocnemius, and 4.2 for the soleus muscle.[19,21] During the same period, the muscle weight rose 21 times in the gastrocnemius and 9.5 times in the biceps brachii, with the average diameter of the muscle fibers increasing over 3-fold, from 16 to 50 μ (10-fold increase in area), while the number of fibers remained constant. In subsequent experiments, Gordon, Kowalski, and Fritts [22] found that after 80 days of life the DNA content of the quadriceps muscle of the rat was constant but muscle weight, total myofibrillar protein, and total sarcoplasmic protein doubled between 80 and 160 days of age. Similar observations have been noted in developing muscles of humans,[23] chicken,[24] pig,[23] and cattle.[25] Thus, the marked increase in muscle protein after cessation of DNA accumulation could be related to Allison's concept of protein reserves. Since muscle tissue represents 60–70% of the total body protein in adult animals (unpublished data), it seemed important to study the relation between the protein metabolism of muscle tissue and Allison's concept of protein reserves.

Tissue Contribution to the Nitrogen Loss During Malnutrition or Stress

The liver is the major single contributor of the nitrogen that is lost during periods of mild starvation or protein-deprivation (Table I). Previous observations [1,26,27] have emphasized that in protein-depleted rats, the liver has undergone the largest percentage decrease in protein content and weight of all organs in the body. This loss in liver protein is associated with a decrease in activity of urea cycle enzymes; [28] amino acid transaminases; [29,30,31] enzymes that initiate catabolism of certain amino acids; [32,33,34] many oxidative enzymes; [12,35] certain enzymes involved in glycolysis, glucogenesis and the hexosemonphosphate shunt,[36,37] and *in vivo* [38] and *in vitro* [39] rate of protein synthesis. Since many of these liver enzymes have relatively short half lives—tryptophan pyrrolase, 2–3 hours,[40] tyrosine α-ketoglutorate transaminase, 3–4 hours,[41] and arginase, 4–8 days,[42] a decreased rate of synthesis concomitant with an unchanged time constant for degradation results in a rapid loss of these proteins. Thus, the change in liver protein content following depletion may not represent a loss of a so-called "stored protein" but can be related to an adaptive mechanism whereby protein synthesis is reduced by a diminished supply of amino acids. This may be an over-simplification of the problem because certain liver enzymes such as cathepsin,[43] alkaline phosphate,[43] or ribonuclease [44] remain constant or are increased during protein depletion. In the liver of fasting rats, hepatic glucose-6-phosphatase and, to a smaller extent, fructose-1, 6-diphosphatase are preferentially maintained as compared to the protein level of the cell.[45] Other carbohydrate enzymes either parallel the loss in hepatic cell nitrogen; e.g., phosphohexose isomerase and lactic dehydrogenase; or undergo a sharp decrease early in the starvation process and reach very low values, and decrease more rapidly than the liver nitrogen; *e.g.*, glucose-6-phosphate dehydrogenase.[45] Drysdale and Munro [46] have demonstrated that following an injection of ferric chloride, the synthesis of ferritin protein can be stimulated in the liver of protein-depleted rats. Under other physiological stresses—such as tumor cachexia,[47] hypertension,[48] pregnancy,[49] radiation,[50] and ardenal corticoid administration [45,51]—certain liver enzymes and total cellular proteins increase even though the animal is losing body nitrogen. Further, the susceptibility of an enzyme response to protein-depletion varies in different tissues with brain enzymes being unaffected, mild effects in kidney, skeletal muscle and

Table I. Percent Contribution of Various Tissues to the Total Amount of Body Protein That Is Lost When Rats are Starved or Fed a Protein-Free-Diet (PFD)

	Starvation (days)			PFD (days)		
Tissues	1	7	14	1	21	100
	% Contribution to total protein loss					
Liver	50.0 ± 2.3	13.9 ± 0.8	9.6 ± 0.6	47.1 ± 3.2	4.3 ± 0.1	3.6 ± 0.2
Viscera	3.8 ± 0.2	19.2 ± 1.5	15.0 ± 1.0	8.8 ± 0.4	12.1 ± 0.8	7.2 ± 0.3
Carcass	30.8 ± 3.2	48.7 ± 4.2	56.6 ± 5.1	32.4 ± 4.0	50.4 ± 2.3	61.2 ± 5.1
Skin	15.4 ± 1.8	18.2 ± 1.5	18.8 ± 1.6	11.7 ± 0.9	33.2 ± 2.6	28.0 ± 3.4
% loss in total body protein	2.4 ± 0.1	14.1 ± 0.7	28.9 ± 1.3	1.3 ± 0.2	17.8 ± 0.6	49.2 ± 4.2
% loss in total body nitrogen	2.5 ± 0.2	14.5 ± 1.3	29.2 ± 1.8	1.5 ± 0.3	18.5 ± 0.9	50.9 ± 3.1

Rats weighing 200–225 g were starved or fed a PFD for various lengths of time. The 14 day starved and 100 day PFD groups were housed in metabolism cages and daily urine and fecal nitrogen determinations were performed on each animal. Percent loss in body nitrogen was calculated by assuming that the rats initially contained 2.54% nitrogen. Liver, viscera, carcass, skin, and total body protein were measured by the previously described biuret procedure.[17] In order to calculate the loss in body protein, the protein content of the various tissues was compared to the zero day concentrations. Each value is the mean ± S.E.M. of six animals.

spleen and a marked loss in liver enzymes.[14] Thus, the contribution that a particular protein will make to the body nitrogen loss is influenced by the physiological stress and the tissue with which the protein is associated. In addition, certain body proteins which have a relatively short half-life, such as alpha- and beta-globulins,[52] are not reduced even during severe protein-depletion.[13] From these data, it is not probable, as suggested previously by Munro,[53] that the disappearance of body protein during protein deprivation simply represents the adaptation of proteins with rapid turnover rates to the diminished supply of amino acids.

The mechanism by which an amino acid deficiency could influence the cellular content of various tissue proteins is under investigation in many laboratories, with most of these efforts limited to the liver cell. It has been demonstrated that deletion of certain essential amino acids or fasting rapidly alters the liver polysomal profiles and the rate of protein synthesis.[54–57] In addition these changes take place even if RNA synthesis has been inhibited by previously treating the animal with actinomycin D. The results suggest that amino acids *per se* can affect the translational phase of protein synthesis. In rats starved for 5 days[58] and in weanling animals fed a low protein diet,[59] the cellular concentration of liver ribosomes is significantly reduced as compared to controls fed an adequate protein diet. Thus, it is possible to postulate two mechanisms for the control of protein synthesis: (a) a rapid but transient effect on the ability of existing ribosomes to incorporate amino acids into protein, and (b) a slower but longer term effect on the number of ribosomal units per cell. The former mechanism could be related to the direct action of amino acids on the cell and the latter may be mediated through other metabolic controls, such as hormones. Also, it has been demonstrated that approximately 40% of the liver messenger RNA (m-RNA) is stable with a half-life equivalent to or greater than the ribosomal half-life of 110–127 hours[58] and that the liver cells from protein-deprived rats have a markedly reduced ability to synthesize cytoplasmic ribosomes and m-RNA.[59]

From these data it is possible to theorize three kinetic modes to explain protein loss and the significance of protein reserves during amino acid deprivation. Model one: the half-life of a particular protein and its m-RNA is very short. Thus, the removal of dietary amino acids would result in an inhibition of synthesis of new m-RNA, a dissolution of the polyribosomal complex and a decreased rate of synthesis of the protein. If the catabolic rate remains constant,[60] then the cellular concentration of this protein will be rapidly reduced until a new equilibrium can be established to correspond to the decreased rate of synthesis. The rapid loss in body nitrogen when an animal is fed an amino acid deficient diet (labile protein reserve) can be characterized as due to this type of kinetic response. Model two: The half-life of the m-RNA for a particular protein is equal to or greater than that of the ribosome. Under these conditions the synthetic rate for the protein is regulated by the ribosomal content of the cell. If the catabolic rate remains constant, then the loss in cellular content of a particular protein is a function of its turnover rate. As the severity of the protein-depletion increases, the proteins that comprise the cellular population of a tissue such as liver are mainly those with a slower turnover rate. This conclusion is supported by findings of Stephens and Waterlow,[61] which indicate that the rate of catabolism of total liver proteins decreases in proportion to the degree of protein-depletion. This second kinetic model could explain the effect of amino acid deprivation on protein metabolism in most nonhepatic tissues of the body. Model three: The particular protein has a rapid half-life but its m-RNA in the cellular polysomes is maintained by external stimuli, such as hormones or ferric chloride. This third model then could explain how serum globulins are maintained or synthesis of ferritin protein could be stimulated in a protein-depleted animal. Thus, the mobilization of the so-called "protein reserve" is merely an adaptive process whereby the body proteins obtain a steady state after the synthetic rates have been altered by either a diminished amino acid supply or other physiological stresses.

Even during mild fasting or protein-deprivation the carcass (mostly skeletal muscle protein) contributes 30% of the nitrogen which is lost (Table I). This represents such a small quantity of nitrogen relative to the total protein mass of skeletal muscle that the difference in nitrogen content is barely detectable. However, as the animal continues to lose body nitrogen, a greater percentage is being contributed by carcass while liver and viscera diminish in their contribution (Table I). In severely depleted animals, skeletal muscle contributes approximately 60% of the nitrogen that is lost. Under these conditions the body is losing 40–50% of its body nitrogen and the reduction in protein content of skeletal muscle is highly significant. Similar results have been noted by Waterlow and Stephens.[27] The labile nature of liver protein is illustrated by the fact that after the first day of fasting or protein-free-diet the contribution of liver protein to the total body nitrogen loss rapidly diminishes until in the severely depleted animal it represents only 4–5% of the total protein that is lost from the body. Since most physiological stresses are correlated with a decrease in muscle protein content,[18,47,51] it may be concluded that this tissue is one of the major contributors to Allison's so-called "protein reserve." This decrease in muscle nitrogen represents a true diminution in muscle protein content as illustrated by fact that the nonprotein nitrogen per cell is not significantly reduced when rats have lost 20% of their body nitrogen by either starvation or protein-deprivation (Table II). Under these conditions the concentration of free amino acids (especially essential amino acids) in the muscles of these depleted rats is reduced below control levels. However, during periods of milder protein-deprivation these values are elevated above control levels.[15,38] Thus, the hypothesis of Basak[6] and Fisher *et al*[7] that the labile protein reserves represents the loss of nonprotein nitrogen is not supported by these data or by the fact that the percentage decrease in body protein is very similar to total nitrogen loss (Table I). As concluded by Munro[4] from other experimental data, the major portion of the nitrogen which is lost even during conditions of mild protein depletion must come from the catabolism of body proteins.

Table II. Non-Protein-Nitrogen (NPN) and Free Amino Acid Content of Paired Gastrocnemii Muscles of Rats That Were Starved or Fed a PFD

Treatment	NPN (mg N/mg DNA)	Essential Free Amino Acids (μmoles/mg DNA)	Nonessential Free Amino Acids (μmoles/mg DNA)
Control	0.73 ± 0.09	4.67 ± 0.11	8.02 ± 0.12
21 day PFD	0.69 ± 0.05	3.73 ± 0.31	8.26 ± 0.68
10 day starved	0.69 ± 0.09	2.55 ± 0.08	5.15 ± 0.42

Rats weighing 200–225 g were starved for 10 days or fed PFD for 21 days. The results are compared to zero day controls that had been maintained on a 18% casein diet. Non-protein-nitrogen (NPN) was determined on perchloric acid extract of the muscle homogenates. Free amino acids were determined on the same extracts by a modification of the technique of Piez and Morris.[62] Muscle DNA was measured by procedure of Wannemacher *et al.*[59,63] Essential amino acid calculations include histidine and arginine. Each value is the mean ± S.E.M. of six animals.

Protein Metabolism in Skeletal Muscles from Starved and Protein-deprived Rats

Studies on the mechanisms which regulate the protein content of muscle cells are not as extensive as in liver cells. However, the data in Figure 1 do indicate a good agreement between the protein content of

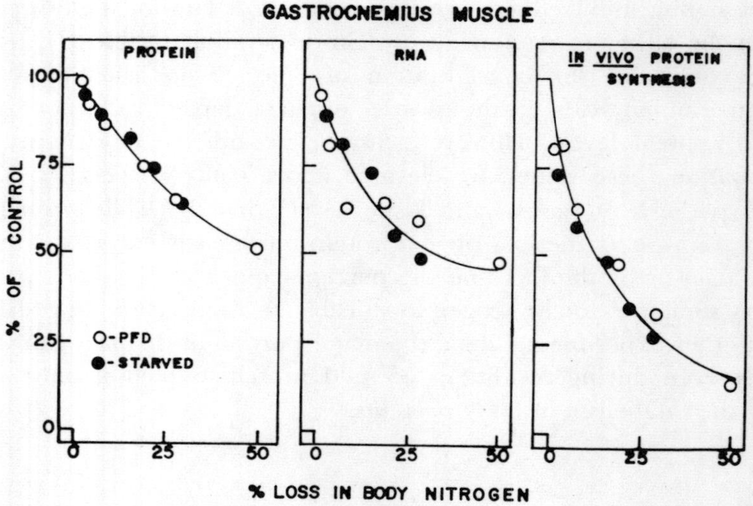

Figure 1. Protein and RNA content and rates of *in vivo* protein synthesis in a pair of gastrocnemii muscles from rats that had lost varying amounts of body nitrogen by starvation or feeding a PFD. The results are expressed as a percent of the values found in the zero time controls which weighed 200–225 g and had been maintained on a 18% casein diet. Protein and RNA content was determined by procedure of Wannemacher *et al.*,[59,63] while the rate of *in vivo* protein synthesis was measured by the radioactive amino acid uptake procedure as outlined by Wannemacher.[64] Each value is the mean of six animals.

the gastrocnemius muscle and the amount of body nitrogen which is lost during either starvation or protein-deprivation. These changes in protein concentration are correlated with a decrease in RNA content and *in vivo* rate of protein synthesis. Similar relationships are noted in adult dogs which are fed deficient levels of dietary protein.[18]

Skeletal muscle is composed of an outer connective tissue sheath which has septa at irregular intervals that surround the bundles of muscle fibers and each of their muscle fibers.[65] Muscle fibers are composed of three major protein fractions: [66] (a) The proteins of the sarcoplasma and grana, which are composed of the soluble proteins from nuclei, mitochondria and microsomes, can be readily extracted with water or neutral salt solution of low ionic strength ($\Gamma/2 < 0.2$). In early stages of development the soluble fraction can comprise nearly 70% of the total muscle proteins but in the adult it represents approximately 30%.[24,67,68] (b) Myofibrillar proteins, which represent the filamentous organization of muscle tissue, are extracted in neutral salt solutions of high ionic strength ($\Gamma/2 > 0.5$).[66] Morphologically these myofibrils are composed of structural units called "sarcomeres."[65] In striated muscles 80–90% of the myofibrillar proteins have been identified as three components: myosin (the major constituent), actin and tropomyosin.[69] (c) Stromal proteins, which are constituted by material of a collagenous nature that contribute to the structure of the sarcolemma and possibly to the Z membrane of the sarcomere, are retained in the residue after prolonged extraction with strong salt solutions.[66]

During growth the number of fibers per muscle remains constant but they increase markedly in diameter, length and myofibril content.[21] These increases in turn are correlated with a marked elevation in concentration of myofibrillar, sarcoplasmic and stromal protein content.[22,70] These same proteins are reduced when the rats are fed a protein-deficient diet or are starved until they have lost 20% of their body nitrogen (open bars, Figure 2). The percentage decrease in sarcoplasmic protein is greater than that noted for total cellular protein content; in contrast, the reduction is less than average in the fibrillar and extra-cellular fractions (proteins of stroma and connective tissue sheath). Similar results have been reported previously for muscles of protein-deprived chickens.[71] As stated earlier, the reduction in muscle protein is associated with a decreased rate of *in vivo* protein synthesis (crosshatched bars, Figure 2A). The synthesis of sarcoplasmic protein is not inhibited to as great an extent as the average protein content of the cell, while fibrillar and extra-cellular proteins are formed at a slower rate. In order to explain these deviant results it is necessary to consider the turnover rates of the proteins from the various muscle fractions. Funabiki and Makato[72] found that the specific activity or mobility was greatest for globular proteins of the sarcoplasma followed in order by sarcoplasmic albumin (sol. 50% $(NH_4)_2SO_4$), myosin, actin and collagen. They also reported half-lifes of 6.5 days for sarcoplasmic globulin, 10 days for sarcoplasmic albumin, 25–80 days for myosin and actin and greater than 100 days for muscle collagen. Thus, the cellular concentration of the sarcoplasmic proteins, which have a rapid turnover rate, quickly reaches a new steady state which corresponds to the reduced rate of pro-

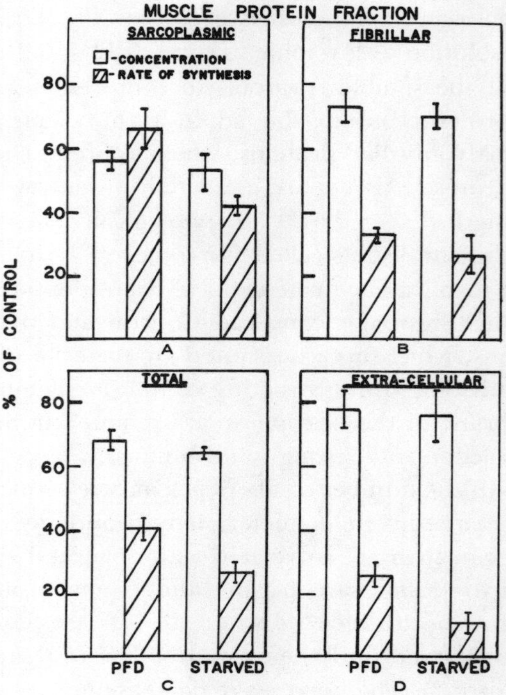

Figure 2. Protein concentration and rate of *in vivo* protein synthesis in various protein fractions of a pair of gastrocnemii muscles from rats that had been either starved for 10 days or fed a PFD for 21 days. The results are expressed as a percent of the values found in the zero time controls which weighed 200–225 g and had been maintained on a 18% casein diet. Muscle proteins were separated into sarcoplasmic, fibrillar and extra-cellular fractions by the technique of Dickerson.[67] Protein concentration and rate of *in vivo* protein synthesis were determined by previously described procedures.[64] Each value is the mean ±S.E.M. of six animals.

tein synthesis. This conclusion is supported by similar percentage reductions in protein concentration and synthetic rate for the proteins in the sarcoplasmic fraction. The proteins of the fibrillar and extra-cellular fractions, which have a relatively inert turnover rate, very slowly reach equilibrium as indicated by the marked difference in the reduction in protein content as compared to that in synthetic rate (Figures 2 C and D). Under these conditions the magnitude of the reduction in protein synthesis is not reflected until steady state conditions are obtained.

Therefore, the observation that the protein content of the fibrillar fraction of depleted muscle cells is above cellular average even though the protein synthesis rate has been reduced below average is not a contradiction when the half-lifes of these proteins are taken into consideration.

The data presented in Figures 1 and 2 indicate that the loss in muscle protein is related to a decreased rate of protein anabolism. The mechanisms which regulate the rate of protein biosynthesis in muscle tissue are currently under investigation in our laboratory. As indicated in Figure 1, the decreased rate of *in vivo* incorporation of amino acids is correlated with a reduction in total cellular RNA from muscles of protein-depleted or starved rats. These changes in RNA content are related to a significant ($P < .01$) decrease in ribosomal RNA (r-RNA) content of muscle cells from the hind leg of protein-deprived rats as compared to pair-fed or *ad libitum*-fed 18% casein controls (Table III). Restricting the food intake of 18% casein control animals results in a 10% decrease in muscle r-RNA, which can be correlated with a reduced growth rate.[14] Ribosomes from the muscles of protein-deprived rats synthesize protein at a slightly lower rate than pair-fed or *ad libitum* controls. While these results are not statistically significant, the differences are of the same magnitude as those noted by Sidransky and Verney[75] for muscle ribosomes of threonine-deficient animals as compared to casein controls. Thus, both the quality and quantity of muscle ribosomes are reduced by protein-deprivation. When both of these factors are multiplied together the so-called "protein synthetic capacity" (Table III) of muscle cells from protein-deprived rats is markedly reduced. In rats that have been depleted of 20% of their body nitrogen, the data from both *in vivo* and *in vitro* experiments indicate that protein biosynthesis in muscle cells is decreased by 30–40% as compared to the synthetic rate in same tissue of nondepleted controls (Figures 2 and 3, Table III).

Since very little muscle protein is lost during the very early (1–2 days) stage of protein deprivation,[27] it is assumed that the half-life of most of its m-RNA is equal to or greater than the 5 day half-life of r-RNA[58] and that the changes in protein synthetic rate are correlated with the ribosomal content of the cell. If the protein catabolic rate remains constant then its cellular loss is a function of the turnover rates of individual protein. This conclusion is supported by the fact that sarcoplasmic proteins which have a relatively shorter half-life are lost at a much faster rate than fibrillar proteins which have a much longer half-life (Figure 2). The fact that protein-deprivation inhibits the synthesis of

Table III. Amino Acid Incorporating Ability and Concentration of Ribosomes from Rear Leg Muscles of Protein-Depleted and Normal Rats

Treatment	mg r-RNA / Muscle Mass	% of Control	Incorporation of ^{14}C-Leucine DPM/μg r-RNA	% of Control	Protein Synthetic Capacity DPM/Muscle Mass	% of Control
PFD pair-fed	0.55 ± 0.05	35	9.51 ± 1.19	86	5.31×10^3 ± 0.64	32
Control pair-fed	1.59 ± 0.21	–	10.92 ± 1.08	–	16.64×10^3 ± 1.33	–
PFD *ad lib.*-fed	0.62 ± 0.05	35	9.46 ± 1.34	81	5.90×10^3 ± 1.01	28
Control *ad lib.*-fed	1.77 ± 0.09	–	11.75 ± 1.21	–	20.71×10^3 ± 2.23	–

Rats weighing 190–210 g were pair-fed or *ad lib.*-fed either a protein-free diet or 18% casein (control) diet for 28 days. The following muscles: gastrocnemius, posterior gracilis, soleus, semitendinosus, semimembranosus, adductor brevis, and posterior biceps femoris, were removed from both rear legs and ribosomes were prepared by a modification of the procedure of Florini and Breuer.[73] Ribosomal-RNA (r-RNA) was determined by a modification of the procedure of Wannemacher *et al.*[59,63] The protein synthetic system, *in vitro,* contains 15 Ug of r-RNA, 300 μg of cell sap protein, 0.14 m moles of sucrose, 0.065 m moles KCl, 0.01 m moles $MgCl_2$, 0.05 m moles tris-HCl buffer, pH 7.6 at 25°, 5 μmoles ATP, 0.1 μmoles GTP, 19 naturally occurring L-amino acids (range: 13-0.2 nmole)[74] and 0.9 nmole of L-U-14C Leucine (0.2 μc) in 1 ml. Incubation of the reaction and isolation of the radioactive protein were described previously.[59] Protein synthetic capacity is the product of the rate of incorporation per unit of r-RNA times the ribosomal content per muscle mass. Each value is the mean ± S.E.M. of six animals.

fibrillar protein to a greater extent than that of the sarcoplasmic fraction might be explained by their difference in molecular weight. Myosin, the major component of fibrillar fractions, has a molecular weight of approximately 420,000,[76,77] contains 3400–5050 amino acid residues[78] and requires a m-RNA with a minimum of 10,000 nucleotides. In contrast, two of the major protein constituents of the sarcoplasma, myoglobin, and cytochome *c*,[66] have molecular weights of 17,000[77] and 12,400,[80] contain 154[81] and 107[80] amino acid residues per molecule and require a m-RNA with a minimum of 465 and 325 nucleotides, respectively. Since larger m-RNA molecules require more ribosomes for stabilization,[82] a reduction in cellular r-RNA content might have a greater effect on a polyribosomal complex which is concerned with the synthesis of myosin than one which is involved in the formation of a smaller protein such as myoglobin or cytochrome *c*.

Conclusion

Figure 3 summarizes the changes that take place in liver and muscle cells of protein-deprived animals. This model is a modification of the one presented previously by Munro.[83] His data and that of others [54–57] indicate that dietary amino acid deficiency affects the liver cell by causing disaggregation of the polyribosomes, an accumulation of ribosomal

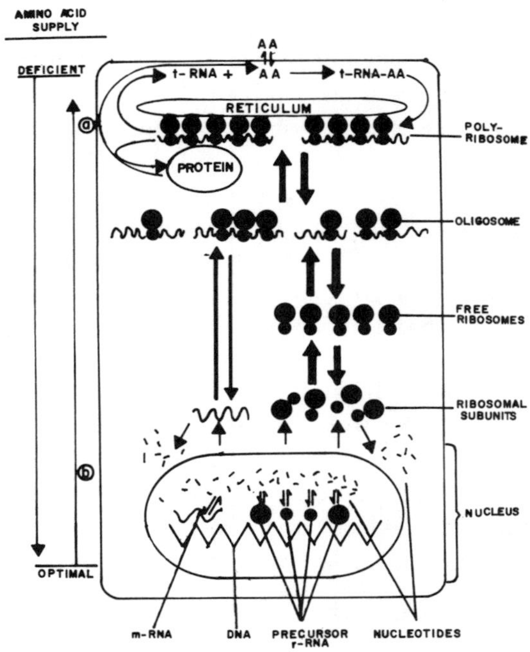

Figure 3. A schematic model of the subcellular structure of a liver or muscle cell to show the effects of amino acid deficiency on structure which regulate the rate of protein synthesis.

degradation products and a marked decrease in the ability of the polyribosomal fraction to synthesize protein (as indicated by (a) in Figure 3). These changes in cellular integrity take place rather rapidly and are independent of nuclear synthesis of new m-RNA or r-RNA.

Changes in cellular protein concentration $\left(\frac{dp}{dt}\right)$ can be expressed by the following kinetic formula:

$$\frac{dp}{dt} = K - k(P)$$

where K is the rate of protein synthesis, k is the first order rate constant for catabolism, and P is the concentration of protein at time zero. The rate of catabolism for an individual protein molecule of a cell is a function of its turnover rate. Thus, for proteins which have a short half-life, a reduction in synthesis (K) results in a rapid cellular loss of these proteins. If k remains constant the cellular concentration of the proteins continues to decrease until a new steady state is reached for the reduced rate of synthesis (K^1), which takes place rather quickly for proteins with rapid rates of turnover. The loss of these proteins corresponds with the rapid decrease in daily urinary nitrogen of animal fed a diet devoid of protein. In the rat and dog [17] this equilibrium is reached in 5–7 days and the nitrogen that is lost during this time period, 6–8% of total body nitrogen,[4] corresponds to the so-called "labile protein reserve" of many investigators.[1–4] The data in Table I indicate that proteins from cells of liver and viscera are the major contributors to the nitrogen which is lost during conditions of mild protein-deprivation. These tissues also contain many proteins with relatively short (5 days or less) half-lifes.[84]

As the animal continues to lose body nitrogen there is a gradual decrease in r-RNA content of liver and muscle [58,59] (Table III) that is correlated with a decreased amount of newly synthesized ribosomal subunits or free ribosomes, which are released from the nucleus into the cytoplasm of the cell [59] (also see (b) Figure 3). The change in r-RNA, $\left(\frac{d\ \text{r-RNA}}{dt}\right)$, can be expressed by the following equation:

$$\frac{d\ \text{r-RNA}}{dt} = K - k(\text{r-RNA})$$

where K is the rate of appearance of new cytoplasmic r-RNA and k is the first order rate constant for the breakdown of r-RNA. Since this r-RNA has a half-life of 5 days, a reduction in the appearance of new cytoplasmic subunits results in a gradual decrease in r-RNA until a new steady state has been obtained, which in turn affects the number of ribosomes that are available for formation of the polyribosomal complexes that are necessary for protein synthesis (Figure 3). These changes prevent the synthesis of proteins that have stable m-RNA (half-life greater than 5 days) as well as those with unstable m-RNA. Under these conditions the rate of protein loss is a function of the half-life of the individual proteins. Those proteins which have a long half-life slowly reach equilibrium with the new rate of protein synthesis. Since the amount of protein which is catabolized per day is a product of the rate constant times the protein concentration, the daily nitrogen loss should gradually decrease as the severity of the protein-depletion progresses. If the above conclusions are correct then most of the nitrogen which is lost during the later stages of protein deprivation is contributed by the so-called "inert proteins" which have relatively long half-lifes. This assumption is supported by the fact that after dogs or rats have been fed a diet devoid of protein for 5 or 6 days, the daily urinary nitrogen execretion continues to decrease at a very slow rate and by the data in Table IV which indicates that the myofibrillar fraction[17] of muscle tissue contributes 30% of the nitrogen which is lost from all the tissues of animals that have been depleted of 20% of their body nitrogen by either protein-deprivation or starvation. This particular muscle fraction is mainly composed of three inert proteins—myosin, actin and tropomyosin—that have relatively long half-lifes.

Table IV. Percent Contribution of the Various Muscle Protein Fractions to the Total Amount of Body Protein That Is Lost When Rats Are Starved or Fed a PFD

Muscle Fraction	10 Day Starved %	21 Day PFD %
Fibrilla	29.9 ± 1.8	28.2 ± 2.3
Sarcoplasmic	16.8 ± 1.5	16.9 ± 2.0
Extracellular	3.7 ± 0.7	3.7 ± 0.7

By assuming that most of the protein that is lost from the carcass (Table I) of the starved or protein-depleted animals is muscle protein, then data from Figure 2 can be utilized to calculate the amount of protein that the various muscle protein fractions contribute to the total body protein loss when rats were starved for 10 days or fed a PFD for 21 days. Protein loss was based on the concentrations found in zero day controls. Each value is the mean ± S.E.M. of six animals.

Therefore, the labile protein reserves of Voit[1] and others[2–4] represents nitrogen which has been mainly contributed by proteins with rapid turnover rates, while Allison's concept of "protein reserves" includes the nitrogen which is lost from these proteins plus that which is contributed by the so-called inert proteins of the body which by our calculations can represent 50–60% of the total protein content of an adult animal. Another control mechanism must exist, especially in liver and brain cells, whereby certain proteins are maintained or stimulated by different physiological stresses, even though most of the other proteins of the body are decreasing. The biological reactions which control this mechanism still have to be elucidated.

References

1. Voit, C., *Ztschr. Biol.,* **2,** 307 (1866).
2. Rubner, M., *Arch. Hyg.,* **66,** 1 (1908).
3. Borsook, H., and Keighley, G. L., *Proc. Roy. Soc.* (*London*), *Ser. B,* **118,** 488 (1935).
4. Munro, H. N., in "Mammalian Protein Metabolism," Vol. 1, H. N. Munro and J. B. Allison, Eds., Academic Press, Inc., New York, 1964, p. 382.
5. Holt, L. E., Jr., Halac, E. H., Jr., and Kajdi, C. N., *J. Am. Med. Assoc.,* **181,** 699 (1962).
6. Basak, M. N., *Indian J. Med. Res.,* **46,** 307 (1958).
7. Fisher, H., Brush, M. K., Griminger, P., and Sostman, E. R., *Am. J. Clin. Nutr.,* **20,** 927 (1967).
8. Allison, J. B., in "Symposia on Nutrition, Plasma Proteins," Vol. 2, Charles C. Thomas, Springfield, 1950, p. 123.
9. McCoy, J. R., Allison, J. B., Crossley, M. L., and Wannemacher, R. W., Jr., *Am. J. Vet. Res.,* **17,** 90 (1956).
10. Allison, J. B., Wannemacher, R. W., Jr., and Miglearese, J. F., *J. Nutr.,* **52,** 415 (1954).
11. Allison, J. B., and Wannemacher, R. W., Jr., in "Amino Acid Malnutrition," W. H. Cole, Ed., Rutgers University Press, New Brunswick, N. J., 1957, p. 1.
12. Wainio, W. W., Eichel, B., Eichel, H. J., Person, P., Estes, F. L., and Allison, J. B., *J. Nutr.,* **49,** 465 (1953).
13. Wannemacher, R. W., Jr., Russell, T. J., and Allison, J. B., *J. Nutr.,* **80,** 315 (1963).
14. Wainio, W. W., Allison, J. B., Kremzner, L. T., Bernstein, E., and Arnoff, M., *J. Nutr.,* **67,** 197 (1959).
15. Allison, J. B., Wannemacher, R. W., Sr., and Banks, W. L., Jr., *Fed Proc.,* **22,** 1126 (1963).
16. Allison, J. B., *Fed. Proc.,* **10,** 676 (1951).
17. Allison, J. B., and Wannemacher, R. W., Jr., *Am. J. Clin. Nutr.,* **16,** 445 (1965).
18. Wannemacher, R. W., Jr., and McCoy, J. R., *J. Nutr.,* **88,** 66 (1966).
19. Enesco, M., and LeBlond, C. P., *J. Embryol. Expt. Morphol.,* **10,** 530 (1962).
20. Winick, M., and Noble, A., *Develop. Biol.,* **12,** 451 (1965).
21. Enesco, M., *Anat. Record,* **139,** 225 (1961).
22. Gordon, E. E., Kowalski, K., and Fritts, M., *Am. J. Physical,* **210,** 1033 (1966).

23. Dickerson, J. W. J., and Widdowson, E. M., *Biochem. J.,* **74,** 247 (1960).
24. Robinson, D. S., *Biochem. J.,* **52,** 621 (1952).
25. Helander, E. A. S., *Acta Physiol. Scand. Suppl.,* **141,** 9 (1957).
26. Addis, T., Poo, L. J., and Lew, W., *J. Biol. Chem.,* **115,** 111, 117 (1936).
27. Waterlow, J. C., and Stephens, J. M. L., *Brit. J. Nutr.,* **20,** 461 (1966).
28. Shimke, R. T., *J. Biol. Chem.,* **237,** 459 (1962).
29. Muramatsu, K., and Ashida, K., *J. Agr. Chem. Soc. Japan,* **31,** 607 (1957).
30. Lin, E. C. C., and Knox, W. E., *Biochim. Biophys. Acta,* **26,** 85 (1957).
31. Auerbach, V. H., and Waisman, H. A., *Proc. Soc. Expt. Biol. Med.,* **98,** 123 (1958).
32. Knox, W. E., *Brit. J. Expt. Path,* **32,** 462 (1956).
33. Sayre, F. W., Jensen, D., and Greenburg, D. M., *J. Biol. Chem.,* **219,** 111 (1956).
34. Pitot, H. C., Potter, U. R., and Morris, H. P., *Cancer Res.,* **21,** 1001 (1961).
35. Muramatsu, K., and Ashida, J., *J. Nutr.,* **76,** 143 (1962).
36. Weber, G., and MacDonald, H., *Expt. Cell. Res.,* **22,** 292 (1961).
37. Soberón, G., and Sánchez, Q. E., *J. Biol. Chem.,* **236,** 1602 (1961).
38. Wannemacher, R. W., Jr., and Allison, J. B., in "Significance of Change in Plasma Amino Acid Patterns for Evaluation of Protein Nutrition," J. H. Leathem, Ed., Rutgers University Press, New Brunswick, N. J. (1968), p. 206.
39. Mendel, P., Quirin, C., Block, M., and Jacob, M., *Life Sciences,* **5,** 325 (1966).
40. Feigeleson, P., Dashman, T., and Margolis, F., *Arch. Biochem. & Biophys.,* **85,** 478 (1959).
41. Kenny, F. T., *J. Biol. Chem.,* **237,** 3495 (1962).
42. Schimke, R. T., Brown, M. B., and Smallman, E. T., *Annals New York Acad. Sci.,* **102,** 587 (1963).
43. Ashida, K., in "New Methods in Nutritional Biochemistry," A. A. Albanese, Ed., Academic Press, Inc., New York, 1963, p. 159.
44. Allison, J. B., Wannemacher, R. W., Jr., Banks, W. L., Jr., Wunner, W. H., and Gomez-Brenes, R. A., *J. Nutr.,* **78** 333 (1962).
45. Weber, G., in "Advances in Enzyme Regulation," Vol. 1, G. Weber, Ed., The Macmillan Co., New York, 1963, p. 1.
46. Drysdale, J. W., and Munro, H. N., *Fed. Proc.,* **25,** 240 (1966).
47. Wannemacher, R. W., Jr., and Yatvin, M. B., *J. Nutr.,* **85,** 393 (1965).
48. Allison, J. B., Brande, P. F., Wannemacher, R. W., Jr., Parmer, L. P., and Leathem, J. R., *J. Nutr.,* **73,** 321 (1961).
49. Blaxter, K. L., in "Mammalian Protein Metabolism," Vol. 2, H. N. Munro and J. B. Allison, Eds., Academic Press, Inc., New York, 1964, p. 195.
50. Hopper, A. F., Yatvin, M. B., and Wannemacher, R. W., Jr., *J. Nutr.,* **88,** 202 (1966).
51. Yatvin, M. B., and Wannemacher, R. W., Jr., *Endocrinology,* **76,** 418 (1965).
52. McFarlane, A. S., in "Mammalian Protein Metabolism," Vol. 1, H. N. Munro and J. B. Allison, Eds., Academic Press, Inc., New York, 1964, p. 298.
53. Munro, H. N., *Proc. Nutr. Soc. (Eng., Scot.),* **13,** 115 (1954).
54. Wunner, W. H., Bell, J., and Munro, H. N., *Biochem. J.,* **101,** 417 (1966).

55. Sox, H. C., and Hoagland, M. B., *J. Molec. Biol.,* **20,** 113 (1966).
56. Webb, T. E., Blobel, G., and Potter, V. R., *Cancer Res.* **26,** 353 (1966).
57. Staehelin, T., Verney, E., and Sidransky, H., *Biochim. Biophys. Acta,* **145,** 105 (1967).
58. Wilson, S. H., and Hoagland, M. B., *Biochem. J.,* **103,** 556 (1967).
59. Wannemacher, R. W., Jr., Cooper, W. K., and Yatvin, M. B., *Biochem. J.,* **107,** 615 (1968).
60. Rechcigl, J., Jr., and Price, V. E., in "New Methods of Nutritional Biochemistry," A. A. Albanese, Ed., Academic Press, Inc., New York, 1963, p. 185.
61. Stephens, J. M. L., and Waterlow, J. C., *Nature,* **211,** 978 (1966).
62. Piez, K. A., and Morris, L., *Analyt. Biochem.,* **1,** 187 (1960).
63. Wannemacher, R. W., Jr., Banks, W. L., Jr., and Wunner, W. H., *Analyt. Biochem.,* **11,** 320 (1965).
64. Wannemacher, R. W., Jr., in "Proc. 7th Int. Congr. Nutr.," Vol. 5, J. Kuhnau, Ed., Vieweg & Sohne, Braunschweig, 1967, p. 205.
65. Wall, E. W., in "The Structure and Function of Muscle," Vol. 1, G. H. Bourne, Ed., Academic Press, Inc., New York, 1960, p. 21.
66. Szent-Gyorgyi, A. G., in "The Structure and Function of Muscle," Vol. 2, G. H. Bourne, Ed., Academic Press, Inc., New York, 1960, p. 1.
67. Dickerson, J. W. T., *Biochem. J.,* **75,** 33 (1960).
68. Hasselback, W., and Schneider, G., *Biochem. Z.,* **321,** 462 (1951).
69. Perry, S. V., *Physiol. Revs.,* **36,** 1 (1956).
70. Mendes, C. B., and Waterlow, J. C., *Brit. J. Nutr.,* **12,** 74 (1958).
71. Dickerson, J. W. T., and McCance, R. A., *Brit. J. Nutr.,* **14,** 331 (1960).
72. Funabiki, R., and Kandatsu, M., *Nippon Nogei Kagakie Kaishi,* **39,** 501 (1965).
73. Florini, J. R., and Breuer, C. B., *Biochemistry,* **5,** 1870 (1966).
74. Cooper, W. K., Muramatsu, K., and Wannemacher, R. W., Jr., *Biochem. Biophys. Acta,* **169,** 269 (1968).
75. Sidransky, H., and Verney, E., *Biochim. Biophys. Acta,* **138,** 426 (1967).
76. Mommaerts, W. F. H. M., and Aldrick, B. B., *Biochim. Biophys. Acta,* **28,** 627 (1958).
77. Kielley, W. W., and Arrington, W. F. H., *Biochim. Biophys. Acta,* **41,** 401 (1960).
78. Kominz, D. R., Hough, A., Symonds, P., and Laki, K., *Arch. Biochem. Biophys.,* **50,** 148 (1954).
79. Edsall, J. T., in "The Proteins," Vol. 1, Part A, H. Neurath and H. Bailey, Eds., Academic Press, Inc., New York, 1952, p. 634.
80. Takahashi, K., *J. Biochem.* (*Tokyo*), **46,** 1323 (1959).
81. Perkoff, T. R., Hill, R. L., Brown, D. M., and Taylor, F. H., *J. Biol. Chem.,* **237,** 2820 (1962).
82. Rich, A., Becker, S. P. Y., Darnell, J., and Hall, C., *Science,* **142,** 1658 (1963).
83. Munro, H. N., Nutr. Dieta, **8,** 179 (1966).
84. Buchanan, D. L., *Arch. Biochem. & Biophys.,* **94,** 489 (1961).

V. Biochemistry

The Phosphorylation of Thymidine as a Factor Controlling Cellular Proliferation *

JAY S. ROTH
Division of Biological Sciences
Section of Biochemistry and Biophysics
University of Connecticut
Storrs, Connecticut

I. Introduction

The intracellular concentration of thymidine triphosphate (TTP) may function as one of the most important controlling mechanisms in cellular proliferation. Some known and possible relationships are shown in Figure 1.

The action of TTP is not simple. Its behavior may be likened to that of a computer control in a commercial chemical plant in which a series of complicated reactions is being carried out automatically. Temperature, pressures, and concentrations are fed into the computer which then makes the necessary adjustments in the flow of material to give optimum production of product.

Each of the arrows in Figure 1 is in itself a delicately balanced system containing from several to many components. (For example, refer to Figure 2 which represents synthesis and allosteric regulation.)

* The author is most grateful to the National Institutes of Health, National Cancer Institute for Career Fellowship 31-63 and for research support from grants 07824 and 08384 and to the American Cancer Society for grant P-303.

I should like to express my appreciation also to Mr. Burr Atkinson and Mrs. Kathy Ellison who compiled the tables in the Appendix and made many useful suggestions concerning the manuscript.

Each component, in turn, may be influenced or controlled by a number of diverse factors.

A simplistic concept of TTP regulatory control from Figure 1 suggests that a high TTP concentration would favor DNA synthesis. High TTP concentration would also shut off further TTP synthesis by allosteric and possibly gene control and might promote more rapid breakdown of the nucleotide. As long as DNA synthesis is proceeding rapidly, TTP might be drawn from the pool rapidly enough to prevent a buildup and the shutting off of synthesis. Many other interpretations or relationships may be derived from the diagram in Figure 1.

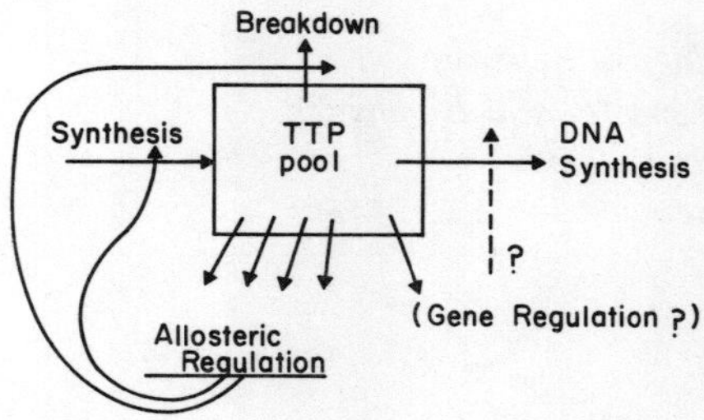

Figure 1. Some known and possible relationships of TTP metabolism and its metabolic control actions.

It is obvious that a thorough comprehension of all of the components of Figure 1 would be necessary to understand the relationships of TTP synthesis to cellular proliferation. Although we are far from this understanding it is often useful to summarize the present state of our knowledge, and this review attempts to do this. Shortly after the completion of this review, a book by Cleaver on thymidine metabolism and cell kinetics was published.[209] This book contains useful and interesting material to supplement this review. Also, Sugino[213] has recently published a review on the metabolism of deoxyribonucleotides.

The main reactions which lead to the synthesis of TTP are depicted in Figure 2 and also some allosteric control reactions. The pathways enclosed in boxes I and II represent important rate-controlling ones for DNA synthesis. Since the number of enzymes involved is very large and their control complex, this report will be primarily confined to the most proximal steps leading to the formation of TTP, namely phosphorylation of TdR, TMP, and TDP, reactions 1, 2, 3 (Figure 2), and

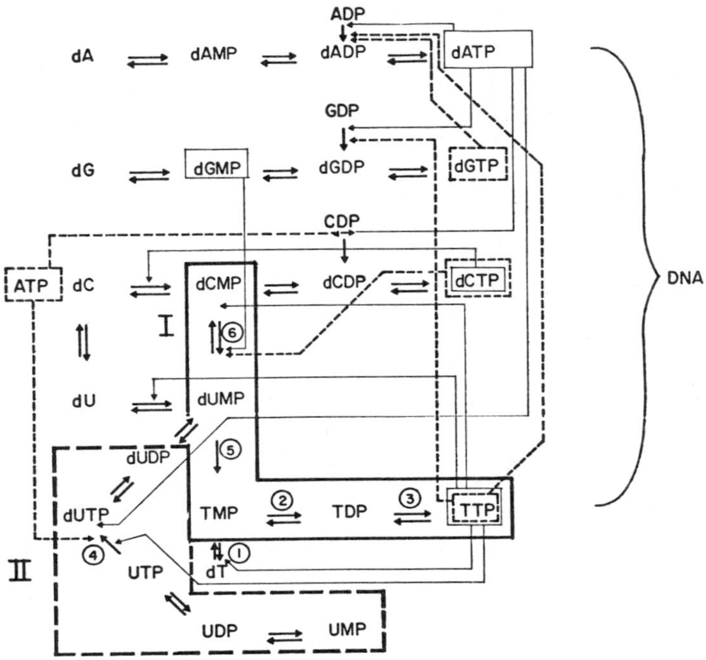

Figure 2. Some reactions and feedback controls involved in the synthesis of precursors of deoxynucleoside triphosphates required for DNA synthesis.

The light dotted lines represent positive (accelerating) feedback control or modulators (ATP) which facilitate the reactions indicated. The light solid lines represent negative feedback controls by the nucleotides that are in the boxes. The reactions within the heavy solid line (I) represent what is believed to be the major pathway to the synthesis of TTP in higher animals, while the reactions within the heavy dotted line (II) represent an alternate pathway to TTP synthesis the importance of which has not been determined. The numbers in circles refer to the following enzymes which catalyze major reactions: 1, thymidine kinase; 2, thymidylate kinase; 3, thymidine di(pyro)phosphate kinase; 4, uridine tri- or diphosphate reductase; 5, thymidylate synthetase and; 6, deoxycytidylate deaminase.

in particular, reactions catalyzed by enzymes 1 and 2 since enzyme 3 not only is present in large amounts in the cell at all times but also has a broad specificity.

Considering enzymes 1 and 2, thymidine kinase and TMP kinase, respectively, the evidence presently available suggests that TMP kinase is *the critical rate controlling enzyme* for TTP synthesis while thymidine kinase, although important to rapidly growing abnormal systems such as tumors and virus infected cells, is simply a "salvage" enzyme not on the main pathway for TTP synthesis (see Figure 2) in normal systems.

The importance of thymidine kinase and especially TMP kinase in relation to proliferation is manifest from the following series of observations which will be documented in detail later on.

1. Although both enzyme activities are generally low or absent in nonproliferating tissues, they are almost always present in moderate to high activities in rapidly proliferating tissues, for example:

a) Tumors: There is often a rough correlation between enzyme activity and rate of proliferation of the tumor. TMP kinase activity is *always* elevated to some degree, but in some tumors thymidine kinase activity may not be appreciably above levels in corresponding normal tissues.

(b) Regenerating liver: The activities of both thymidine kinase and TMP kinase increase several-fold in regenerating liver starting about 18 hours after the removal of two-thirds of the liver and this increase is closely correlated with an increase in DNA synthesis observed about this time.

(c) Embryonic tissues: In the developing chick embryo TMP kinase activity is low initially but rises to maximum levels after 4 days of growth. Thymidine kinase is found only in very low amounts and does not change during the entire course of development. Similar results are found in the embryo of the frog *Rana pipiens.*

(d) Viral infected systems: An early prelude to manufacture of virus DNA is invariably an increase in TMP kinase and thymidine kinase activities.

2. The levels of TMP and thymidine kinase are sensitive to the physiological state of the cell. Thus, injury such as by X-radiation and chemicals, or cessation of growth due to a variety of causes, leads to a rapid decline in the activities of TMP kinase and thymidine kinase.

3. The levels of the kinases for dCMP, dGMP and dAMP are high even in resting tissues and do not change significantly when rapid growth ensues.

In the discussion that follows, the enzymes thymidine kinase and TMP kinase are, for the most part, separately described although their close relationship at times requires them to be studied and considered together.

II. Thymidine Kinase

A. Distribution

Thymidine kinase has been reported to be present in a wide variety of animal, plant, and microbiological forms. The amount, or activity of the enzyme may vary considerably depending on the source and physiological state of the tissue. Although a limited number of higher animals have been examined, the enzyme is probably widespread if not universal.

In the field of microorganisms, the enzyme in *E. coli* has been purified and its properties investigated by Okazaki and Kornberg.[1,2] It has also been found in *Tetrahymena pyriformis,*[3] *Chlorella pyrenoidosa,*[4] and the slime mold *Physarum polycephalum.*[5] Thymidine kinase (and TMP kinase) was found in low amounts in extracts of the wing epidermis of the silkworm *Antheraea pernyi* during diapause.[6,7] Mammalian tissues, particularly the rat, have been examined most extensively. Adult liver contains only small to trace amounts with activity increasing with decreasing age. The enzyme is present also in thymus and kidney. Many tumors appear to contain elevated amounts of thymidine kinase[8–18] as do most cells in tissue culture[11,17–21] with the exception of some mutants.[22] Regenerating liver contains considerable activity particularly compared to normal adult liver which has only trace amounts.[11,16,23–37] Thymidine kinase is present also in rabbit bone marrow.[27]

Only small amounts of thymidine kinase have been found in the whole chick embryo,[38,39] chick embryo muscle,[40,44] or embryo of *Rana pipiens.*[42] Its distribution in mammalian embryo has not yet been studied.

In plants it has been found in *Zea mays,*[43,44] in roots of sprouting seedlings,[45] in microspores of *Lilium longiflorum*[46] and *Trillium.*[47]

Morse and Potter[17] have studied thymidine kinase in several cell cultures derived from the Novikoff hepatoma. Indirect evidence for high levels of thymidine kinase activity was obtained by complete inhibition of cell growth by 10^{-3} M thymidine or 10^{-8} M 5-fluorodeoxyuridine and complete reversal of amethopterin and 5-fluorodeoxyuridine inhibition by 10^{-5} M thymidine. Thymidine toxicity was readily reversed by 10^{-5} M deoxycytidine. Two variant strains of cells were found, one of which apparently lacks thymidine kinase since it was irreversibly inhibited by the presence of amethopterin and was resistant to high levels of thymidine and 5-fluorodeoxyuridine. The second variant strain possessed thymidine kinase but was resistant to high levels of thymidine.

1. INTRACELLULAR LOCATION

It would be expected that enzymes concerned with the synthesis of immediate precursors of DNA should be found in close conjunction to the source of DNA, that is, in the nucleus. However, in a homogenate of tumor, or regenerating liver, prepared in 0.25 M sucrose solution, almost all thymidine kinase activity is obtained in the high speed supernatant fraction, and very little in the sediment. In our laboratory we have observed[48] that a considerable portion of the enzyme may be retained in the nucleus by using sucrose solutions of higher concentration.

In the way of indirect evidence for localization of thymidine kinase, Stone, Miller and Prescott[49] using a dry emulsion film were able to show that in Tetrahymena, the soluble thymidine derivative pool exists exclusively in the nucleus. This is in agreement with other studies which indicate that thymidine nucleotides are localized mainly in the nucleus.[50,51]

Hacker has reported[52] that although purified intact mitochondria from normal liver, regenerating liver, and chemical carcinogen-induced hepatoma showed little thymidine kinase activity, alternate freezing and thawing released considerable amounts of activity which could then be recovered in a high speed supernatant fraction. Similar effects were noticed with mitochondria from the Morris 5123 hepatoma but in this case the intact mitochondria already had significant activity.

The finding that mitochondria contain thymidine kinase (and TMP kinase) is of interest in that it adds additional weight to the theory that these particulates are autonomous.

B. Thymidine kinase—purification

Bresnick and Thompson have purified the enzyme from Novikoff ascites and McCoy MDAB hepatomas as well as the Walker carcinosarcoma.[8] Specific activities ranged from 160 to 1320, representing a concentration of enzyme activity of about 90-fold. It should be noted that specific activities of the order 1000 have been obtained from crude high-speed supernatant fractions of the Novikoff (solid) and MDAB hepatomas assayed by a different procedure.[15] The purification procedure involved ammonium sulfate precipitation, heating at 60°, use of calcium phosphate gel and a final ammonium sulfate precipitation. Yields ranged from 25 to 35% for the three tumors. There were no significant differences in the enzymes from the three sources.

Ives, Morse, and Potter[9] separated thymidine kinase from thymidylate kinase by adjusting a high-speed supernatant fraction to pH 4.5 with 1 N acetic acid, centrifuging and neutralizing the supernatant fraction to pH 7.5 with KOH. From 66 to 80% of the thymidine kinase activity was retained, but practically no TMP kinase.

Okazaki and Kornberg [7] purified the enzyme approximately 1200-fold from *E. coli.* The steps involved heat treatment, streptomycin precipitation, two ammonium sulfate precipitations, and column chromatography on DEAE cellulose.

Grav and Smellie have separated and purified thymidine and TMP kinases from Landschutz ascites tumor cells.[53] The procedures used for separation were:

(a) A high speed supernatant fraction was made up in 0.01 M phosphate buffer pH 7.6 to a protein concentration of 6–8 mg/ml.

(b) The pH was adjusted to 4.5 and held there for 30 minutes. TMP and TDP kinases were destroyed.

(c) If, however, 80 mμ moles of TMP per 1.5 mg of protein was added before adjustment of the pH to 4.5 and this was followed by centrifugation and return of the pH to 7.6, this allowed recovery of 80–120% of the original TMP kinase activity. Much of TDP kinase was destroyed as before.

(d) Alumina Cγ gel absorbed most of the TMP kinase activity. In the presence of TMP at the same concentration as above, 40–80% of the TMP kinase was not absorbed and could be further purified by gradient elution from Sephadex G-100 with 0.01 M phosphate buffer, pH 7.6 to give TMP kinase purified about 1000-fold.

(e) Chromatography of the pH 4.5 fraction directly on the gel led to the separation of three kinases. The stability in increasing order was thymidine kinase, TMP kinase, TDP kinase.

Wanka and co-workers purified thymidine kinase from germinating seeds by heat-treatment and fractionation on Sephadex G-50 and carboxymethyl cellulose-25.[45] About a 40-fold purification was achieved.

C. Properties of thymidine kinase

1. STABILITY

Activity in extracts from wing epidermis of the silkworm *Antheraea peruyi* is quite stable.[7] No loss could be detected after storage for a month in the frozen state or after incubation for 2 hours at 37°. The enzyme was also stable to dialysis in the absence of thymidine for 16 hours at 2° against sucrose-Tris-HCl, pH 7.7.

Partly purified enzyme from hepatomas is rapidly inactivated at 60°. Within 5 minutes only 20% of the activity remains. The addition of thymidine greatly stabilized the enzyme under these conditions. The optimum concentrations of ATP, Mg^{++}, and thymidine have been determined (see appendix) for enzyme from many sources. As reported by Bianchi *et al.*,[10] Bojarski and Hiatt[54] and Morse, Ives and Potter[9] the enzyme is relatively stable to dialysis.

The enzyme from *E. coli* also appears to be quite stable. In the purification, heating at 70–72° in a crude extract results in only a 10% loss of activity.[7]

Bukovsky and Roth studied the stability to storage at −17° of thymidine phosphorylation in several tumors.[13] Formation of TMP was not affected in the McCoy MDAB or Dunning hepatoma after storage up to 90–100 days but in the Morris 5123 hepatoma it dropped to half the control value after 4 days. Formation of TDP + TTP was not affected for 27–36 days in the first two tumors but again decreased to half the control value in the 5123 hepatoma after 4 days.

Purified thymidine kinase from germinating seeds[45] was quite stable to 40° (20 hours) and to 5 minutes at 50° but after heating for 2 minutes at 60° only 25% of the activity was recovered.

2. EFFECT OF METAL IONS

There is fairly general agreement that added Mg^{++} is not required for thymidine kinase activity (see references 9 and 13); Mn^{++} can substitute in part for Mg^{++} and all activity is lost by addition of EDTA. Determination of metal requirements is complicated by the fact that ATP is an excellent chelating agent.

Bresnick *et al.*[55] working with thymidine kinase from tumor reported than Mn^{++}, Co^{++}, Ca^{++}, and Fe^{++} could replace Mg^{++} to a small extent, but none were as effective and higher concentrations were inhibitory in the presence of Mg^{++}. Thymidine kinase from *E. coli* required Mg^{++}, Mn^{++}, ATP, and bovine serum albumin. Omission of Mg^{++} or Mn^{++} gave only a slight reduction of activity. Omission of both led to complete loss of activity.[1]

3. OCCURRENCE OF TWO COMPONENTS OF THYMIDINE KINASE IN PLANTS

Wanka and co-workers[43–45] studied thymidine kinase in roots of germinating corn and wheat seedlings. They found that the enzyme could be separated by precipitation with ammonium sulfate into two components designated P and T, both enzymatically inactive. However, high activity was restored by combining the two components either from the same or different sources. The components, except for component P from wheat, were stable for several months if stored in a deep-freezer. The functional enzyme was found in tissues in which mitosis or polyploidization was occurring. Component P was present in all plant materials studied. From the data it appeared that enzyme activity is controlled via formation and degradation of component T. Similar results were observed for the phosphorylation of uridine.

4. MOLECULAR WEIGHT

The molecular weight of the aggregated form of the enzyme is in excess of 600,000 as determined by Sephadex G-200.[55]

5. INHIBITORS OTHER THAN TTP

Thiol reagents profoundly reduced enzymatic activity of the enzymes from tumors with *p*-hydroxymercuribenzoate being the most effective. Mammalian thymidine kinase was also inhibited by sodium dodecyl sulfate, methyl guanidine sulfate, dioxan, ethylene glycol and urea. It was not affected by dCDP[55] in contrast to the enzyme from *E. coli.*[2]

Utilizing a new spectrophotometric assay for thymidine kinase, Baker, Schwan and Santi[56] studied the inhibition of thymidine kinase by nucleoside derivatives. The method depends on the use of 2′-deoxy-5-fluorouridine as the substrate. The concentration of the resultant F-dUMP was assayed by measuring the inhibition of TMP synthetase. The assay has a magnification of 1000-fold since F-dUMP is known to bind to TMP synthetase 1000-fold better than the normal substrate dUMP. *E. coli* thymidine kinase was used. The order of binding to the enzyme was found to be thymidine > FUdR > UdR. Table I shows some of the results.

Table I. Some Inhibitors of Thymidine Kinase

Thymidine Analog	Conc.	Loss in Binding	% Inhibition
3′-deoxythymidine	3 mM	80-fold	64
5′-deoxythymidine	3 mM	30	84
5-cyclopentane	2 mM	120	44
5-ethoxy	2 mM	600	21
N^3-C_5H_{11}	3 mM		70
4-thiothymidine			74
5,6-dihydrothymidine			0

Large groups on the 5′-hydroxyl had no effect. (From Baker *et al.*[56])

Gray *et al.*[27] have reported that when supernatant fraction from regenerating rat liver, Ehrlich ascites cells or bone marrow cells, all of which contain thymidine kinase, are mixed with supernatant fractions from resting tissue, such as rat liver or kidney, the total ability to phosphorylate thymidine was considerably reduced. They suggested that an inhibitor might be present in the nonproliferating tissues. The nature of this inhibitor was not studied in this paper, although it was probably protein since boiled or acid-precipitated extracts had no effect. The activity of extracts in normal liver was also increased by ammonium sulfate fractionation. The most likely explanation of these results is that the resting tissues contained high phosphatase activity which hydrolysed TMP as it was formed. There appears to be very little evidence to date for specific thymidine kinase or TMP kinase inhibitors in tissues.

Bresnick and Williams have studied the action of 5-trifluoromethyl-2′-deoxyuridine on thymidine kinase.[57] The analog can be phosphorylated as readily as thymidine or 5-bromodeoxyuridine, but inhibits the formation of TMP. K_i was 3.7×10^{-6} M. The inhibition was a complex function of ATP concentration.

Azaserine (o-diazoacetyl-L-serine) was found by Sartorelli and Booth[58] to depress the levels of thymidine kinase and TMP kinase in sarcoma 180 ascites cells in mice. There was a 60% decrease in the amount of adenine nucleotides in the inhibited cells. Addition of adenine reversed the decrease in kinase activity suggesting a possible role for the adenine nucleotides in regulating the activities of the kinases.

6. SUBSTRATE SPECIFICITY

Relatively few studies of substrate specificity with purified enzymes have been carried out. Partly purified enzyme from hepatomas phosphorylated deoxyuridine, thymidine, 5-halogendeoxyuridines and azathymidine (30% of thymidine).[55] The *E. coli* enzyme phosphorylates deoxyuridine, thymidine and the 5-substituted halogen derivatives equally well. dGTP could serve as a phosphate donor almost as well as ATP; dATP and dCTP were somewhat less active, while GTP and ITP were relatively poor donors. Other nucleoside tri- and diphosphates were inactive. Very high levels of ATP were necessary to saturate the enzyme.[2]

The "apparent" Michaelis constants for thymidine have been reported to be 2.7 mM for Novikoff hepatoma, 3.2 for McCoy MDAB hepatoma and 3.7 for Walker tumor thymidine kinases.[55]

There are many observations in the literature on the toxicity of thymidine or deoxyuridine derivatives. It is generally assumed that toxicity can only be exhibited if (1) the compound passes through the cell membrane, and (2) is transformed by thymidine kinase and other kinases to a phosphorylated derivative (TTP) which then exerts an inhibitory effect.

7. AGGREGATION AND DISAGGREGATION

Thymidine kinase from hepatomas has been shown by Bresnick, Thompson and Lyman [55] to exist in two forms: In dilute solution it is an aggregate with M.W. around 600,000. If the ionic concentration (KCl) is increased to around 0.2 M or a sucrose concentration greater than 0.15 M is used, the enzyme disaggregates into particles with molecular weight around 1.13×10^5. Bresnick and collaborators observed that with aggregated enzyme, TTP inhibited competitively while with the disaggregated form, the inhibition was noncompetitive. Some constants are compared below:

	M.W.	K_m thymidine	K_i TTP	K_m/K_i
Aggregated	6×10^5	3.3 uM	4.8 uM	1.5 (compet.)
Disaggregated	1.13×10^5	6.9 uM	1.3 uM	5.3 (noncompet.)

The ratios of K_m/K_i suggest that the disaggregated form is more effectively inhibited with TTP.

	K_m ATP
Aggregated	$1.8 - 2.1 \times 10^{-4}$ M
Disaggregated	2.1×10^{-4} M

The pH optimum for both forms was the same and the heat stabilities varied only slightly. The catalytic activity was not changed appreciably by aggregation.

8. MECHANISM OF THYMIDINE PHOSPHORYLATION

In early studies on the phosphorylation of thymidine by supernatant fractions from lymph nodes and spleen of leukemic mice, Bianchi and collaborators [10] concluded that TTP was formed by pyrophosphate addition to TMP. This erroneous conclusion was reached probably because their techniques were not exact enough to detect small quantities of TDP formed initially. It was recognized later, that TDP kinase is so active that any TDP formed would almost instantly be converted to

TTP. Although Bianchi and collaborators diluted their supernatant fractions 17-fold, the concentration of TDP kinase was probably still high enough to convert all TDP to TTP in the first few seconds of the reaction. In a later paper Grav and Smellie corrected this mistake.[59] McAuslan and Joklik using HeLa 53 cells also proved that TDP was an intermediate in the formation of TTP.[60]

Ives has provided conclusive evidence that TDP is the precursor of TTP in crude extracts prepared from the Novikoff hepatoma.[61] The proof depended on showing that TTP produced from TMP by ATPγP^{32} was doubly labeled rather than singly labeled. Use of βP^{32} labeled ATP gave no label in the TTP. The level of TDP kinase was found to be about 1000 times that of thymidine or TMP kinase and the TDP kinase was quite nonspecific in its requirements. Bianchi and co-workers[62] also showed high levels of TDP kinase in spleens of leukemic mice, rat liver, and rat and human erythrocytes and thus explained their earlier inability to detect TDP as an intermediate in the phosphorylation of TTP.

9. ALLOSTERIC EFFECTS OF TTP AND OTHER NUCLEOTIDES: FEEDBACK INHIBITION BY TTP

One of the most interesting and possibly important properties of thymidine kinase is its marked feedback inhibition by TTP. Enzyme from all sources examined thus far, with the exception of *Tetrahymena pyriformis* (see below) is inhibited to some degree by small concentrations of TTP.

Ives, Morse, and Potter were one of the first groups to study feedback inhibition of thymidine kinase by TTP.[9] They used Novikoff hepatoma cells in tissue culture. The reaction was quite specific for TTP and they concluded that the inhibition was reversible or competitive in contrast to the results of Bresnick.[8] This discrepancy may possibly be explained on the basis of a modulating-effect of the ATP-Mg^{++} complex (Figure 3). As the concentration of the latter increased, the inhibition by a given quantity of TTP decreased markedly. Another possibility is that the two groups were working with different aggregate forms of the enzyme, since as indicated above by Bresnick and co-workers[55] one form is competitively inhibited and the other noncompetitively.

Ives, Morse, and Potter[9] also reported feedback inhibition by TTP of thymidine kinase activity from rat spleen, thymus, regenerating liver, Morris 7800 hepatoma, *E. coli,* and *Physarum polycephalum.*

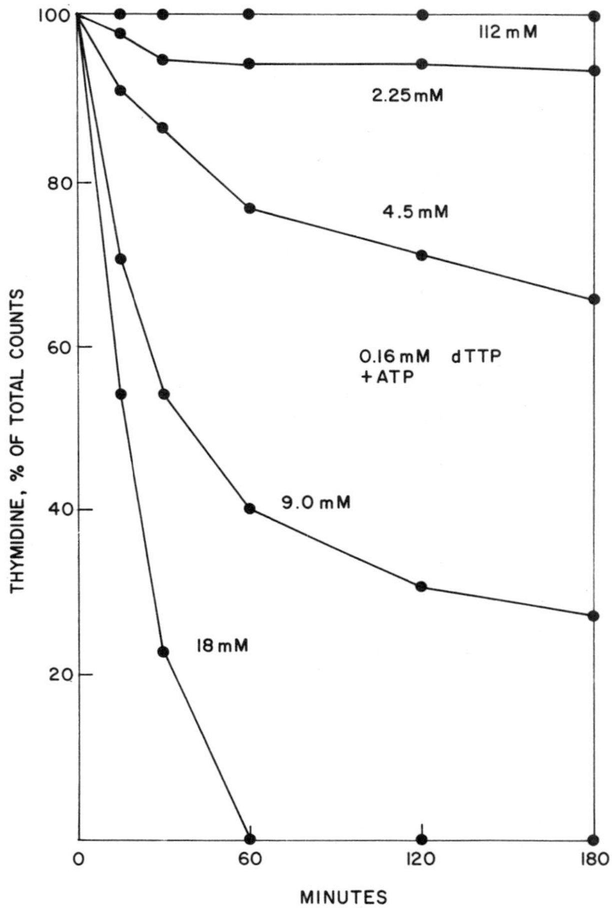

Figure 3. The modulating effect of ATP-Mg^{++} on the inhibition of thymidine kinase by exogenous TTP (0.16 mM). The disappearance of thymidine was strongly inhibited at all time points when the ATP-Mg^{++} concentration was low, but weakly inhibited at high concentrations of ATP-Mg^{++}. From Bresnick *et al.*[8]

Bukovsky and Roth[13] studied the feedback inhibition of thymidine kinase in a series of transplantable hepatomas in the hope of finding differences that could be correlated with the growth rate of the tumors. All the tumors examined showed feedback control and although significant quantitative differences were observed, these could be due to different properties of the crude homogenates rather than to intrinsic differences in the allosteric properties of the enzyme. They noted that inhibition was considerably more complete at lower levels of TTP in the absence of Mg^{++}.

Bresnick and collaborators found inhibition of thymidine kinase by both TTP and dCTP.[14] Their results are briefly summarized in the table below. Eight other tumors of varying growth rate, mostly slow, gave similar results. (Only one sample of each was tested.) On the other hand, the Novikoff ascites and Dunning LC 18 while inhibited strongly by TTP were not at all affected by dCTP. Induced and spontaneous mouse hepatomas fell into the class of normal liver. The authors suggest the possibility of two thymidine kinase enzymes or isozymes.

Table II. The Negative Feedback Effect of TTP and dCTP on Some Normal Tissue and Tumors

Tissue	TTP	dCTP
Adult liver	++	++
Embryonic liver	+++	0
Regenerating liver	+++	0
Morris 7288C (5)	++	++
Morris 7800 (5)	++	++
Morris 5123D (2)	++	++
Novikoff ascites (3)	+++	0
Dunning LC 18 (3)	++	0

++ = moderate.
+++ = strong.
From Bresnick *et al.*[14]

Using a partly purified enzyme from hepatoma, Bresnick and Thompson studied feedback inhibition as a function of TTP and ATP concentration.[8] The kinetics were complex and appeared to be noncompetitive with thymidine. A more pronounced inhibition was obtained at lower concentrations of ATP. The inhibition by TTP was not affected by $HgCl_2$, thiol reagents, duponol, urea, or other agents which disrupt hydrogen and hydrophobic bonds.

Inhibition by TTP appeared to be competitive with thymidine but not with ATP or dCDP and actually increased as the concentration of the latter increased. Aging of the enzyme changed its kinetic properties somewhat suggesting that conformational changes were taking place on storage. In physiological terms, the kinase is activated by those deoxynucleoside di- and triphosphates which accumulate when the supply of TTP is insufficient. This activation operates by reducing the K_m for deoxythymidine thus making for its more efficient utilization at low concentrations.

The enzyme from silkworm[6] is inhibited about 45% with equimolar TTP and about 50% by a 10-fold higher concentration of dCTP; dGTP and dATP had no effect.

Bresnick and Karjala[16] also demonstrated that normal and leukemic human leukocyte thymidine kinases were readily susceptible to feedback inhibition by TTP. Crude extracts of cells were used which probably explains the almost equal effectiveness of TDP; dCTP and dCMP also inhibited the kinase with dCTP being slightly more effective than TTP and dCMP considerably less active than dCTP. Bresnick and Karjala suggested that the low levels of thymidine kinase activity reported earlier by Bianchi[34] may have been due to feedback inhibition by TTP formed in the incubation mixture. This may also be the reason for low incorporation of added thymidine ^3H observed in some rapidly growing tissues.

E. coli thymidine kinase[2] was markedly stimulated by dCDP, hydroxymethyl dCDP, dADP, dGDP, and GDP which are inert as phosphate donors. The lower the concentration of ATP, the greater was the stimulation. Stimulation by dCDP was also observed when dATP, dGTP, and GTP were the phosphate donors.

The reaction was inhibited 50 to 90% by TTP at 2 to 8×10^{-4} M when the concentration of thymidine was 0.57×10^{-3} M and dCDP was 2.86×10^{-3} M. At 7.1×10^{-4} M, BrdUTP was as inhibitory as TTP while the effect of FdUTP was marginal.

Kit and Dubbs have compared the properties of thymidine kinase isolated from noninfected and virus (vaccinia, Herpes simplex) infected cells[63] by a modification of the procedures of Weissman *et al.*[11] and Ives *et al.*[9] Enzyme purification was only 3–5-fold in both cases. Both enzymes responded to about the same degree to feedback inhibition by TTP. The half life to incubation (37°) of enzyme from vaccinia virus infected cells was greater than 400 minutes, while from the controls it was 30 minutes, even in the absence of all small molecules (Sephadex treated). The enzyme from Herpes simplex infected cells, however, was only slightly more stable than that from the noninfected cells. Antiserum to the vaccinia induced enzyme inhibited the enzyme from infected cells but did not affect the activity from control cells. Growing normal cells in the presence of thymidine, prevented the decline in enzyme activity usually observed at about 72 hours after subculture.

Feedback inhibition by TTP has also been reported in the chick embryo by Maley and Maley[39] and Roth and Askew[38] as well as for regenerating liver by Breitman[64] who was one of the first to observe it.

Shoup, Prescott, and Wykes[3] have studied thymidine kinase activity in *Tetrahymena pyriformis* strain HSM. Tetrahymena can produce TTP from thymidine or uracil. Cells were lysed in the presence of 0.2% Triton X-100. Maximal enzyme activity was obtained at about 1.5 mM ATP, considerably lower than for other systems. TTP did not exert any feedback inhibition under a variety of conditions of TTP and ATP concentrations. (It appears possible that Triton X-100 may alter the conformation of the enzyme abolishing the allosteric effect.) No inhibition or activation was noted with dCTP and several other nucleotides. This is the only report of lack of feedback inhibition and since, in reality, Tetrahymena may be a plant[197] this unique property may be common to plant thymidine kinases. It would be of considerable interest to test thymidine kinases from other plants for feedback inhibition. On the basis of their results, Shoup *et al.* tentatively assume that thymidine kinase may be a constitutive enzyme in Tetrahymena. This is based on the observations that enzyme levels are the same in extracts of stationary and log phase cells. Also no differences are noted in the activity of the enzyme in cells grown in the presence of thymidine compared to those without it. Thymidine kinase is still present in cells that have ceased to make DNA.

10. FEEDBACK INHIBITION BY TTP AFFECTING GROWTH

Several investigators[65] have shown that TTP formed intracellularly at high levels will inhibit cell growth. This inhibition can be released by addition of small amounts of CdR,[66] indicating that the block exists in the reduction CDP → dCDP. Deoxyadenosine and deoxyguanosine also inhibit reduction of ribonucleotides and the inhibitions have been investigated *in vitro* with partly purified reductase by Moore and Hurlbert.[67]

III. Thymidylate Kinase

A. Distribution and properties

Although by virtue of its great importance TMP kinase is an extremely interesting enzyme, relatively little is known about it. It is probably found in all proliferating tissues and occurs in at least small amounts in nonproliferating tissues as well. The enzyme has been found in regenerating rat liver,[10,25,28,30,31,32,68] normal rat liver,[32,52,70] kidney,[21,25] mouse liver,[71,72] and all tumors studied so far.[10–13,15,53]

Brookes and Williams [7] and Brookes [6] have reported the enzyme to be present in high-speed supernatant fractions prepared from the silkworm *Antheraea pernyi* wing epidermis. The activity is low during diapause and increases markedly during development. The maximum activity during development is uncertain since thymidine was used, for the most part, as substrate and thymidine kinase is rate-limiting. It has been studied in *Chlorella pyroidenosa* [4] and many viral infected systems.[74] It is high in the chick embryo [15,38,41] and frog embryo [42] and is found, surprisingly, in human erythrocytes.[212]

1. INTRACELLULAR DISTRIBUTION

TMP kinase, like thymidine kinase is found almost entirely in the high-speed supernatant fractions of cell homogenates or extracts. It would be expected that its original location would be the cell nucleus and possibly mitochondria. Repeated attempts to demonstrate nuclear TMP kinase activity in nuclei from hepatomas, using a large variety of preparative methods, have not been successful in our laboratory. Kielley, however, has found particle-bound TMP kinase in mouse liver,[71] and Hacker demonstrated TMP kinase activity in both nuclear and mitochondrial fractions from liver and tumor. Results from his work [52] are illustrated in Table III below. The purity of the fractions has been assessed.

Hacker's results illustrated in Table III raise a number of interesting questions. For example:

1. The amounts of TDP (not illustrated) in many cases are equal to, or a substantial fraction of, the TTP. This is of interest as in most cases previously reported, only small amounts of TDP are recovered.

2. The relatively high specific activities of nuclei, mitochondria and whole homogenate are of interest. The significance of these is somewhat difficult to assess without values for recovery of activity and nitrogen from all the fractions.

3. The changes in mitochondrial specific activity in regenerating liver may indicate a release of activity from the particle to the soluble fraction. This is also true of studies by Fausto and Van Lancker.[69]

Hacker's results, if confirmed, may be quite useful in extending our understanding of TMP kinase functioning. It might be of value to consider the report by Kielley [71] which, in some ways at least, appears to confirm Hacker's observations. Kielley [71] found that the TMP kinase

Table III. Production of TDP + TTP from TMP by Various Tissue Fractions *

Tissue Source	TDP + TTP: Normal Liver	Regenerating Liver: 12 hr	Regenerating Liver: 24 hr	Regenerating Liver: 36 hr	Induced Hepatoma	Morris 5123 Hepatoma
Whole homogenate	0.56	0.92	8.73	5.80	5.34	2.62
Nuclei (Type I)	0.2	0.53	1.24	1.27	4.37	0.28
Mitochondria	4.1	2.61	1.10	0.35	2.20	0.40
Microsomes	0.1	0.1	0.36	0.32	0.10	0.1
Soluble fraction	0.34	0.11	1.34	11.1	0.93	0.24

Activities are in mumoles per 300 μg protein in 25 minutes at 37°. Values reported for lysosomes have been left out as they are uniformly less than 0.1 in all cases. Values reported for Nuclei Type II have also been left out as they do not differ significantly from Type I. TTP + TDP originally reported separately have been added together as a measure of TMP kinase activity.

* The methods used for assay are as follows: The complete system contained in 0.15 ml, 5.0 μmoles tris-HCl buffer pH 7.6; 2 μmoles $MgCl_2$; 1 μmole ATP; 4 μmoles KCl; 1.6 μmoles trisodium phosphoenolpyruvate; 60 μg pyruvate kinase (Sigma Type II); 20 mμmoles $2^{-14}C$–5′–TMP with 306,000 counts/min and an appropriate amount of fraction. After the incubation period (25 min. at 37°) aliquots of each reaction mixture were streaked together with nonradioactive markers on Whatman No. 1 paper and developed 36 hours in isobutyric acid, conc. ammonia, water (66:1:33) v/v/v. The appropriate spots were cut out and counted by liquid scintillation.

level in resting mouse liver is high, possibly higher than in some growing tissues, and that a major part of the enzyme is bound to cell particles from which activity can be released. (Note she used glycylglycine buffer containing TMP in the homogenizing medium.) The activity was released from the residue by alternate freezing and thawing in the presence of TMP as shown in Table IV.

Table IV. Release of TMP Kinase Activity from Particulate Preparations

Tissue Preparation Homogenate	TMP Kinase Activity (μmoles TMP phosphorylated/hr/mg tissue) *: Liver	Hepatoma
Soluble fraction	5.1	16.3
Residue fraction	7.7	14.7
Extract of frozen residue	1.0	
Extract of frozen homogenate	18.9	

* This probably should be mμ moles—Author.

There was no significant difference in the activities of fresh and frozen homogenates. When similar procedures were tried by Bukovsky and Roth [13] with rat liver no evidence of particle-bound TMP kinase was found. This has been confirmed by V. R. Potter (private communication). Furthermore, Hacker reported [52] that freezing and thawing of his particulate extracts (probably in absence of TMP) destroyed most of the TMP kinase activity. Additional studies on the nature of particulate TMP kinase activity are obviously needed.

Fausto and Van Lancker [28] have reported values for TMP kinase in subcellular fractions of regenerating and normal liver of the rat. In normal rats 60% of the TMP kinase activity was recovered in the supernatant fraction, 23% in mitochondria, 10% in nuclei and about 7% in the microsomal fraction. Total recovery was 174% that present in the whole homogenate.

2. STABILITY

The enzyme is generally unstable in the absence of TMP or other nucleotides.[68] Addition of TMP to the homogenization medium can make a profound difference in the amount of TMP recovered.

Kielley [72] reported that 0.05 M glycylglycine buffer pH 7.1 or 0.25 M sucrose as a homogenizing medium gave the highest values of TMP kinase activity. Calcium chloride in the medium (2×10^{-3} M + 2.5×10^{-4} M Mg^{++}) was inhibitory. She pointed out the importance of substrate concentration and found maximum activity at about 1.5×10^{-4} M. Kielley also observed that activities in mouse liver were higher than in rat liver; Av. 11.3 *vs.* 2.9. She has also compared the activities of other kinases in the rat and mouse.[72]

Atkinson and Roth [42] have noted also that in *Rana pipiens* embryos glycylglycine gives a 3-fold activation (or stabilizes) TMP kinase. Glycylglycine which resembles TMP in structure (Figure 4) may be able to stabilize the enzyme in the same way as its substrate.

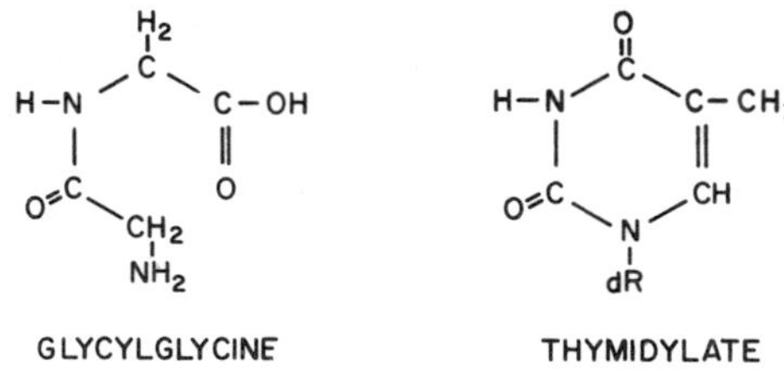

Figure 4. Comparison of the structure of glycylglycine and thymidylate.

The enzyme from extracts of wing epidermis of the silkworm was quite stable to conditions of storage, dialysis and incubation, even in the absence of thymidylate or thymidine.[6,7] Ives, Morse, and Potter[9] reported that thymidine and TMP kinase activities in a high speed supernatant fraction prepared from Novikoff hepatoma cells grown in tissue culture were stable for several months when stored at −20°. Johnson and Schmidt[4] found that freezing and thawing of a high speed supernatant fraction from *Chlorella pyroidenosa* completely inactivated TMP kinase. Freezing and thawing of whole cells led to a loss of about 25% of the activity. The instability of TMP kinase in a crude supernatant fraction was probably first observed by Bianchi and co-workers[10] and these investigators found that the enzyme was stabilized to a considerable extent to dialysis and incubation by addition of thymidine. All kinase activities were reported to be maintained after freeze-drying and storage at −50°.[53] Similar results have been noted for dCMP deaminase.[48] The stabilizing effect of thymidine and thymidylate have been investigated in much detail.[54,68] Kokolis and collaborators have reported that TTP and sulfhydryl groups are effective in protecting the enzyme.[75]

Effect of metal ions: There is fairly good agreement that TMP kinase requires added Mg^{++} for activity. Mn^{++} can replace this in part. Omission of Mg^{++} or addition of EDTA leads to complete loss of activity.[9]

B. Purification of TMP kinase

As already noted Grav and Smellie have purified TMP kinase from Landschutz ascites tumor cells.[53] Kokolis and co-workers[75] have purified the enzyme from normal mouse spleen supernatant fraction using ammonium sulfate precipitation and Sephadex column fractionation. The enzyme by the latter method had a molecular weight of approximately 70,000.

Nelson and Carter[73] have purified TMP kinase 140-fold by acid precipitation, absorption of impurities by alumina $C\gamma$ and fractionation on a hydroxylapatite column. 0.3 mM TMP was required for maximum velocity. Replacement of Mg^{++} with Mn^{++} gave 50% activity.

IV. Regulation of Thymidine and Thymidylate Kinase Activities

A. Thymidine kinase

Thymidine kinase activity varies widely with the age of a culture of mouse fibroblasts[11,19,76] (Figure 5), increasing during the first days of

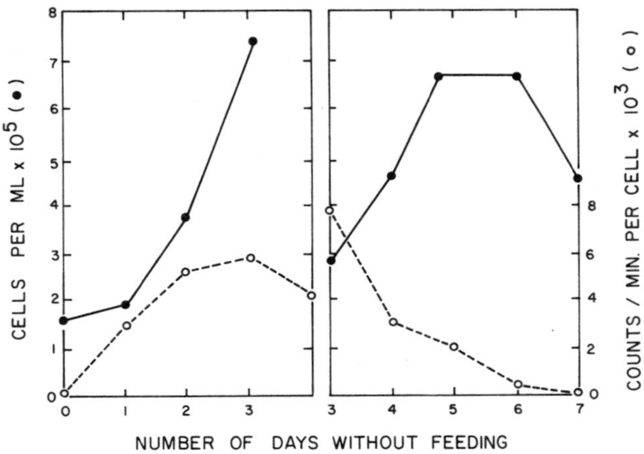

Figure 5. Changes in the concentration of cells per ml (●) and in thymidine kinase activity (o) during growth. To begin the experiment a stationary culture which had not been fed for 5 days was diluted with fresh warm medium. At this time and daily thereafter an extract was made from an aliquot of cells and thymidine kinase activity determined. Another culture treated similarly was used for the latter half of this experiment. In this particular experiment thymidine was not present during the preparation of extracts; entirely similar results have twice been obtained with thymidine present. From Littlefield.[19]

growth and then decreasing, but does not necessarily parallel the number of cells in the culture. Puromycin added to the culture medium prevented the increase in thymidine kinase activity suggesting that synthesis of new enzyme is necessary for the increase. The fall in activity in an ageing culture was prevented by thymidine, presumably by the mechanism of enzyme stabilization.

Sublines of fibroblasts deficient, or completely lacking thymidine kinase can be obtained by growing the cells in the presence of 5BrUdR for a considerable period.[19] Such cells could not survive aminopterin treatment except by a reversion. Thymidine kinase activity in 5BrUdR resistant cells was generally inversely proportional to the degree of resistance. Littlefield interpreted his results as signifying that kinase is continually synthesized in proportion to total protein synthesis which is rapid in recently-fed cultures and slowed in stationary cultures. The enzyme may be stabilized by endogenous or exogenous thymidine.

Hotta and Stern in an interesting series of papers[46,47] have correlated several facets of DNA synthesis and cell division with development of the microspores of *Lilium longiflorum.* In these spores the chain of development is first marked by a pulse-like production of deoxyribosides outside the cells.[77] The abrupt formation of these precursors seems to be due to the induction of deoxyribonuclease activity which persists for an interval of 6–8 hours in the region of the anther adjacent to the microspores.[78] The deoxyribosides, in turn, induce the enzymic phosphorylation of thymidine,[47] the induction taking place shortly before DNA synthesis begins (Figure 6). In lily, the duration of phosphorylation is about 12 hours (in Trillium at 2–4°, about 2 days). Inhibitors were not found in inactive cells. Addition of reagents such as 5-fluorouracil, chloramphenicol, 5-methyl tryptophan and ethionine all blocked the pulse in enzyme activity[46] suggesting that the appearance of the enzyme represented synthesis of RNA and protein. Studies with labeled amino acid showed that the enzyme activity coincided with labeled protein and all these, although not conclusive, are suggestive that thymidine kinase is being synthesized *de novo.* Factors underlying the removal of thymidine kinase were also studied.[79] Disappearance of enzyme required aerobiosis (prevented by dinitrophenol and anaerobic conditions). Fluorouracil but not FUdR, X-radiation, and acridine orange, all disrupted the inductive period of enzyme (both temporally and quantitatively). The authors state that these experiments suggest that interference with regulation is achieved by a complexing or alteration of an already-formed nucleic acid molecule (presumably DNA in the chromosome).

In a different system, Hotta and Stern[80] showed that thymidine may act as an inducer of thymidine kinase activity in cells of higher plants (wheat embryos). Forty-eight hour old wheat seedlings double their thymidine kinase but not total protein within 24 hours.[81] When the embryos were excised from such seedlings and grown in a nutrient medium for 4 to 6 hours, no increase in thymidine kinase activity was noted. However, if the embryos were exposed to various concentrations of thymidine, significant increases in enzyme activity occurred. The inductive effect of thymidine was prevented with actinomycin or chloramphenicol and both drugs caused a marked decline in activity, with or without the presence of inducer, although the thymidine induced group was more sensitive to actinomycin D. With microspores of lily developing synchronously, potentially inducible cells were found to be susceptible to the inductive effect of thymidine only during about 10% of the G-1 period and close to the interval when thymidine kinase activity normally appears.

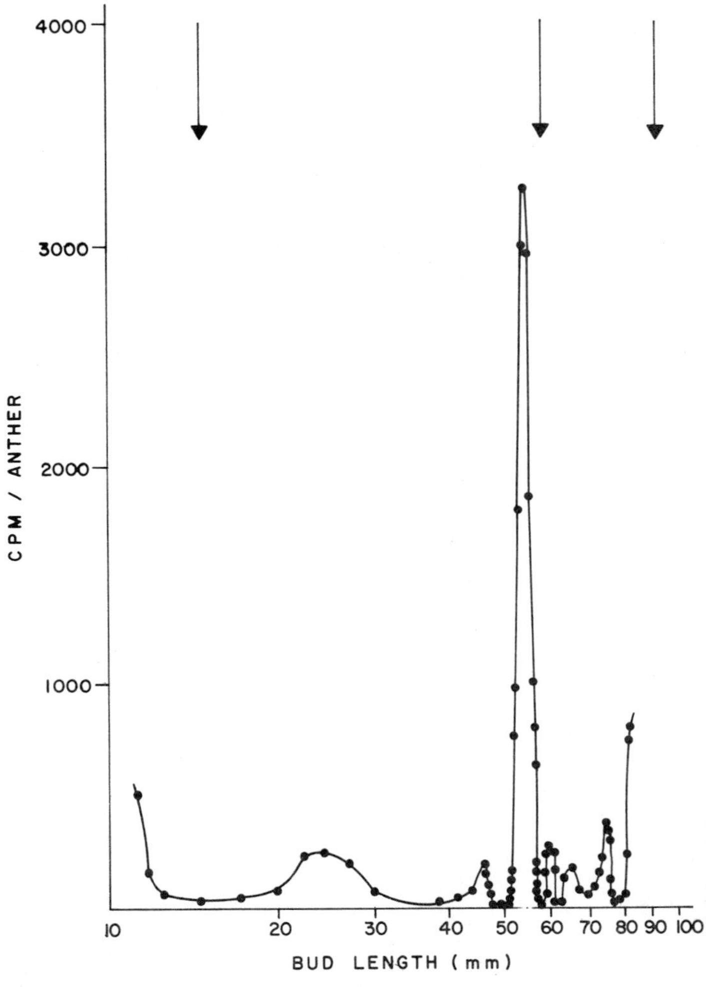

Figure 6. Periodicity of thymidine phosphorylation by suspensions of microsporocytes and microspores of lily. Vertical arrows indicate approximately the intervals of DNA synthesis. Tetrads resulting from meiosis appear at 24 to 25 mm bud lengths. Mitosis occurs in the region of 62 to 63 mm. From Hotta and Stern.[47]

Brent, Butler, and Crathorn[18] have studied variations in phosphokinase activities during the cell cycle in synchronous populations of HeLa cells. The results obtained are depicted in Figure 7. Thymidine kinase in the high speed supernatant fraction is low immediately after division (G1), rises during the S period and reaches a maximum about 5 hours after the peak of DNA synthesis, after which it falls rapidly during subsequent division. TMP kinase activity follows a similar pattern. The peak in its enzyme activity occurs somewhat later than the peak of DNA synthesis, as is the case for regenerating liver.[26] The authors conclude that since these enzymes must have been present in sufficient quantity at the beginning of the S period in order to produce the precursors required for DNA synthesis, it must be assumed that they are present in the cell at such times in a form which is not apparent on assay of the enzyme activity in the supernatant fraction. Two possible explanations of this finding can be suggested: (a) the enzymes are present before and during DNA synthesis in the form of a complex with the nuclear particulate fraction of the cell, and the ob-

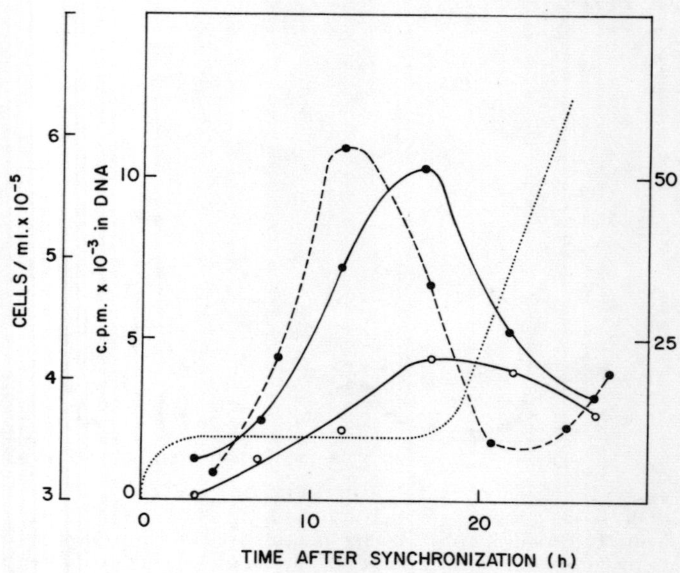

Figure 7. Synchronous growth of HeLa cells through one division cycle after collection of mitotic cells at zero time. Cell number (.........) was measured by haemocytometer. The rate of DNA synthesis (●----●), given as c.p.m. in DNA was determined by the incorporation of tritiated thymidine at each time by an aliquot of cells during a 15-min pulse. Thymidine kinase activity (●——●), given as $\mu\mu$moles TdR phosphorylated/hr/10^5 cell equivalent, and TMP kinase activity (o——o), given as $\mu\mu$moles TMP phosphorylated/hr/10^5 cell equivalent, were determined in the supernatant fractions. From Brent, Butler, and Crathorn.[18]

served increase in activity represents their release to the supernatant fraction when their function has been completed and, (b) the observed increase in activity is due to actual enzyme synthesis and the subsequent decrease due to the binding of this enzyme to a nuclear particulate fraction preparatory to its functioning in the next cell cycle. It has been shown, however, that the lifetimes of these two phosphokinases in the cell are considerably less than one cell cycle. It therefore seems more probable that these enzymes will be formed shortly before they are to be used by the cell. In order to distinguish between the two hypotheses it will be necessary to determine whether the rise

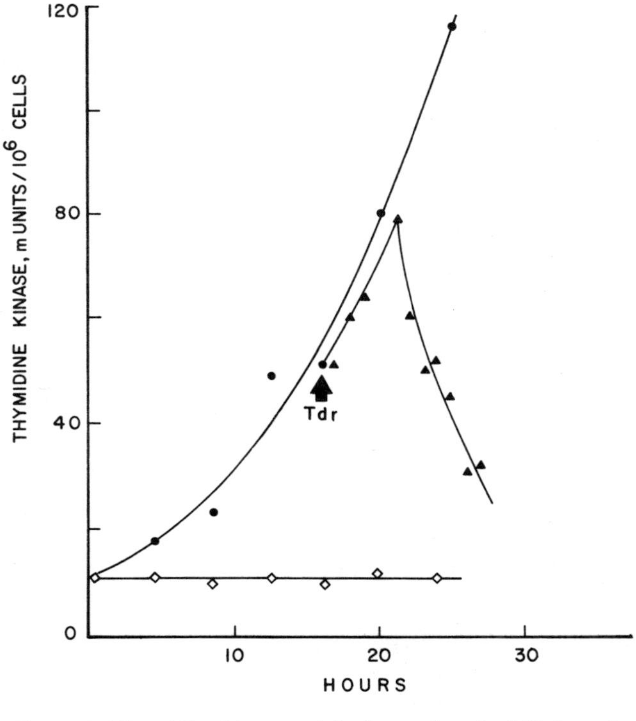

Figure 8. Thymidine kinase activity in synchronized HeLa cells. Cultures treated with amethopterin and thymidine (◇) maintained a constant level of enzyme as they grew logarithmically. In the absence of thymidine (●) the enzyme level rose sharply. In cultures given thymidine at 16 hours (▲) enzyme activity continued to increase until the cells began to divide 8 hours later. From Stubblefield and Mueller.[20]

in enzyme levels observed is due to new protein synthesis at this time or some earlier stage.

Stubblefield and co-workers[20,194,195] have investigated thymidine kinase activity in HeLa cells in tissue culture synchronized by means of a thymidineless state, induced by aminopterin (Figure 8). In the absence of thymidine, the enzyme level rose sharply relative to control cultures supplemented with thymidine. When the blocked cultures received thymidine at 16 hours, the enzyme level was almost 5 times that of controls. After the thymidine reversal the enzyme level continued to rise until DNA synthesis was complete and then decreased rapidly as the cells entered mitosis.

Eker has studied factors affecting the thymidine kinase activity of Chang human liver cells in culture.[215] Deoxythymidine was inhibitory at concentrations higher than 0.03 M and the inhibition could be prevented by deoxycytidine. Addition of amethopterin or 5-fluorodeoxyuridine led to a marked increase in the specific activity of thymidine kinase which could be prevented or reversed by TdR. The amethopterin induced activity could be prevented by puromycin. Russel[217] has pointed out that infection of tissue cultures by mycoplasmas can lead to marked increase in thymidine kinase and other enzyme activities which could be mistaken for enzyme induction in the tissue culture.

1. MESSENGER STABILITY

The significance of differences in the stability of messenger RNA for important enzymes as a mechanism of controlling enzyme activity has been considered by Pitot and co-workers[29] and Roth.[15] Pitot investigated the period during which induction of a given enzyme was insensitive to actinomycin D. This was assumed to be the lifetime of the RNA template for that enzyme. Thymidine kinase template half life under these conditions was less than 3 hours. Differences were found in the half lives of a number of enzymes in tumors as compared to rat liver.[69]

Baserga and collaborators,[83] Roth,[15] and Bresnick[28] all reported that under some conditions, actinomycin D had no effect on the thymidine kinase activity of Ehrlich ascites cells, Novikoff hepatoma or McCoy MDAB hepatoma. This surprising finding suggests that there is a fundamental difference in the stability of thymidine kinase in the tumors mentioned, compared to normal liver (and a small increase in stability in regenerating liver). Roth has suggested that the difference may reside in the location of the m-RNA. In liver it may be produced and utilized in the nucleus and there it may be susceptible to degradation and turnover. In tumor it may be found in the cytoplasm possibly associated with endoplasmic reticulum and not be subject either to inhibition by actinomycin D or normal turnover. Puromycin (2.5 mg I.P. per hour) also did not affect thymidine kinase activity in Ehrlich ascites cells (up to 450 minutes) although Roth obtained opposite results for the action of this inhibitor of protein synthesis with respect to this and other enzymes.[48]

In an important report, Bresnick and colloborators[23] studied the rates of turnover of thymidine kinase and its template RNA in regenerating and control liver. The turnover was measured by administering actinomycin D and cycloheximide at various times after partial hepatectomy (Figures 9 and 10) and measuring the enzyme activity at 24 hours after hepatectomy. It is assumed that actinomycin D prevents the synthesis of further template and the decay of enzyme activity is then due to limitations of synthesis due to decrease in availability of template (there are some problems with this concept). Similar conclusions are drawn with respect to use of cycloheximide. Template is available but protein synthesis is reduced. The results obtained were:

		Half-life hrs
	Control	Regenerating liver
1. Thymidine kinase template	3.0	7.5
2. Thymidine kinase protein	2.6	3.7

There was no change in the stability of the template for aspartate transcarbamylase or for its protein, measured under similar conditions. Bresnick suggests that immediately following partial hepatectomy, a substance is elaborated which may prolong the life of the thymidine kinase messenger RNA by inactivating a catabolic mechanism and in this way ultimately lead to a greater intracellular concentration of enzyme.

Mayfield and collaborators have correlated increases in enzymes of DNA synthesis in the surviving kidney of the rat after uninephrectomy with mitotic activity measured by radioautographic techniques.[84] Thymidine kinase, TMP kinase, and dCMP deaminase increased in ac-

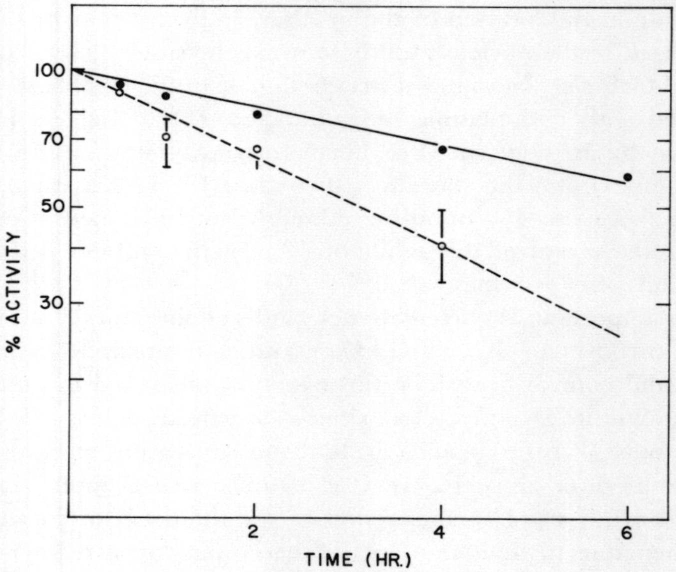

Figure 9. Stability of thymidine kinase template in control and regenerating liver. Actinomycin D, 500 μg/kg body weight was injected i.v. into rats at 19, 21, 23, 24, and 24.5 hours after partial hepatectomy. The animals were sacrificed at 25 hours after partial hepatectomy. Thymidine kinase activity was determined (solid lines). The time indicated on the *abscissa* represents the duration of actinomycin D treatment prior to sacrifice of the animal. In a similar set of experiments, control animals were injected with actinomycin D and periodically thereafter were sacrificed; thymidine kinase activity in the liver was determined (dashed line). The specific activity of thymidine kinase in control and 24-hour regenerating liver is 0.6 and 7.5 mμmoles of thymidine-5′-monophosphate/mg protein/15 min. The average of 6 results (6 individual rats) is presented in the figure. From Bresnick *et al.*[28]

tivity just prior to the burst in mitotic activity while dCMP kinase did not show any consistent change.

B. Thymidylate kinase—regulation

It has been suggested that control of growth may be regulated by control of the activity of TMP kinase.[38] There is much circumstantial evidence for this. Thus, TMP kinase activity is high in rapidly proliferating tissues, regenerating liver, embryonic tissues of all types and most tumors while it is low or absent in most mature, nongrowing tissues, such as adult liver, spleen, etc. Thymidine kinase as already noted is eliminated from consideration as an enzyme controlling growth rate

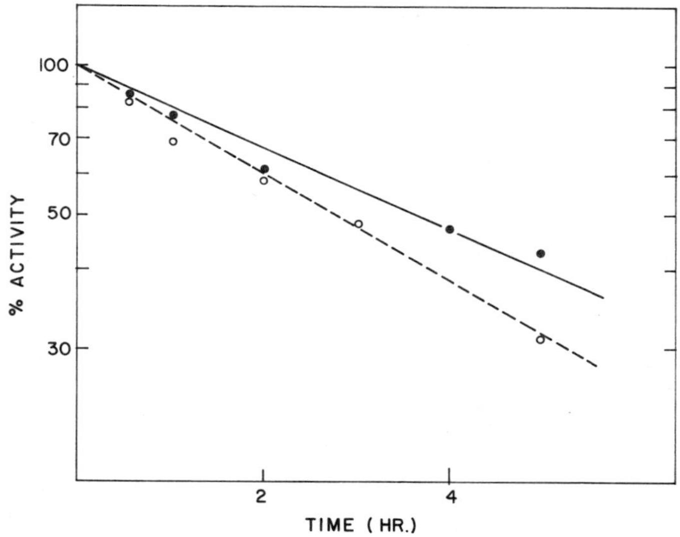

Figure 10. Effect of cycloheximide upon thymidine kinase activity in control and regenerating liver. At 20, 21, 23, 24, and 24.5 hours after partial hepatectomy, cycloheximide, 1.5 mg/kg body weight, was injected i.p. into rats. The rats were sacrificed at 25 hours after partial hepatectomy, and thymidine kinase activity in the liver was determined (solid line). The specific activity of the enzyme in regenerating liver was 7.5 mμmoles of thymidine-5′-monophosphate (dTMP)/mg protein/15 min. Similarly, cycloheximide, 1/5 mg/kg body weight, was injected into control animals (dashed line). The rats were sacrificed at 1, 2, 3, and 5 hours after administration of the drug, and the thymidine kinase activity of the liver was ascertained. The specific activity of the enzyme in control liver was 0.6 mμmole dTMP/mg protein/15 min. The duration of the treatment with the antibiotic is indicated on the abscissa. The average of 6 results (6 rats) is presented in the figure. From Bresnick *et al.*[28]

because it is not present in appreciable amounts in some rapidly growing embryonic tissues and the phosphorylation of thymidine represents, in a sense, a salvage pathway, or reclamation reaction not on the main pathway to TTP synthesis.

TMP kinase unlike thymidine kinase does not appear to be subject to feedback control. Its activity must therefore be regulated by other means among which the following are some possibilities either separately or in combination:

(a) Regulation of rate of enzyme synthesis by:

(1) Nucleotide concentration directly, or

(2) Indirectly through repressors.

(b) Regulation of the state of aggregation-disaggregation of other changes in conformation.

(c) Activation and inactivation by means of:

(1) Inhibitors or activators (protein or nucleic acid),

(2) With or without the mediation of hormones or other substances.

The evidence for these or other possible regulatory mechanisms will be examined and a working hypothesis is presented in the recapitulation. It is obvious that different systems (plant, animal, and bacterial) may use different control mechanisms, and this, indeed, appears to be the case in some plant systems that have been already described.

1. TMP KINASE ACTIVITY DURING SYNCHRONOUS GROWTH

The use of synchronized cells to study changes in enzyme activity is of great value as it enables one to relate activity to specific phases in the growth cycle. Johnson and Schmidt [4] found that the level of TMP kinase exhibited a marked periodism during synchronous growth of *Chlorella pyrenoidosa*. The periodism was almost identical to that observed for the accumulation of DNA (Figure 11) except that the DNA curve was dis-

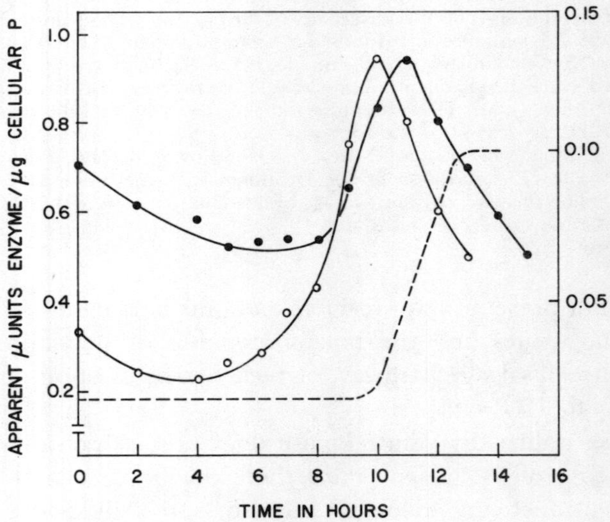

Figure 11. Relationship between the level of cellular DNA and the apparent level of TMP kinase during synchronous growth of *Chlorella pyrenoidosa* (strain 7-11-05). ●——● DNA phosphorus; o——o dTMP kinase; ----, cell number per ml; initial $12 \cdot 10^6$ terminal $94 \cdot 10^6$. From Johnson and Schmidt.[4]

placed to one hour later suggesting, according to the authors, that TMP kinase may indeed be limiting DNA synthesis. Similar fluctuations were observed in the level of dCMP deaminase by Shen and Schmidt.[85] Johnson and Schmidt could not demonstrate any inhibition or activation by adding heated supernates from different time periods, suggesting that no small, relatively stable molecules are involved in the changes observed. Atkinson and Roth have also examined TMP kinase activity during synchronous division of fertilized *Rana pipiens* eggs (see reference 42 and the section on kinase activity during development).

In a series of transplantable hepatomas of different growth rates, Bukovsky and Roth reported a correlation of high thymidine kinase activity with high growth rate of the tumor but TMP kinase did not appear to be related in this way.[12] In a later investigation these authors found that there were a number of important factors affecting activity, particularly the growth stage of the tumors.[13] Both thymidine and TMP kinase activities were low during early stages of growth when the tumor tissue contained a high ratio of connective tissue cells to tumor cells, and increased rapidly to a maximum as the tumor tissue became preponderant. Then the activity declined to very low levels when the tumor became large and many of the cells stopped proliferating probably due to deficiencies in food or oxygen.

Bessman has reported that TMP kinase is relatively constant during the growth cycle of *E. coli.*[86]

C. Thymidine phosphorylation in development

Sachsenmaier and Ives[5] have followed thymidine kinase activity during development of the slime mold *Physarum polycephalum.* Activity increases sharply at the end of interphase and maximum activity is reached at the end of mitosis which coincides with the onset of DNA replication (Figure 12).

Sachsenmaier and collaborators also found that actinomycin C and cycloheximide prevented the increase when administered at appropriate times.[206] Although the results suggest that an increase in the synthesis of new protein molecules is prevented, the possibility that the antibiotics may function by blocking enzyme turnover and stabilization of the enzyme molecule is mentioned. A third possibility is that stabilization or activation of m-RNA for thymidine kinase is prevented.

Weinstock and Dju have studied thymidine phosphorylation in developing breast muscle of the normal and dystrophic chicken.[40] Phosphorylation was at equally high levels in both dystrophic and control

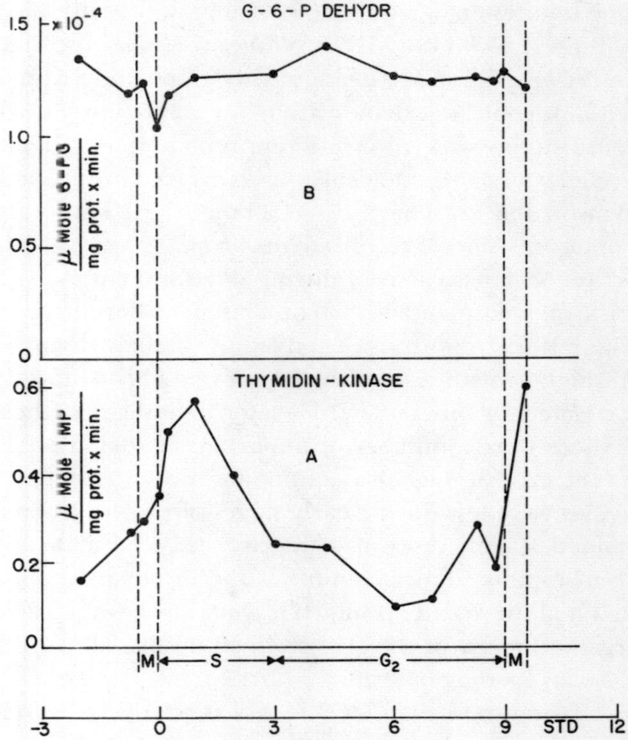

Figure 12. Enzyme activities in various stages of the division cycle of synchronous macroplodia of slime mold *Physarum polycephalum.* (A) thymidine kinase, (B) Glucose-6-phosphate dehydrogenase. From Sachsenmaier and Ives.[5]

breast muscle in the 12-day embryo, a period of development when DNA synthesis is rapid. The level of enzyme activity in control muscle was found to have dropped markedly by the 18th day. On the contrary, in dystrophic muscle the level remained high out to at least 14 weeks. (Although a sharp drop followed by a rebound was observed at two weeks this point may be artifactual.)

Scholl, Herrmann, and Roth [41] studied TMP kinase activity in developing leg muscle of the chick from the 7th day of development to hatching and correlated enzyme activity with mitotic activity. Enzyme activity remained at about the same high level from the 7th to the 16th day and then by the 18th day fell markedly to about 25% of its initial activity. The decrease in enzyme activity did not correlate with proliferative activity which by the 12th day had fallen to quite low levels. The change in enzyme activity did occur, however, at the time when there was an active change from DNA synthesis to myosin synthesis, in other words it did correlate with the process of differentiation. Thymidine kinase activity was at very low levels throughout and could not be increased by a number of techniques that were tried.

Roth and Askew[38] studied the phosphorylation of thymidine by the developing chick embryo from 3 days to hatching. Thymidine kinase activity was low (3–6 mμm/mg prot./hr) and did not change significantly during the entire course of development. TMP kinase reached a maximum value of about 100 mμmoles/mg prot./hr at 3–4 days, declined fairly rapidly by about the 12th–14th day to a value one-quarter of this, and then remained fairly constant until hatching time. A comparison of the activities of several kinases was made on days 2, 3, and 4 of development and the results are given in Table V.

Table V. Kinase and Deoxycytidylate Deaminase Activities in the Early Embryo—Average of 3 Experiments
(mμmoles of product/mg prot./hr)

	Day 2	Day 3	Day 4
Thymidine kinase	0.3 (0–0.5)	1.8 (0.4–3.5)	2.1 (0–3.2)
TMP kinase	18.0 (9.5–23.7)	32.1 (26.0–36.3)	102.2 (67.7–139)
dCMP deaminase	2060 (1665–2700)	3420 (2600–4800)	3930 (3390–4450)
dG kinase	12.4	22.7	30.5
dGMP kinase	263	455	690

From Roth and Askew.[38]

From the second to the fourth day there was a marked increase in TMP kinase specific activity (5-fold by day 4). Extrapolation to younger embryos might give an even more marked rise. Although the other enzymes measured did increase in activity (approximately 2-fold) the changes are not nearly as marked or significant, particularly since it is clear that the activities of dCMP and dGMP kinase were at all times from 6 to 40 times greater than those of TMP kinase. The latter would appear, therefore, to be the *significant rate-limiting enzyme* in the production of TTP and DNA. Although activities of dAMP and dCMP kinases were not determined, studies by Beltz in normal and regenerating rat liver[31] and by many other investigators[27,70,87,88] using different systems, have indicated that the activity of these enzymes are high, even in adult liver and do not change significantly during regeneration.

To look further into the initial thymidine and TMP kinase activities, Atkinson and Roth [42] have studied thymidine phosphorylation in the developing embryo of *Rana pipiens.* Use of the frog egg offers two advantages: (a) It is a simple matter to compare the unfertilized egg with the fertilized at any time even short intervals after fertilization; (b) The first few divisions of the fertilized egg are in good synchrony.

Changes in TMP kinase activity during the first few divisions are depicted in Figure 13. Values for whole homogenate and high speed supernatant fraction are compared. The following points are of interest:

1. TMP kinase activity is present in the unfertilized egg. After fertilization it rises to a maximum in both homogenate and supernatant fraction approximately half way to the first cleavage and then declines to a minimum at the time of the first cleavage.

2 The activity in the supernatant fraction follows this cyclical pattern for at least one and probably two more cleavages. The activity

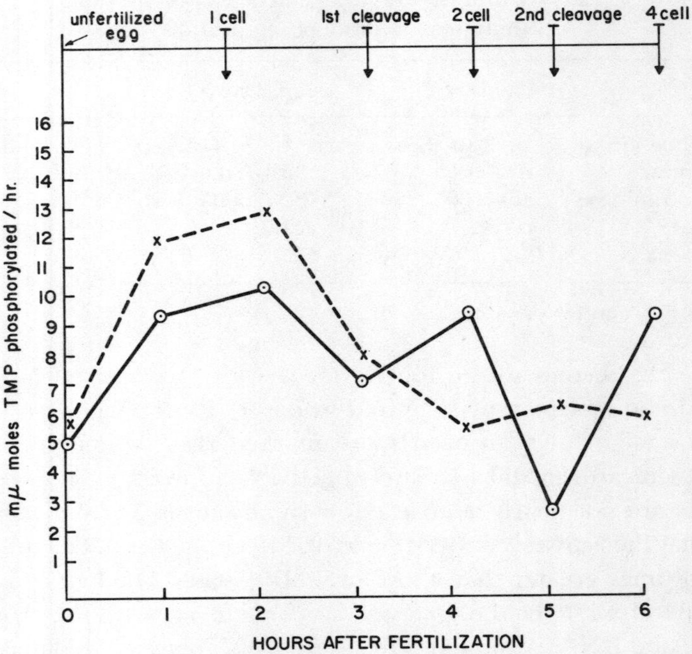

Figure 13. TMP kinase activity in the developing eggs of *Rana pipiens.* Eighty eggs were hand stripped of jelly coats for determination of each experimental point. Three separate experiments were run all giving the same result. From studies by Atkinson (unpublished).

changes in the whole homogenate fall out of phase, however, with that in the supernatant fraction after the first cleavage.

These results suggest that in keeping with other systems, already described, TMP kinase activity in the amphibian embryo reaches a maximum at or near the period of maximum synthesis of DNA and then decreases as the cell prepares for division.

Duspiva and Hansen-Delkeskamp have followed thymidine phosphorylation in *Bufo bufo*[208] eggs but since they did not employ short time intervals they were not able to detect the kinds of changes illustrated in Figure 13.

Changes in TMP kinase activity in supernatant fraction prepared from whole embryos of *R. pipiens* up to 240 hours are shown in Figure 14. Specific activity remained relatively constant or increased slightly up to about 120–144 hours (gastrula stage) and then increased rapidly up to 192 hours when it appeared to level off. Several other enzymes related to DNA metabolism were found to behave in the same way as TMP kinase.[42] Recently, Klemperer and Haynes followed thymidine kinase activity in rat livers during development.[198]

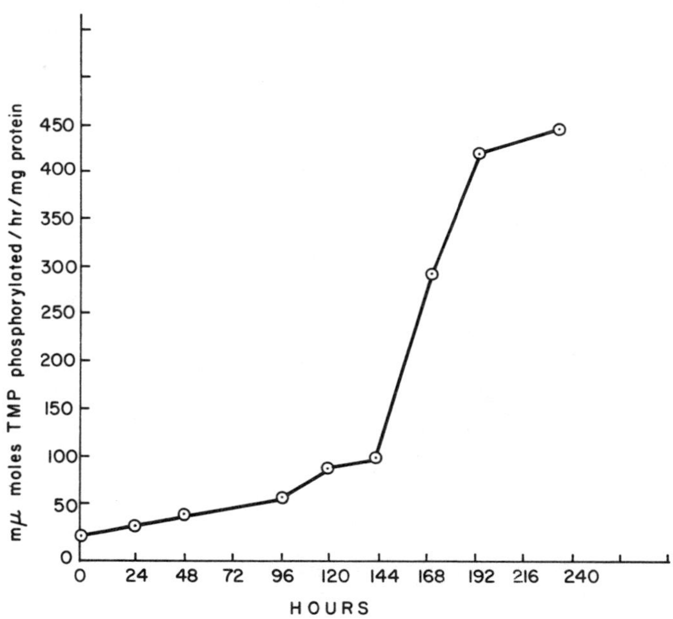

Figure 14. TMP kinase during early development of the frog *Rana pipiens*. Whole embryo. From studies by Atkinson (unpublished).

D. Miscellaneous studies on control of phosphorylation

Kielley[33] has compared the patterns of triphosphate synthesis in mouse liver with mouse ascites hepatoma. Broken cell preparations were used so that interactions from other cell components which might have a regulatory function were not excluded and continuous time assays were made so better overall comparisons could be made. The results indicated that mouse liver and mouse ascites hepatoma differed markedly in patterns of synthesis of TTP. In mouse liver synthesis of TTP is almost completely prevented. In hepatoma synthesis of dCTP was restricted. The limitation of TTP synthesis in mouse liver homogenate was not due to a deficiency of TMP kinase but to interfering actions. TMP alone among all the deoxyribonucleotides studied was rapidly dephosphorylated under kinase assay conditions (Figure 15). Addition of TMP to the homogenizing medium increased the activity moderately (2×) in liver but not in hepatoma but did not otherwise alter the characteristic pattern of the activities. The falling off in activity illustrated in Figure 15 was not a property of the whole homogenate since when a high speed supernatant fraction was used (liver S_2 fraction) activity continued for at least 40 minutes with little decline in rate.

Cannon and Breitman have studied the control of deoxynucleotide synthesis in *E. coli,* specifically the effect of exogenous TMP in the medium.[89] Exogenous nucleotides are utilized in *E. coli* after dephosphorylation to nucleosides.[90–92] They tested the two possibilities: (a) that thymine deoxynucleotide synthesis *de novo* is regulated by thymidylate synthetase, ribonucleotide reductase or another enzyme in the pyrimidine synthetic cycle, or (b) there is no regulation *de novo;* simply a dilution of thymine deoxynucleotides synthesized *de novo* by those obtained from an exogenous source. The decision between the two possibilities was made by comparing the steady state level of pyrimidine nucleotide pools in *E. coli* grown with and without exogenous TMP. In *E. coli* ^{14}C from exogenous uracil is incorporated into all nucleotides (illustrated in the pathways in Figure 3, Reference 89) whereas ^{3}H from exogenous TMP is incorporated only into thymine compounds. 52% of DNA thymine was shown to be derived from the exogenous TMP. If there is no regulation it would be expected that the ^{3}H TMP would be diluted by the *de novo* pools and excess ^{14}C thymine compounds would be excreted into the medium. There was little excretion of ^{14}C thymine compounds into the medium and there was a 50% expansion of the total thymine deoxyribonucleotide pool, accompanied by a reduction in both cytosine and uracil deoxynucleotide pools and increases in the ribo-

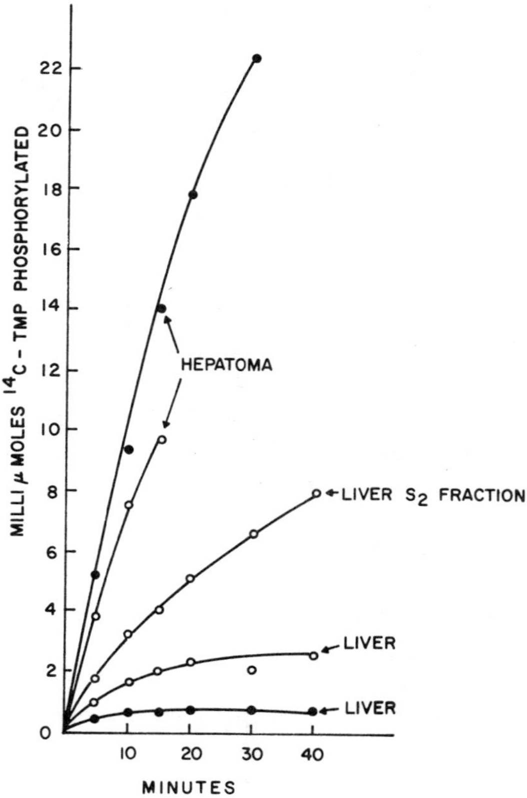

Figure 15. Time course of TMP phosphorylation by mouse liver and hepatoma. 40 mμmoles of TMP-C^{14} in assay system. ●——● Exp. 1 tissue concentration. 4.2 μg. DNA-P. o——o Exp. 2, TMP added to the homogenizing medium. Tissue concentration 2.8 μg. DNA-P. From Kielley.[33]

nucleotide pools along with a reduction of the contribution of synthesis *de novo* to the thymine deoxynucleotide pool. All these results are consistent with a regulatory effect at the ribonucleotide reductase level and not at the thymidylate synthetase step. A distinction cannot be made between feedback control or enzyme repression.

A report of some significance in relation to regulation of DNA synthesis is that of Powell[200] who found that when strain L fibroblasts were treated with small doses of UV light or puromycin, DNA synthesis was inhibited 5–8 hours after treatment but there was virtually no change in the phosphorylation of thymidine to TTP. The activity of

DNA polymerase was also not affected. A more valid test of the importance of phosphorylation would be to use synchronized cells and determine the effect of these treatments on the cyclical increase in thymidine and TMP kinase activities that are observed in synchronized cells.

Scholtissek has studied nucleotide phosphorylation in chick embryo fibroblasts cultured at low temperatures.[202–204] Triphosphates were found to accumulate except for thymidine and deoxyuridine in which case the monophosphates accumulated.[202] At 37° in the presence of cytosine arabinoside and actinomycin D, triphosphates of most of the ribonucleosides or deoxynucleosides accumulated, with the exception of UTP or TTP.[203] It was suggested that at low temperatures (4°) a release of negative feedback inhibition of TMP kinase by TTP might be achieved. Scholtissek's conclusions are[204] that lowering of the incubation temperature by a few degrees shifts the overall pattern of cell synthesis from mainly macromolecular synthesis to the synthesis of energy-rich phosphates which accumulate.

E. Effect of hormones on thymidine phosphorylation

Little work has yet been reported in this field.

Weiss and co-workers[93] have noted that testosterone (up to 2 mg/mouse) increased thymidine kinase activity in seminal vesicle. At low doses (but not high doses) the effect could be counteracted by intraperitoneal injection of 125 mg/kg of 5-fluorouracil. Thymidylate synthetase behaved similarly.

Nelson and Carter[73] found that diethylstilbestrol was a strong inhibitor and 17 α-ethynyl-β-estradiol a less effective inhibitor of TMP kinase in extracts of L5178Y mouse leukemic cells. Preparations of TMP kinase from mouse spleen, calf thymus, three human leukemias, and *E. coli* were not inhibited. The inhibition was irreversible and a high concentration of TMP protected the enzyme from inhibition but could not reactivate the inhibited enzyme. The inhibition of TMP kinase was prevented by several sulfhydryl compounds of which dithiothreitol was most effective, but addition of this compound to the inhibited enzyme did not restore activity. Diethylstilbestrol ^{14}C did not specifically associate with the enzyme protein in a partially purified TMP kinase preparation and the addition of the protective agents dithiothreitol and TMP did not alter the degree of binding. Diethylstilbestrol caused no inhibition of thymidylate metabolism in L5178Y cells incubated *in vitro* nor in the development of the ascitic tumor in the mouse. The estrogens were added in methanol solution. The final concentration of methanol (1%) had no effect on the reaction.

Brookes and Williams [7] found that the activity of thymidine kinase and TMP kinase increased in the wing epidermis of the silkworm *Antheraea pernyi* during early adult development under the influence of ecdysone. Ol'Shanetskaya and Novikova [199] reported that injection of 6 mg/100 g of cortisone acetate into hepatectomized rats at the time of maximum kinase activity markedly inhibited both thymidine and TMP kinase activity. Sakuma and Terayama also showed [201] that cortisone injection daily for 5 days prior to partial hepatectomy retarded the increase in DNA synthesis and induction of thymidine and TMP kinase activity.

V. Induction of Thymidine and Thymidylate Kinase in Kidney and Liver

Subtotal hepatectomy and uninephrectomy initiate a series of responses in the remaining organ which lead to regeneration or compensatory hyperplasia. One of the earliest events is an increase in the enzymes that convert thymidine to thymidine phosphate.[84,94,95] The increase in enzyme activity can be prevented or delayed by administration of actinomycin D or by X-rays given shortly after hepatectomy.[95,96,160,207] A radioprotective dose of s-2-aminoethylisothiuronium chloride injected into rats before partial hepatectomy also delays the onset of kinase activity.[207] An increase in liver and kidney TMP kinase activity is also observed when a large dose of thymidine is administered intraperitoneally to rats.[70,97] This effect may be due to stabilization of the existing enzyme by virtue of increased cell levels of TMP. Evidence to support this latter possibility was presented by Adelstein and Kohn [160] who found that prior X-ray and actinomycin D treatment did not affect the increased level of TMP kinase activity found 1 hour after thymidine was administered. Starvation up to 72 hours also had little effect. It is of some interest to note, as Adelstein and Kohn have pointed out [160] that TMP kinase differs in this response from a number of other enzymes induced by substrate. The differences may be seen at a glance in Table VI.

TMP kinase differs from the others, as well, in being related to nucleic acid metabolism while the other enzymes are all concerned with the metabolism of amino acids.

Main, Cole, and Walwick [94] studied the increase in TMP kinase activity after unilateral nephrectomy in the rat. When radiation was applied

Table VI. Response of Substrate-Induced Enzymes to Various Treatments

Enzyme	Stimulus	X-rays	Actino-mycin	Ref.
Serine dehydrase	Casein hydrolysate	Prevents	Prevents	98
Ornithine-δ-transaminase	Casein hydrolysate	Prevents	Prevents	98, 99
Tryptophan pyrrolase	Tryptophan	Prevents	None	100
Tyrosine transaminase	α-methyl tryptophan	Prevents	None	101
TMP kinase	Thymidine	None	None	160

From Adelstein and Kohn.[160]

after uninephrectomy, a sharp cut off time was noted, between 20 and 24 hours. Before 20 hours the increase in activity was prevented After 24 hours there was little or no effect. Similar results were observed with DNA polymerase. Uninephrectomy was also carried out after irradiation with 850 r. From 0 hours to about 3 days, the response was either abolished or greatly reduced. At intervals greater than 3 days some ability to respond was restored.

In regenerating liver, DNA synthesis starts 18 hours after partial hepatectomy, rises to a maximum between 24 and 30 hours after operation and decreases abruptly during the following hours. TMP kinase and DNA polymerase are known to increase in the supernatant fraction of cell extracts.[11,23,69,98] Fausto and Van Lancker[28] have tried to decide among several possibilities: (a) activation of latent enzyme, (b) elimination of inhibitors, or (c) synthesis *de novo*. Regenerating liver was separated into 4 subcellular fractions and the amount of TMP kinase in each fraction and the total recovery determined. In normal liver, recovery in the fractions was 174% of that in the homogenate, while 12, 24, and 36 hours after hepatectomy it was 127, 100, and 87%, respectively. There was no evidence of any inhibitors. After hepatectomy, the total percentage of enzyme in the supernatant fraction increased sharply between 18 and 24 hours and it declined in the other fractions. There was no change in either the amount or distribution of TMP phosphatase during the course of liver regeneration. Microsomes added to the supernatant fractions had no uniform effect on the enzyme activity of the latter. TDP kinase was found to be associated to the extent of 64% with particulate material. No changes in activity or distribution of this enzyme were found during regeneration and actinomycin injected 3 hours after hepatectomy had no effect on it.

The increases in DNA polymerase activity in 24 hour regenerating liver were quite small. The average of 6 animals was: normal 175 ± 20, regenerating 231 ± 41 and this would appear to eliminate this enzyme from an important rate-controlling role in growth. One point to note is that TMP kinase activity (as well as DNA polymerase, total activity) increased most 24–36 hours *after* hepatectomy, well after the largest part of the DNA is synthesized (Figure 16). At present there is no explanation for this finding which was also noted some time ago by Bollum and Potter [101] and these investigators in a general way, came to the same conclusions as Fausto and Van Lancker.[28]

Fausto *et al.* have shown [95] that 600 r of X-radiation given 24 hours prior to hepatectomy does not affect the subsequent appearance of TMP kinase; 1000 r does, however, decrease the response. Neither dose of radiation has any effect on DNA polymerase activity in regenerating liver.

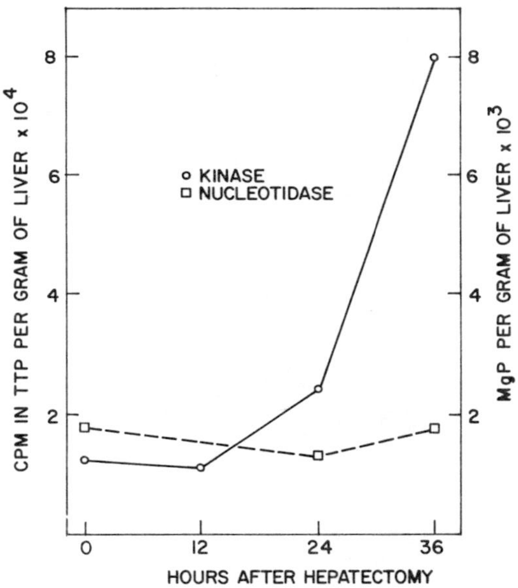

Figure 16. Comparison of total TMP kinase activities (o) and total TMP nucleotidase activities (□) in normal and regenerating livers. The results are expressed in counts per minute per g of liver in TTP (left) for TMP kinase activities, and mg of phosphorus liberated per g of liver (right) for nucleotidase activities. Each point represents the total kinase or nucleotidase activity recovered (obtained by adding the enzyme activity of each cell fraction) after tissue fractionation. From Fausto and van Lancker.[28]

Beltz has compared the effect of X-radiation on the elevation of thymidine kinase and TMP synthetase during liver regeneration.[24] As reported by many others, elevation of thymidine kinase was completely prevented by irradiation with 1500 r 6 hours after hepatectomy, while the activity of TMP synthetase was elevated approximately 3-fold in spite of early exposure to X-ray. Irradiation of the animals (1500 r) 15–16 hours after hepatectomy gave essentially the same results. Although thymidylate synthetase activity in the irradiated rats was only 80% of the controls, this difference may not be significant. Activity of dCMP deaminase was the same in supernatant fractions prepared from 6, 16, and 24.5 hour regenerating liver of rats irradiated 6–7 hours after partial hepatectomy. Stevens and Stocken[102] and Myers *et al.*[103,104] have demonstrated that exposure of partially hepatectomized rats to X-radiation shortly before or after surgery, blocks the elevation of dCMP deaminase usually observed in regenerating liver.

Bianchi, Crathorn, and Shooter[54] studied the appearance of thymidine kinase in regenerating rat liver in relation to the onset of DNA synthesis. Both the phosphorylation of thymidine and DNA synthesis were found to occur *in vivo,* before the kinases could be detected *in vitro* using particle free fractions. Using ^{3}H thymidine as a precursor, the extent of incorporation 24 to 26 hours after partial hepatectomy, increased to 2–3 times that of normal liver. Little activity was seen at 20 hours. The appearance of TMP kinase was minimal at 24 hours and increased up to 26 hours. Thymidine kinase appeared between 20 and 24 hours. On the other hand, when thymidine was injected *in vivo* a considerable portion was phosphorylated to TMP at all times after hepatectomy and even in normal liver. Phosphorylation to TTP was found to take place in normal rats, but not in regenerating liver at 18–20 hours after hepatectomy; but again at 24–26 hours it occurred to a smaller extent than in normal liver. The authors suggest that one is measuring competing reactions

$$\text{(1) TdR} \rightarrow \text{TMP} \rightarrow \text{TTP}$$
$$\text{(2) dUMP} \rightarrow \text{TMP} \rightarrow \text{TTP}$$

and that (1) is relatively high in normal liver but (2) is low. In regenerating liver the reverse is true. The delay in appearance of TMP kinase is suggested as being a consequence of the enzyme being bound to some particulate structure (nucleus?) initially followed by release after the major portion of synthesis to TDP has occurred.

Beltz has compared five enzymes related to DNA metabolism in normal and regenerating liver.[31] The content of each enzyme in a 24.5 hour regenerating liver was compared with its content in the same tissue immediately after surgery. Four- and 6-fold increases in the level of thymidylate synthetase and thymidine kinase were found before the level of dCMP deaminase had changed appreciably. No significant change in the content of dCMP or dAMP kinase was observed. Fluoride at a concentration that inhibited 80% of the phosphatase activity did not affect kinase activity. Behki and Morgan confirmed this[32] but reported that TMP kinase in regenerating rat liver was fluoride sensitive. The latter investigators were concerned with possible interfering reactions of UdR → dUMP or CdR → dCMP which would compete with thymidine → TMP, phosphorylation. As might be expected, UdR present in the reaction medium reduced the formation of TTP in both whole homogenate and supernatant fractions. It, however, increased TMP formation in the supernatant fraction but not whole homogenate. UdR appeared to strongly stimulate the dephosphorylation of TMP.

Maley, Lorenson, and Maley reported[105] that the increase in thymidine kinase activity observed in regenerating rat liver at 24 hours could be prevented by administration of *p*-fluorophenylalanine, puromycin or actinomycin D immediately after hepatectomy. Bresnick has followed early changes in pyrimidine biosynthesis after partial hepatectomy.[36]

In an early study, Canellakis and collaborators[37] compared phosphorylation of deoxynucleotides by extracts of normal and regenerating liver and the results are illustrated in the following table.

Table VII. Phosphorylation of Deoxyribonucleotides by Extracts of Normal and Regenerating Rat Liver

	Percent Distribution of Deoxynucleotides					
	Normal			Regenerating		
^{14}C Substrate	Mono	Di	Tri	Mono	Di	Tri
dAMP	10	10	80	10	10	80
dGMP	10	30	60	5	10	85
dCMP	10	5	90	10	10	80
TMP	95	5	5	5	5	90

The time period was not given (probably 24 hours). The rate of uracil breakdown in the liver decreased immediately after hepatectomy.

VI. Induction of Kinases in Other Systems

A. Animal

Lieberman and co-workers using trypsinized cells from rabbit kidney cortex and cultured directly, observed that for the first 12 hours of culture the rate of turnover of nuclear RNA remained at the *in vivo* level although the low initial rate of synthesis of stable ribosomal RNA increased progressively.[21,106,107] Between 12 and 22 hours a sharp increase in the rate of turnover of nuclear RNA occurred and this rate was maintained until the onset of DNA synthesis. Addition of actinomycin D or *p*-fluorophenylalanine prevented these changes suggesting that early synthesis of RNA and protein is essential for later DNA replication. This forms a useful system in which to study changes in enzymes concerned with RNA and DNA synthesis. Some of the changes in enzyme activity are depicted in Table VIII.

Table VIII. Changes in Enzyme Activity in Kidney Cells Cultured Directly from the Animal

Age of Culture	DNA Polymerase	Thymidine Kinase	DNA Synthesis
0 hrs	3.6	1.3	
21	3.1	1.4	180
36	4.4		310
46	11.	2.1	1230
72	47.	7.5	8500

Relative enzyme activity is given. From Lieberman *et al.*[21]

Actinomycin D and *p*-fluorophenylalanine prevented the increases measured above, as did EDTA applied at zero time. A dose of 3000 r of X-radiation also largely prevented the increases. (Of interest is a decrease in RNase activity). The system of Lieberman is similar in its responses to parenchymal cells of hepatectomized rat liver.

Gebert and Potter[108] have shown a transient increase in thymidine kinase activity in rats fed a protein-free diet for 4–12 days which were then intubated with 1 g of protein hydrolysate (casein) every 12 hours for 60 hours. Thymidine kinase activity did not change on the protein-free diet (30 mμmoles/g tissue/hr) but at 24 and 36 hours after feeding the casein hydrolysate there was an increase to double the control values and a return to control levels at 48 and 60 hours. The phenomenon was sensitive to actinomycin D. A similar effect was observed in the Reuber hepatoma and liver of tumor-bearing animals but in this case, the increase in the tumor was 7-fold at 24 hours.

Thymidine kinase activity can be induced in several systems by inhibiting the formation of TMP.[26,194–196,214]

B. Viral

Bessman[86] was one of the first to measure the effect of phage infection on the activity of bacterial kinases. Using *E. coli* and the different T phages he found:

(a) T2r+, T4r+ and T6r+ gave 10–20-fold increases in TMP and dGMP kinases. No change was observed in dAMP or dCMP kinase activity.

(b) T5 gave a 6–10-fold increase in TMP, dGMP and dCMP kinases. There was no change in dAMP kinase.

(c) T3 and T7; no change in any of the kinases.

Chloramphenicol (1×10^{-3} M) added at the time of infection prevented the increase in TMP kinase activity during T2 infection. With Bellow and Van Bibber, Bessman showed[74] that there were two forms of TMP kinase in infected *E. coli*. The two forms were separated on DEAE columns. The enzyme formed on infection was inhibited by dGMP while the enzyme from infected cells was not. The viral induced dGMP kinase was also shown to be different from the normal cell enzymes.[111,112]

McAuslan and Joklik studied phosphorylation of various nucleosides and nucleotides in a HeLa cell system in tissue culture infected with pox virus (vaccinia).[60] Their results may be briefly summarized:

(1) Thymidine kinase activity increased markedly (10–15-fold) between 4 and 6 hours after infection, reaching a plateau at about 9 hours.

(2) There was no increase in the level of TMP kinase or TDP kinase during the first 9 hours post-infection. The increase in thymidine kinase activity was not obtained in the presence of 10^{-3} M *p*-fluorophenylalanine, which inhibited incorporation of amino acids into protein by 90%. The authors pointed out that this does not necessarily prove *de novo* synthesis of enzyme since an "uncoating" enzyme is required and this may be necessary for induction of enzyme activity. Infection with UV inactivated virus (uncoated) gave increases, but not infection by heat-inactivated reactivable virus (coated) suggesting that viral DNA is required for the appearance of activity. Infection with rabbit pox and ectromelia also gave increases.

(3) No change was found in the level of dAMP and dCMP kinases.

(4) dUMP and dUDP formation from CdR was increased 3–5-fold in infected cells. DNA polymerase levels increased by less than 50%.

In a later report Duckworth and Bessman [109] purified a deoxynucleotide-kinase from *E. coli* infected with T4 bacteriophage 400-fold. The enzyme phosphorylated TMP, dGMP and d-hydroxymethylCMP to diphosphates. Biochemical evidence showed it to be a single enzyme.

Nohara and Kaplan [110] working with rabbit kidney cells infected with pseudo-rabies virus (cells were in the stationary phase) found that there was a 100-fold increase in thymidine incorporation, at a rate comparable to logarithmically growing noninfected cells. They also observed that the TMP kinase of infected cells was considerably more thermostable than the similar enzyme from noninfected cells. The increased stability of the enzyme from infected cells did not appear to be due to changes in intracellular TMP concentration or inhibitors.

Kit, Piekarski, and Dubbs [76,113] have also studied the induction of thymidine kinase by vaccinia infected mouse fibroblasts and they obtained results similar to those already mentioned. They were able to show, also, that the inhibition of induction by puromycin was not the result of interference with the uncoating process [114] but an actual inhibition of thymidine kinase synthesis. The magnitude of the induced thymidine kinase activity was also shown to be a function of the virus multiplicity.[115]

These investigators investigated, in addition, the effect of actinomycin D, 5-fluorouracil, 5-fluorodeoxyuridine and mitomycin C on the viral infected fibroblasts. When cells were treated with high levels of actinomycin D one hour prior to the addition of vaccinia, induction was prevented. Cells pretreated for 8 hours with 1 μg/ml actinomycin D, 24 hours before the incubation of virus showed only about one-half of the amount of thymidine kinase induction although the incorporation of ^{3}H uridine into RNA was inhibited about 95%. Pretreatment of noninfected cells with mitomycin had little effect on the subsequent induction. 5-Fluorouracil caused only moderate reductions in the induction of thymidine kinase activity in vaccinia infected cells.

Dubbs and Kit have developed a method for the isolation of vaccinia mutants deficient in thymidine kinase inducing activity [116,117] and studied the properties of these viruses. The key to the isolation has been the availability of a thymidine kinase-free cell line. The mutant viruses were not affected by bromo or iodouridine when growing in thymidine kinase-free cells but were inhibited by these agents when growing in cells containing thymidine kinase, presumably because the cells phosphorylated the drugs and the phosphorylated derivatives were then used by the virus with subsequent interference with viral growth. Similar conclusions may be reached from the experiments of Centifanto and Kaufman [118] who studied thymidine kinase activities and induction in IUdR resistant and sensitive Herpes simplex virus in normal and thymidine kinase-low mouse fibroblasts.

Simian adenoviruses, simian papovavirus, SV40 and an adenovirus SV40 hybrid all produce increases in thymidine kinase activity following infection of green monkey kidney cells [119] as do many other viral infected cells.

Infection of mouse embryo cells with polyoma virus gave an increase in thymidine kinase activity 8 and 10 hours after infection and 2 hours before the onset of viral DNA synthesis. Some of the properties of the induced virus differed from those present in uninfected cells.[120] As noted in another system, the virus in infected cells was more thermostable.

Polyoma virus infection effects on thymidine kinase activity have also been studied by Littlefield and Basilico [121] as well as Kit, Dubbs, and Frearson.[122] The former investigators infected a thymidine kinase deficient line of hamster BHK cells with polyoma virus. Unlike the other systems described, neither infected cells, nor transformed cells exhibited increased incorporation of 2^{14}C thymidine into DNA, or increase in thymidine kinase activity after infection. Further study by the second group of investigators mentioned above using mouse embryo cells suggested that the reason for this may be that polyoma virus does not induce synthesis of a new viral thymidine kinase enzyme. In the embryo cells an increase in enzyme activity took place 16–48 hours after virus inoculation. The presence of puromycin from 18–28 hours after infection prevented this increase, but if removed at 28 hours, thymidine kinase synthesis was restored. Ara C or 5BrUdR added at 2 hours after infection did not inhibit the appearance of thymidine kinase activity at 28 or 48 hours. Extracts from infected cells did not activate extracts from noninfected cells nor did extracts from noninfected cells inhibit extracts from infected cells. The enzyme from infected and noninfected cells had essentially the same K_m and heat inactivation rate. TMP kinase activity was also increased in infected cells but uridine kinase and TMP phosphatase were not changed. Actually, initial high TMP kinase activity was maintained 30–48 hours in infected cells, while in the control it fell from 252 to 70 units. Thymidine kinase activity increased only 3-fold over control activity at 28 hours, the time of maximum activity. (For review of references on polyoma virus see this paper.)

Rous Sarcoma. Chick chorioallantoic membrane infected with this oncogenic RNA virus showed increases in uridine and cytidine kinases but no significant increases in nucleoside kinases phosphorylating deoxynucleosides (all four were tested).[82] The increase is not due to decreased catabolism of uridine or decrease in the phosphatase hydrolysing UMP.

Ohno has shown that chick embryo cells infected with vaccinia virus and pretreated with interferon did not show an increase in thymidine kinase activity until 6 hours after the infection.[123] This is in accord with the suggested action of interferon in interfering with the reading of the messenger RNA. Takahashi and collaborators[124] have also demonstrated increased thymidine kinase activity in human embryonic kidney cells infected with human adenovirus types 5 and 12. Kaplan and coworkers have summarized their results with pseudorabies virus recently.[125] After infection with pseudorabies virus extracts of rabbit kidney cells exhibit an increased level of activity of TMP kinase and thymidine kinase. The increase in the level of activity of TMP kinase although prevented by inhibitors of protein synthesis, is not due to the synthesis of new enzyme protein. Instead as a result of infection, the enzyme present in noninfected cells which is unstable *in vitro* seems to acquire new properties which confer upon it stability *in vitro.*

TMP kinase is stable *in vivo* in noninfected, as well as infected cells, and since an increase in the stability of the enzyme *in vitro* presents no obvious advantage to the growth of the virus, it seems likely that the virus-induced change may also affect other properties of the enzyme that cause it to possess greater activity *in vivo.* Aside from stability, no other differences in behavior between the enzymes present in infected and noninfected cells was noted. These experiments were done with cell extracts in which the enzymes were stabilized with relatively high concentrations of TMP and it is possible that these conditions may have obliterated the differences between the enzymes.

Experiments, in which cells were exposed to conditions in which normal cut-off of the increase in enzyme synthesis does not occur, revealed that whereas the cut-off of thymidine kinase activity is subject to a control mechanism, the cut-off of TMP kinase activity is not. This finding lends further support to the notion that the mechanism of increase in the level of activity of these two enzymes in the infected cells is different. Although the large increase in the level of thymidine kinase activity seems to have little survival value for pseudorabies virus when the infective process occurs in growing cells, it may be essential when the virus infects stationary phase cells in which the level of activity of phosphatases is high. The same state of affairs would also be true for TMP kinase, if the configuration of TMP kinase present in the infected cells is changed in such a way as to confer upon the enzyme higher activity *in vivo.*

Phosphorylation of thymidine has been investigated with DNA viruses in many other viral-infected systems. In general both thymidine and TMP kinase activity is elevated not long after viral infection. The increased activity may be due to increased synthesis of cell enzymes, increased stability of cell enzymes, or synthesis of new viral-coded enzymes.

VII. Degradation of Thymidine Phosphates

Although the emphasis thus far has been on the synthesis of TTP and the synthesis of kinases, it must be acknowledged that control may be equally well achieved by changes in degradative processes and many authors have suggested this. There is some evidence available to indicate a reciprocal relationship in certain tumors between the synthesis of TTP and its degradation. That is, when anabolic processes increase catabolic processes concomitantly decrease. Such reciprocal relationships probably exist in the normal cell as well.

A. Animal systems

Several investigators have studied the degradation of TMP in a variety of tissues.[12,13,126–129] Bukovsky and Roth investigated the activity and distribution of TMP phosphatase activity in homogenates and subcellular fractions of normal liver, several hepatomas and liver from the tumor-bearing animals.[12] In all the samples activity was concentrated 2–3-fold in the microsomal fraction compared to the homogenate with activity in this fraction being particularly high in the Dunning and Morris 5123 t.c. hepatoma. (A similar concentration in the microsome fraction was reported by Fiala *et al.*[126]) Approximately 90% of the TMP phosphatase activity was recovered in the three particulate fractions and was relatively equally distributed among them. There was no evidence for an inhibitor or any important interactions among the fractions. In liver, but not in hepatoma, enzyme activity was concentrated approximately 2-fold in the nuclear fraction. The decreased nuclear concentration in hepatomas could be an artifact of preparation since the nuclear fractions from the hepatomas contained much more nitrogen than normal or host livers.

There is a rough relationship between growth rate and the amount of phosphatase activity, with the rapidly growing tumors having the lowest amounts. Exceptions to this are noted, however.

Bukovsky and Roth also reported [13] that phosphatase activity in a series of transplantable tumors was greatly influenced by the quantity of ATP and Mg^{++} in the reaction mixture. Stein and Rubini [127] suggested that dephosphorylation of TMP by the alkaline phosphatase of granulocytes, may be used to regulate DNA synthesis. Dephosphorylation was studied in a variety of patients and activity varied considerably depending on the type of patient. Alkaline phosphatase was high in myeloma, acute myologenous leukemia, and carcinomatosis, but low or absent in pernicious anemia, polycythemia and hemochromatosis. TMP was readily dephosphorylated in all cases and the changes in activity against all substrates were in proportion suggesting one common alkaline phosphatase enzyme. Trubowitz [128] using human normal purified granulocytes reported that of 32 substrates TMP was most rapidly split by an alkaline phosphatase preparation.

Eker has studied the hydrolysis of various nucleotides in Chang human liver cells grown in tissue culture.[129,210] dTMP, TDP, TTP and pyrophosphate were used as substrates and the hydrolysis measured by formation of inorganic phosphate. The rate of release of phosphate from TMP, TDP and TTP was not affected by final concentrations of fluoride from 1–200 mM, contrary to a number of other reports which observed an inhibition of phosphatase activity by fluoride. Magnesium was present in the reaction mixture. Sodium azide (5 mM) *p*-chloromercuribenzoate (2 mM) and N-ethyl maleimide (2 mM) were also without significant effect. Under the conditions of assay, a large variety of other substrates were dephosphorylated at about the same rate. In addition to the difference with respect to fluoride, normal liver TMP dephosphorylation was optimal at pH 6.0 (viz. 9.4 for Chang cells) and not affected by Mg^{++}. (Chang cells were stimulated by up to 5 mM Mg^{++}.) The activity of thymidine kinase and TMP phosphatase was also followed by Eker during the growth cycle of Chang liver cells.[164] TTP was, as expected, quite inhibitory to thymidine kinase and TDP equally so, while TMP had a small effect, probably due to synthesis of TTP. The activity of thymidine kinase gradually increased during the log and early log period, reached a maximum at from 2–4 days of growth and then declined at 5 days (stationary period). Dephosphorylation of TMP and TDP declined slightly in the first 24 hours and then rose to a maximum (2-fold increase) by the 4th day. TTP dephosphorylation remained constant (see reference 164 for a review on effects of fluoride).

Fiala and co-workers have investigated deoxynucleotidase activity in rat liver and tumors.[126] They pointed out that 5′-nucleotidase, although it has a broad specificity for ribonucleotides, has little activity against deoxynucleotides. Normal liver homogenate of adult rats rapidly hydrolysed all deoxynucleoside monophosphates. For many experiments acetone-powdered tissue was used. Embryonic liver, normal spleen and thymus had low dCMP phosphatase activities compared to adult liver, while kidney had somewhat more than half the activity of adult liver. Novikoff hepatoma, Ehrlich ascites and mouse sarcoma also had low activities of dCMP phosphatase in comparison to normal liver. In general, there was an inverse correlation of dCMP phosphatase activity with growth rate and with dCMP deaminase activity. The phosphatase was inhibited by fluoride but generally unaffected by Mg^{++} up to 1×10^{-3} M. Separation of liver into subcellular fractions showed that about two-thirds of the total deoxynucleotidase activity was in the cytoplasmic granules (mitochondria and microsomes) with the remainder equally distributed between the supernatant fraction and nuclei. Washing and resedimentation of mitochondria led to an increase in specific activity towards dCMP, dAMP and dGMP and loss of most of the TMP phosphatase activity. The latter was, however, concentrated in the microsomes, suggesting two specific enzymes. The enzyme hydrolysing dCMP, dAMP and dGMP may be a lysosomal acid phosphatase.

In early experiments using leukemic spleen and lymph nodes from mice, Bianchi and collaborators[10,130,131] observed that TTP production was maximal in their crude system at around 1 hour, but by 4 hours all of the TTP had disappeared, degraded, presumably by phosphatase activity. Concomitant increases in thymidine TMP and TDP were observed. Studies were also carried out with regenerating liver and Ehrlich ascites cells. Contrary to the results of Gray[27] these investigators did not observe inhibition of phosphorylation when extracts of normal tissues of the mouse were mixed with mouse leukemic tissues.[130]

Alkaline Phosphatase. Rubini and co-workers demonstrated that a factor present in homogenized normal human blood leucocytes containing alkaline phosphatase activity, depressed the incorporation of tritiated thymidine into DNA.[132] This material when purified 200-fold appeared to be an alkaline phosphatase which readily dephosphorylated TMP and was free of thymidine phosphorylase activity. It probably entered test cells by pinocytosis.

^{3}H thymidine incorporation was measured *in vivo* in the regenerating rat liver by Chang and Looney.[133] The isotopic compound was injected intravenously and within 1 hour most of it had disappeared from the blood stream. Catabolism was very rapid, only 2 minutes after injection, two-thirds of the labeled thymidine present in the acid-soluble supernatant fraction was present as thymine and β-aminoisobutyric acid. Only 1.2% of the injected thymidine was incorporated into newly synthesized DNA.

Fritzson[134] has studied the hydrolysis of UMP, CMP, dUMP, dCMP, and dTMP in the soluble fraction of regenerating and normal rat liver during an 8-day period after partial hepatectomy. He observed that the dephosphorylation of dTMP, dUMP and dCMP in the soluble fraction was due mainly, to nonspecific acid phosphatase activity, whereas the dephosphorylation of the ribonucleotides UMP and CMP is due largely to a specific 5′-nucleotidase. The latter was also capable of hydrolyzing the deoxynucleotides to a small extent. During liver regeneration, the acid and alkaline phosphatase activities showed slight variations which could not be related to the growth rate of the liver. In contrast, 5′-nucleotidase showed distinct cyclic variations which were inversely related to growth rate and were similar to changes obtained with uracil reductase and N-carbamoyl-β-alanine aminohydrolase. All three enzymes are on a common pathway and may be closely associated and regulated by the same mechanism.

B. Viral systems

After infection of *Bacillus subtilis* SB19 with phage SP5C, an enzyme is found which cleaves TMP to thymidine and inorganic phosphate.[135] (See also references 136, 137.) The increase in activity is approximately 10-fold. There is little or no increase in the rate of breakdown of dCMP, dAMP or dGMP. The enzyme was partly purified (15-fold) and found to be extremely labile. Mg^{++} is necessary for its activity. The purified enzyme had small activity (10–20%) against dAMP, dUMP and *d*-hydroxymethyl UMP and the hydrolysis of these compounds may be due to contaminating enzymes. The induced enzyme did not degrade dTTP as reported for an enzyme in extracts of *B. subtilis* infected with phage SP8.[136] A similar TMP phosphatase activity that increased to the same magnitude has been found after infection of *B. subtilis* SB19 with other subtilis phages whose DNAs contain hydroxymethyl uracil, uracil, or thymine. The TMP phosphatase may serve to prevent incorporation of TMP into the DNA of the infected host. The presence of this enzyme in a thymine-containing phage is unexpected and suggests a new pathway for the formation of TTP which may not involve the phosphorylation of TMP or thymidine.

VIII. Pools of Thymidine Derivatives in Cells

The cellular concentrations of thymidine and its phosphorylated derivatives are of profound importance in the regulation of the intermediary metabolism of these components as well as in the regulation of a large variety of enzymes both directly and indirectly related to thymidine metabolism. Some of the factors involved which have already been discussed are:

1. The stabilizing effect of thymidine and its phosphorylated derivatives on thymidine and TMP kinase.

2. The wide variety of feedback inhibitions and modulating controls exhibited by TTP, ATP and other nucleotides not only for thymidine kinase but for many other enzymes (see Figure 2).

A completely unknown factor is the possible action of thymidine and its phosphorylated derivatives in the induction of enzyme synthesis by interactions with genes directly or indirectly through repressors or activators. A second unknown factor is the possible interaction of thymidine and its derivatives with enzyme precursors or activators and inhibitors, either protein or nucleic acids, normally present in the cell.

With the above in mind it is obviously difficult to understand growth control mechanisms involved in thymidine phosphorylation without an intimate knowledge of pool sizes of these components and changes that occur during growth and development. It is unfortunate, for these reasons, that there is only a modest amount of information available on the content and distribution of thymidine and thymidine nucleotides in cells. Although a few studies have been made of deoxyribonucleotide contents of various tissues, the kind of detailed information required is generally not available.

One of the earliest reports of deoxyribosides in animal tissues was that of Schneider.[138] He prepared extracts from a variety of tissues and estimated deoxyribosides microbiologically using *Lactobacillus* R-26. Colorimetric tests and paper chromatography were also carried out. Whole tissues were used; these were generally frozen immediately in liquid nitrogen, powdered and then homogenized with cold TCA. Deoxyribosides were observed in a large variety of mouse and rat tissues. They were mostly pyrimidine deoxyribosides, mainly deoxycytidine, and no significant amount of purines was found. Using a crude particulate fraction prepared in isotonic saline, 81% of the deoxyribosides were found in the supernatant fraction.

Later work showed that in the Novikoff hepatoma, thymidine, deoxyuridine and high concentrations of deoxynucleotides were present in addition to deoxycytidine [139] and large amounts of deoxynucleotides were observed in liver shortly after hepatectomy and in other growing tissues.[140,141] R. L. Potter and co-workers isolated the mono-, di-, and triphosphates of thymidine and deoxycytidine from thymus.[142] Deoxyuridine nucleotides were not found. These experiments although demonstrating the wide occurrence of deoxyribonucleosides and deoxyribonucleotides in a variety of tissues, did not tell anything about the amounts of these components or in any way correlate changes in amount with growth parameters.

Shimabayashi and Iwamoto have determined deoxyribosides in the acid-soluble fraction of chick embryo during development. A bacteriological assay method was used which did not distinguish among the deoxyribosides but ratios of CdR, TdR, and UdR to total pyrimidine deoxyribosides were assayed.[211]

A. Thymidine derivative pool sizes in normal and irradiated animals

The incorporation of isotopically labeled thymidine into DNA has been used by many investigators to study the effect of radiation on the rate of DNA synthesis. The assumption has been made that the endogenous TTP pool is not affected significantly by either the added tracer compound and unlabeled carrier or by the irradiation.

Stewart and collaborators [143] have raised serious objections to this technique and formulated a mathematical model for DNA synthesis which can be used to predict that the change observed in DNA synthesis after irradiation is due to an increase in the TTP pool (see below). Logsdon and Kundel [144] have presented evidence to support this view. Using rat spleen and a variety of pool sizes, they found that over several orders of magnitude, the difference between the curve for the irradiated and nonirradiated animals consists of a shift of the curve to the right (Figure 17). This shift could occur simply from an increase in the size of the endogenous pool of TTP. Nygaard and Potter [146] studied the conversion of thymidine into nucleotides after irradiation using

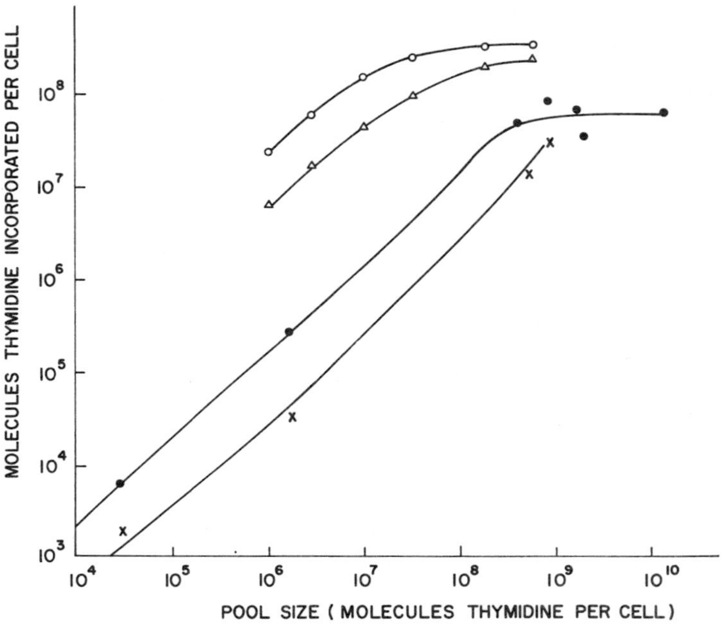

Figure 17. Plot of molecules of thymidine incorporated versus thymidine pool size. o——o Control rat spleen; △——△ rat spleen after irradiation; ●——● control rat spleen; ×——× rat spleen after irradiation. From Logsdon and Kundel.[144]

a pool dilution technique. They found only a slight increase in the TTP pool. Because of this they concluded that there was no significant effect of pool size on incorporation of thymidine into DNA. Huntley and Lajtha [145] studied the effect of pool size on the incorporation of labeled thymidine into mouse bone marrow cells by using thymidine $2C^{14}$ of high and low specific activity. They found no significant changes in the percent incorporation into DNA but they covered only a narrow range of total pool sizes. The four-factor model of Stewart *et al.*[143] is depicted below in Figure 18.

Exogenous thymidine arrives at the pool at a rate which depends on the amount injected, its concentration in the circulation and on permeability and distribution factors. If p_1 molecules of labeled thymidine and p_2 molecules of nonlabeled thymidine are injected together then the amounts in the pool after t times are s_1t and s_2t. All the distribution and permeability factors are lumped into a parameter b_4 assumed to be independent of s. The rate of removal of injected thymidine at the pool can

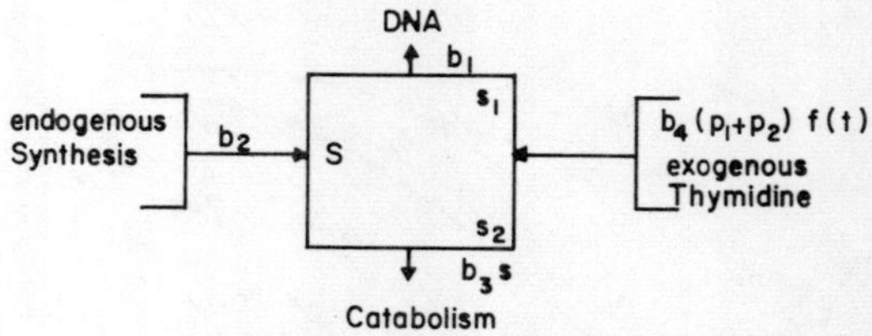

Figure 18. Four factor model of Stewart *et al.*[143] s represents pool size of TTP. b_1 represents rate of TTP incorporation into DNA (assumed to be constant during synthesis and independent of S). b_2 represents rate of synthesis of TTP (also constant and independent of pool size). b_3 equals the rate of catabolism of TTP (first order) so elimination rate is b_3s.

then be written $b_4(p_1 - p_2)f(t)$ molecules/cell/min. $f(t)$ has been determined for a number of systems.[143]

Stewart and collaborators, using squash preparations and grain counts, obtained data in accord with their predictions, namely: (a) a depression of labeled thymidine uptake by exogenous nonlabeled thymidine, and (b) that depression of uptake of thymidine after irradiation in intestinal cell crypts may be due to increasing the effective TTP pool size.

The neglect in the above model of feedback control does not seem warranted. It has been demonstrated by many investigators that exposure of cells to excess thymidine not only stops growth[147–150] but results in toxic effects if continued for a considerable period of time.[151] The explanation of the toxicity is probably the formation of high levels of TTP which repress enzyme activity. Protein and RNA synthesis can continue but cell viability decreases rapidly, perhaps this results in a condition of unbalanced growth similar to that seen in thymineless bacteria. The toxic effect of thymidine can be inhibited or reversed by 2′-deoxycytidine. Cells deficient in thymidine kinase activity are not affected by high concentrations of thymidine suggesting that phosphorylation is necessary for the inhibitory effect.[152] The block is depicted below.

T → TMP → TDP → TTP
CMP → CDP → / dCDP → dCTP → DNA
←┘ ↓
dCMP

Deoxycytidine also markedly increased the utilization of radioactive formate for the formation of TMP, TDP, and TTP.[66] A detailed study of the effects of deficiencies of vitamin B_{12}, folic acid and thymine on intracellular deoxyribosyl compounds has been carried out using several different bacteria.[153] Pool sizes were decreased strikingly by deficiency of vitamin B_{12} and increased several fold by deficiencies of folic acid or thymine.

Biswas and collaborators[154] have suggested that the level of ribonucleotide reductase may be inversely related to the intracellular concentration of TTP with a low concentration resulting in derepression. The results of Cannon and Breitman[89] are consistent with this interpretation.

R. L. Potter and Nygaard[35] studied the conversion of thymidine to thymidine nucleotides and DNA *in vivo* in the spleen and thymus of 45-day old male rats. Thymine nucleotides were recovered by column chromatography on Dowex 1X10 formate. The specific activities of the acid-soluble thymine derivatives was much greater in spleen than in thymus due to considerably larger pool sizes in the latter tissue and possibly a differential uptake of thymidine. The pool sizes of thymidine derivatives were calculated and compared to a few other reports in the literature (Table IX). Soska and Soskova[155] found levels of total

Table IX. Pool Sizes of Thymidine Derivatives in Spleen and Thymus of Rat

	Spleen	Thymus	Pool Sizes in Thymus Relative to Spleen	
	(mμmoles/g of tissue)		(per g of tissue)	(per unit of DNA)*
Thymidine	5.74 ± 0.35 (7.82)	15.63 ± 1.25 (21.4)	2.72	1.41
dTMP	0.62 ± 0.04	4.29 ± 0.31	6.94	3.60
dTDP	2.50 ± 0.18	25.4 ± 0.8	10.15	5.25
dTTP	1.45 ± 0.13	32.3 ± 2.7	22.3	11.6
DNA-thymidylic acid	13,500	26,000	1.93	1.00

The values recorded are means and standard error of the means of 4–10 separate determinations. The values given in parentheses for thymidine were obtained by isotope dilution. The average weight of the organs were spleen 0.765 and thymus 0.578 g.

* These data were derived from those in the previous column by dividing with 1.93 (the ratio between the amounts of DNA thymidylic acid per gram of tissue for the two organs) and presumably represents the ratios between the average pools of each thymidine derivative on a per cell basis.

From Potter and Nygaard.[35]

deoxynucleosides and deoxynucleotides in rat spleen corresponding to about 30 times that reported by Potter and Nygaard for thymine derivatives alone although a statement in their text suggests that the true values may be ten times lower. In the case of rat thymus, Ord and Stocken[156] have reported values for dCMP, dCTP, dUMP and dUTP. Bettendorf, Künkel, and Maass[157] have reported pools of the thymidine nucleotides, roughly one order of magnitude greater than those in Table IX.

Calculations by Potter and Nygaard suggest that the radioactive TTP pool in spleen and thymus turns over twice every minute. These data appear to contradict the idea that there should be an accumulation of deoxynucleotides in the cell before the onset of DNA synthesis and suggest a rapid and continuous formation of these compounds throughout the period of DNA synthesis.

Gentry and co-workers[158] studied thymidine phosphorylation in Novikoff tissue culture cells made totally dependent on thymidine supplied via the medium by blocking thymidylate synthetase with aminopterin. They summarized their observations as follows:

(a) The cells rapidly concentrated thymidine from the medium into the acid-soluble fraction, reaching equilibrium in less than 10 minutes.

(b) Radiochromatographic studies showed that the intracellular thymidine was largely in the form of TTP (80–90%) with some TDP (10–20%) and little if any TMP or thymidine.

(c) Increasing the concentration of thymidine in the medium caused an increase in the intracellular concentration of thymidine compounds.

(d) For cell multiplication the optimal intracellular concentration of thymidine was roughly 5×10^{-5} M.

(e) While the overall rate (but not onset) of DNA synthesis of Novikoff tissue culture cells can be modified by limiting the intracellular TTP concentration the importance of this mechanism *in vivo* remains to be demonstrated.

The authors point out that some evidence is available that alterations in the tissue of intracellular levels of thymidine compounds may have an effect on the rate of DNA synthesis, *e.g.*, Greulich and collaborators[159] found that within 6 hours following an intraperitoneal injection of 10 μg of thymidine per adult male mouse, there was a 29% increase in the number of metaphase mitotic figures in the duodenal epithelium suggesting alterations in the mitotic cycle and perhaps the rate of DNA synthesis.

Some evidence that TMP concentration is not important in initiating

DNA synthesis was obtained by Rueckert and Mueller (unpublished observations [160]) who found that in aminopterin induced TMP starvation of mammalian cells, the onset of DNA synthesis occurs on schedule.

O'Brien has discussed the relationship of the folic acid coenzymes to TMP synthesis and suggests that this reaction may be rate-limiting for DNA synthesis [161] and Friedkin [162] has speculated on the significance of TMP synthesis in the regulation of cellular proliferation.

Cleaver and Holford have studied the incorporation of ^3H thymidine into L strain cells in the exponential phase of growth.[163] Their results can be summarized: (a) ^3H thymidine is incorporated into DNA at a constant rate for at least 1 hour, between concentrations of 10^{-8} and 4×10^{-5} M. (b) The rate of the incorporation is proportional to the square root of the ^3H thymidine concentration in the external medium up to 10^{-6} M. At higher concentrations the incorporation mechanism appears saturated and the incorporation rate is independent of the concentration of thymidine in the external medium. (c) A pool of ^3H thymidine phosphates is formed within the cell during the period of labeling. If cells are allowed to continue growth after they have been labeled, the contents of this pool are used as a source of precursors for DNA synthesis and the amount of ^3H thymidine in DNA continues to increase up to 3 hours after the removal of label. (d) At a concentration of 2.5×10^{-6} M the pool of thymidine phosphates reaches an equilibrium level in the cell after 5 minutes incubation. After an incubation time of 10 minutes in 2.5×10^{-6} M thymidine the pool contains approximately twice the amount of ^3H thymidine that has been incorporated into DNA during that period. The pool size is determined by the external thymidine concentrations and is detectable at external concentrations as low as 7.5×10^{-10} M. Almost all the pool consists of TTP and the ratios are: thymidine 480, TMP, 356, TDP 786 and TTP 14,225. Miller [164] has shown that the soluble thymidine pool is located in the nucleus of Tetrahymena and that ^3H thymidine of the medium enters the pool during S but not during G^1 and G^2.

Biswas and collaborators [154] reported a release of repressor control of ribonucleotide reductase by thymine starvation in *E. coli* $15T^-$. Ribonucleotide reductase activity was increased 10–12-fold and the increase could be prevented by prior addition of chloramphenicol.

The situation in mammalian cells appears to be quite different. Cooper, Perry and Breitman [165–166] have studied pyrimidine metabolism in human leucocytes from patients with myelogenous leukemia. The questions asked by Cooper and co-workers were: (a) To what extent does incorporation of exogenous thymidine into DNA decrease incorporation of thymidine synthesized *de novo?* (b) Does the availability of an exogenous source for thymidine nucleotide synthesis affect thymidine nucleotide synthesis *de novo?* It was found that: (a) DNA thymine derived from exogenous thymidine increases from 13 to 87% over a range of thymidine concentrations from 0.03 μM to 300 μM. (b) Thymidine caused expansion of the total TDP and TTP pool but did not influence the contribution of the *de novo* pathway to this pool. Thus the increasing contribution of exogenous thymidine to the formation of DNA thymine occurred because of progressive dilution of TTP synthesized *de novo* with TTP derived from exogenous thymidine. Also, above 0.3 μM, thymidine inhibited DNA synthesis but not RNA synthesis. This inhibition could be reversed by deoxycytidine. This might be cited as evidence that thymidine and possibly its phosphorylated derivatives do not regulate synthesis of kinase enzymes. It is clear from the work of Cooper and co-workers that in cells with a normal complement of enzymes, TTP is not limiting DNA synthesis. (See discussions of mechanisms, this paper.)

In a second paper, Cooper, Perry, and Breitman studied the sizes and some kinetics of the pools of thymidine nucleotides in normal and leukemic leukocytes.[166] At a concentration of 4 μM exogenous ^{3}H thymidine, using leukemic leucocytes, labeling of TDP and TTP pools reached a maximum within 5 minutes and remained constant for 20 minutes. Similar results were obtained at all thymidine concentrations examined between 0.17 μM and 30 μM. Equilibration of TMP pool with 4 μM ^{3}H thymidine, however, was not obtained within a 25 minute experimental period. In normal leucocytes equilibration was obtained with the TMP pool at 18 minutes while maximal labeling of the TDP and TTP pool again occurred at 5 minutes. Lowered temperatures slowed equilibration of the TDP and TTP pool. Gentry *et al.*[158] reported similar results with Novikoff hepatoma cells in tissue culture suspension.

With increasing concentration of thymidine, expansion of the TMP pool derived from thymidine was greater than expansion of the TDP and TTP pools. In leukemic leucocytes the TDP and TTP pools were approximately equal in size and accounted for 65% of the thymidine nucleotides in the presence of 1 μM thymidine but only 10% in the presence of 300 μM thymidine. In normal leucocytes at these concentrations, the TDP and TTP pools accounted for 35% and 5% of the total thymidine nucleotides at the two concentrations, respectively.

Over the range of thymidine concentrations studied, incorporation of ^{3}H thymidine into the TMP pool was 2–5-fold greater, and incorporation into the TDP and TTP pool was 6–15-fold greater in leukemic than in normal leucocytes. Incorporation of 3 μM ^{3}H thymidine into DNA was 30-fold greater in leukemic than in normal leucocytes. In the presence of thymidine at a concentration of 14 μM, the turnover rate of the thymidine nucleotide pool was 3 minutes. Approximately 35% of the TTP turnover was due to its incorporation into DNA. The finding that the radioactivity of TDP + TTP reached a maximum more rapidly than TMP is not consistent with the sequence ^{3}H thymidine → TMP → TDP→TTP but could be explained on the basis that there are two intracellular pools, one equilibrating rapidly and one slowly. Since approximately 75% of the chronic myologenous leucocytes used were morphologically mature and were not undergoing DNA synthesis, if these, like normal leucocytes accumulated excess of TMP over TDP + TTP, the findings can be explained on the basis of one rapidly equilibrating nucleotide pool ^{3}H thymidine → nucleotide pool and one in which ^{3}H thymidine → TMP → TDP + TTP (slowly). There are other explanations for these data, however (see recapitulation).

B. Pools—Tetrahymena

The situation is complicated in Tetrahymena by the presence of a micronucleus and a macronucleus which synthesize DNA at different times in the cell cycle. Thymidine is also incorporated into the cytoplasm.[167] In spite of this, Stone and collaborators[49] have concluded from direct measurements that a soluble pool of ^{3}H thymidine derivative is present from the end of one DNA synthetic period to the next, but does not turn over during G-2 or G-1. This persisting pool is utilized in DNA synthesis during the next S period and is turned over completely during one S period or less. They visualize the sequence of events in Tetrahymena as follows: (a) DNA synthesis is initiated and continues for a short time at the expense of a pool formed during the preceding S period. (b) The turnover of the pool stimulates the synthesis of additional phosphorylated precursors thus producing the uptake of endogenous thymidine into the intracellular pool. According to this interpretation, the immediate precursor pool would have no influence on the initiation of DNA synthesis but rather DNA synthesis would be the stimulus for the turnover of the thymidine derivative pool. The evidence that Tetrahymena utilizes a pre-existing pool of thymidine derivatives for some DNA synthesis in each cell cycle raises the possibility that S may actually begin at an earlier point in the cell cycle than is indicated by the uptake of ^{3}H thymidine.

Previous experiments with single Tetrahymena [49] have shown that macronuclear DNA synthesis can be initiated in the absence of essential amino acids but that the DNA increase is limited to 20%. For this 20%, ^{3}H thymidine of the medium is not used. This suggests that some of the enzymes of DNA synthesis are present even in the absence of new enzyme synthesis. Another interpretation is that there is a critical level of TTP or some other precursor(s) which turns on synthesis of kinases and other related enzymes (except possibly not DNA polymerase). Some of these ideas are expressed in Figure 19.

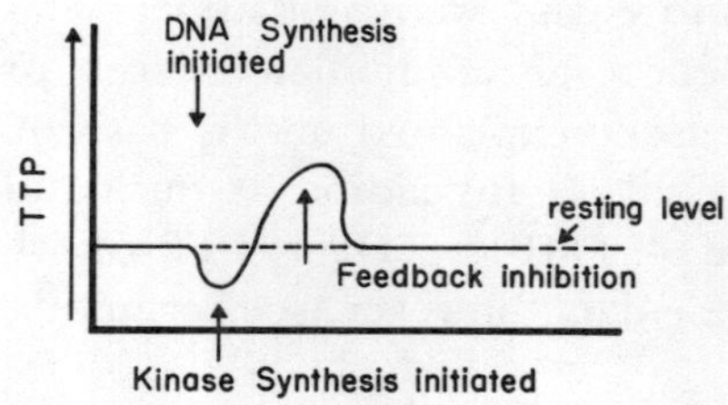

Figure 19. Suggested relationship of intracellular concentration of TTP to events in the cell cycle in tetrahymena.

IX. Other Enzymes Related to Thymidine Phosphorylation

A. Thymidine phosphorylase

The enzyme thymidine phosphorylase catalyzes the following reaction:

$$\text{deoxythymidine} + P_i \rightleftharpoons \text{thymine} + \text{deoxyribose-1-phosphate}$$

As such it is only indirectly related to thymidine phosphorylation. However, the presence of large quantities of phosphorylase in a tissue can reduce the possibility of the reaction TdR → TMP. Friedkin and Roberts [168] have studied the distribution of thymidine phosphorylase in mammalian tissue. In the rabbit activity was highest in intestinal mucosa (31-9 units), liver (13-6), and bone marrow (3.2-0.2) and lower in kidney (2.3-1.0), spleen (3.4-0.5), and heart (0.55-0.05), and lung (0.58-0.13), and essentially absent in skeletal muscle and brain. The phosphorylase was inhibited by isolated mitochondria. In chick embryo (5–18 days) most of the activity occurs in the liver, the rest of the embryo having much lower activity. (This may be why thymidine does not cause induction of thymidine kinase activity.) The specificity of the enzyme was broad, uracil, 5-halogen uracils, aminouracil and thiouracils were all substrates.[169]

The pyrimidine nucleoside phosphorylases have been reviewed[170] and thymidine phosphorylase of *E. coli* has been studied in some detail.[171] Inhibitors of the enzyme[172–176] and its specificity[177] have been investigated recently, while thymidine phosphorylase and nucleoside deoxyribosyltransferase in normal and malignant tissues has been studied by Zimmerman and Seidenberg[178,179] and by Jacquez.[180]

B. Deoxyribose-5′-phosphate aldolase
This enzyme catalyzes the following reaction:

$$\text{Deoxyribose-5}'\text{-phosphate} \rightleftharpoons \text{acetaldehyde} + \text{glyceraldehyde-3-phosphate}$$

Reversal of this reaction provides a possible source of deoxyribose other than the reduction of ribose (see reference 181).

C. Pyrimidine deoxyribosyltransferase
This enzyme catalyzes the following reaction:

$$B_1 + B_2dR \rightleftharpoons B_2 + B_1dR \qquad \text{(see reference 182)}$$

D. Nucleoside diphosphokinases
Nakamuro and Sugino[183] have obtained a nucleoside diphosphokinase from calf thymus. The enzyme was purified 2800-fold and found to have a very general specificity in that all nucleoside di- and triphosphates related to nucleic acids were active as phosphate acceptors and donors, respectively. (For review, see references 1–15 in reference 183.) The pH optimum was 9.0 and the reaction absolutely dependent on a divalent cation and ATP. Mg^{++}, Mn^{++}, and Co^{++} gave about equal activity. A ratio, Mg^{++}/ATP of 4:1 was optimal. The diphosphokinase separated into two peaks on DEAE column chromatography and two forms of the enzyme with different sedimentation rates were detected by sucrose density gradient centrifugation. The presence of a phosphorylated protein as an active intermediate in this type of transphosphorylation reaction has been reported by Mourad and Parks[184] and Norman *et al.*[185]

Ratliff and co-workers have isolated a crystalline nucleoside diphosphokinase from yeast [186] which also appeared to have a wide specificity. These nucleoside dephosphokinases may be the same enzymes that phosphorylate TDP. (See references 53 and 61.)

E. Deoxythymidine phosphotransferase

Seifert and Riman [205] have reported a deoxythymidine phosphotransferase occurring in the high-speed supernatant fraction of fowl leukemic myeloblasts. The reaction catalyzed is:

$$\text{TdR} + \text{nucleoside-5}'\text{-monophosphate} \rightleftharpoons 5'\text{-TMP} + \text{nucleoside}$$

Nucleoside diphosphates and triphosphates may also be the donor. UDP and CDP are the most efficient donors and the triphosphates are generally the poorest donors. The TdR phosphotransferase activity of the supernatant fraction is competitively inhibited by ATP which acts presumably to regulate the reaction.

F. Phosphorylation of thymidine-3'-pyrophosphate

Coutsogeorgopoulus and co-workers have shown [187] that a high-speed supernatant fraction prepared from regenerating rat liver will form thymidine-3'-triphosphate from added ^{14}C-thymidine-3'-pyrophosphate. Hacker has also demonstrated that 3'-TMP is converted to a mixture of the 3' and 5' di- and triphosphates.[52] An identical reaction using extracts of *Bacillus subtilis* was described by Canellakis and collaborators.[188] The latter investigators showed that the reaction requires ATP; that 3'-TMP is not phosphorylated and that the synethesis is not inhibited by a 50-fold excess of 5'-TDP. The enzyme was not purified from either source and its identity and specificity are presently unknown as is the significance of this reaction.

X. Conversion of Thymine and Azathymine to Ribonucleosides and Deoxyribonucleosides

Thymidine [189] but not free thymine is extensively utilized by animal cells.[190–192] Nevertheless, Gaito and Prusoff [193] found some conversion of both thymine and azathymine to the deoxyriboside both *in vivo* and *in vitro.* Radioactive 2'-deoxyuridine was also isolated on incubation of a particle-free supernatant fraction (100,000 × *g*) of a homogenate of mouse liver with 0.05 M Tris buffer pH 7.4. Following incubation for 1 hour, no deoxyuridine was found. When unlabeled 2'-deoxyuridine was added, radioactive 2'-deoxyuridine was obtained. Since the radioactivity of the nucleoside and uracil prepared from it were the same, the deoxyuridine must have been derived from uracil, possibly by the following equilibrium:

C^{14} uridine + 2′-deoxyuridine ⇌ C^{14} deoxyuridine + unlabeled uridine

XI. Recapitulation

Fact

1. The control of TMP kinase activity could be one of the major mechanisms for the control of cell proliferation.

2. TMP kinase is found in significant amounts in all rapidly proliferating cells. Its activity appears to be closely related to the proliferative activity of the cells. High activities are found in tumors, embryonic tissues, virus infected cells, regenerating liver, and in the hypertrophic kidney after uninephrectomy. In the latter two cases maximum enzyme activity is observed at or near the time of maximum DNA synthesis.

3. The enzyme is recovered principally in high speed supernatant fractions of cell extracts or homogenates. During division of synchronous cells it undergoes cyclical variations in activity reaching a maximum at or near the time of maximum DNA synthesis.

4. Thymidine kinase, although important for rapidly growing abnormal systems such as tumors or virus infected cells, is not found in significant amounts in some types of embryos (chick and frog) and is therefore relegated to a supporting role, that of a salvage enzyme off the main pathway of TTP synthesis.

5. Significant quantities of TMP kinase are found in mitochondria and there may be some present in nuclei. The nuclear location of TMP kinase has not been adequately demonstrated.

6. TMP kinase is not subject to feedback control by TTP or other nucleotides at the enzyme level.

7. It is, however, a very unstable enzyme which is remarkably stabilized by its substrate TMP, by TTP and, possibly, by other nucleotides.

Fancy (but consistent with available data)

1. The activity of TMP kinase may be controlled at the enzyme level by a repressor system which in turn is controlled by the intracellular concentration of a critical nucleotide. The nucleotide may be TMP (or it might be TTP) and the system might work in the following way (see Figure 20).

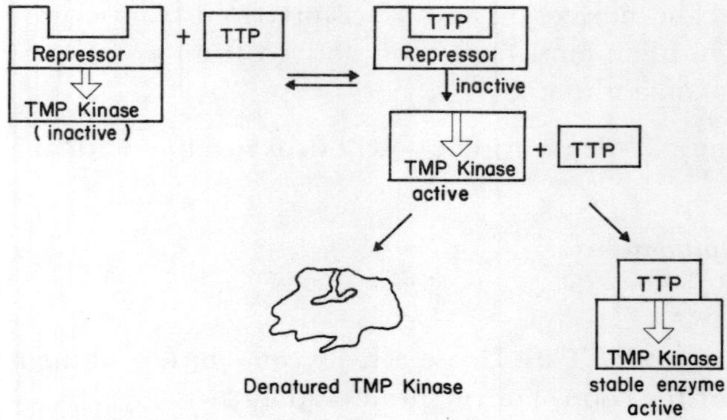

Figure 20. Hypothetical representation of a scheme for control of TMP kinase activity. It is assumed that TMP kinase is manufactured at a constant rate by the cell. When TTP (or TMP) levels are low (a) repressor binds and inactivates TMP kinase, and (b) additional amounts of TMP kinase which are synthesized or which are released by the repressor are rapidly denatured. As TTP levels rise, the TTP not only stabilizes newly synthesized enzyme but provides for a burst of activity due to release from the repressor. The only function of the repressor would be to provide for rapid surges of activity. The stabilization phenomenon alone could account for the observations that have been described. If the stability of the repressor was high, this system could account very well for the cyclical changes in activity that are observed in synchronized cells. The control would, of course, be conditioned by the amount of TMP fed into the system.

It is assumed that TMP kinase is manufactured at a constant rate by the cell. When TTP (or TMP) levels are low (a) repressor binds and inactivates TMP kinase, and (b) additional amounts of TMP kinase which are synthesized or which are released by the repressor are rapidly denatured. As TTP levels rise, the TTP not only stabilizes newly synthesized enzyme but provides for a burst of activity due to release from the repressor. The only function of the repressor would be to provide for rapid surges of activity. The stabilization phenomenon alone could account for the observations that have been described. If the stability of the repressor was high, this system could account very well for the cyclical changes in activity that are observed in synchronized cells. The control would, of course, be conditioned by the amount of TMP fed into the system.

This system allows one to make the following predictions which could be experimentally tested.

1. TTP (or TMP, whichever is the repressor nucleotide component)

levels would be low in nonproliferating tissues and TTP levels would decrease during development or in any situation in which TMP kinase activities decrease. The reverse would be true for situations in which TMP kinase activity increased.

2. The increases in TMP kinase activity found in regenerating liver, viral infected systems (in which TMP kinase is of cell and not viral origin) should not represent an increased rate of synthesis of TMP kinase protein.

3. Since the source of TTP is either

$$\text{(1) TdR} \rightarrow \text{TMP (thymidine kinase)} \rightarrow \text{TTP}$$

or

$$\text{(2) UTP} \rightarrow \text{dUMP} \rightarrow \text{TMP} \rightarrow \text{TTP}$$

one or more of these pathways should be increased in all situations in which TMP kinase activity increases. Concomitantly the reaction TMP$\rightarrow$ TdR + P_i may decrease.

4. If a repressor for TMP kinase exists it should be capable of binding TTP.

5. TMP kinase should exist in the cell bound to a repressor and should be released from this repressor by addition of TTP (or TMP) under the appropriate conditions.

6. In tumors with high TMP kinase activities this may be due to:

(a) High TTP (or TMP) levels.

(b) Failure to produce repressor or faulty repressor.

(c) Increased rate of synthesis of TMP kinase by derepression of the gene.

Viral oncogenesis could be directly concerned with these processes. The virus could derepress the gene for TMP kinase synthesis in which case the rate of TMP kinase synthesis would be considerably higher than normal or the virus might prevent the synthesis of repressor. In the latter case one would expect a lag period to occur during which repressor originally present is being used up.

These predictions should provide a fertile field for experimentation and the knowledge gained in doing so will undoubtedly provide a better understanding of the control mechanisms involved in TTP synthesis.

Appendix 1. Some Characteristics of Assay

Ref.	TdR Kinase Source	pH	Mg^{++}	ATP	ATP/Mg^{+2}
216	Human KB cells and adenovirus 2 or vaccina. 1–2 mg prot. extract per assay mix	7.8	5 mM	5 mM	1
			5 mM/ml	5 mM/ml	

Comments:
1. TdR kinase of infected and uninfected similar (Sp. act. infected, 20% higher).
2. Vaccina infection yield 8× increase in TdR kinase activity over uninfected controls.

Ref.	TdR Kinase Source	pH	Mg^{++}	ATP	ATP/Mg^{+2}
11	*Tetrahymena pyriformis*	8.4	4.8 mM	3.0 mM	0.625
			24 mM/ml	15 mM/ml	

Comments:
1. Less than 1% TDP and TTP formed.
2. TTP does *not* inhibit.
3. Optimal ATP level is relatively low.
4. At suboptimal ATP levels TTP and other deoxypyrimidine nucleotides stimulate the reaction. (TTP and CTP at least function as phosphate donors.)
5. Enzyme activity same in log and stationary phase – appears to be constitutive enzymes.

Ref.	TdR Kinase Source	pH	Mg^{++}	ATP	ATP/Mg^{+2}
55	Walker carcinosarcoma (150× purified enzyme)	8.0	25 mM	50 mM	2
			10 mM/ml	20 mM/ml	
53	Landschutz ascites-tumor cells	8.0	5 μM	5 μM	1
			0.005 mM/ml	0.005 mM/ml	

Comments:
1. TdR kinase purified 3–4-fold.
2. pH optimum 7.8–8.1.
3. Absolute requirement for Mg^{+2} (5 mM optimal).
4. 2-mercaptoethanol gave only marginal enhancement of TdR kinase, BSA had no effect.

Ref.	TdR Kinase Source	pH	Mg^{++}	ATP	ATP/Mg^{+2}
9	Novikoff hepatoma	7.75 (assay)	9 mM	9 mM	1
		optimum 8.25–8.5	18 mM/ml	18 mM/ml	0.32 mM/ml

Comments:
1. TdR kinase exhibited linear and complete utilization of substrate; was independent of ATP-Mg^{++} conc. and exhibited zero order kinetics over a 16× range TdR conc.
2. Excess Mg^{++} stimulated phosphatase activity.
3. TTP was a negative feedback inhibitor. Inhibition was maximal at low ATP-Mg^{++} conc. while high ATP-Mg^{++} conc. reversed the inhibition.

Systems for Thymidine Kinase

TdR	Incub. Time and Temp.	Buffer	Total Assay Vol. ml	Other Reagents	Stop Reaction With	Separation of Product
TdR-C^{14} 16 μM 288 μC 0.016 mM/ml	37° Time varies	40 mM Tris-HCl	1	5 mM 3-phosphoglycerate	Boiling H_2O bath	Paper electrophoresis
TdR-H^3 0.2 mM 1 mM/ml	30 min linear 30°C 5 min preincub.	Tris-HCl 12 mM	0.2	No stimula. by 3-P-glycerate or phosphocreatine	TCA ppt	Paper chrom.
TdR-C^{14} 0.1 μC	37° 15 min	0.05 M Tris-HCl	0.25	(0.2 M KCl for aggregate form only)	Boiling H_2O bath	Paper chrom.
TdR-H^3 2.8 μM 1.0 μC 0.0028 mM/ml	37° 15 min	Tris 100 μM	1.0	0.1 μM mercapto-ethanol	Boiling H_2O	Ecteola-cellulose column chrom.
C^{14}-TdR 0.16 mM (1–2 μC)	2–3 hr. 37°	Tris-HCl 0.05 mM	0.5	3-P-glycerate 7.5 mM KCl	Boiling	Paper chrom.

Appendix 1 (*Continued*)

Ref.	TdR Kinase Source	pH	Mg^{++}	ATP	ATP/Mg^{+2}
16	Leucocytes (normal and leukemic)	8.0	5 mM	5 mM	1
					0.00008 mM/ml
			20 mM/ml	20 mM/ml	0.00007 mM/ml

Comments:
1. Both sources inhibited by 5′-mono-, di-, and triphosphates of TdR, by dCMP and dCTP.
2. TdR kinase gradually inactivated at 50°C.

Ref.	TdR Kinase Source	pH	Mg^{++}	ATP	ATP/Mg^{+2}
215	Chang human liver cells (tissue culture)	8.0	100 mM	100 mM	1
			200 mM/ml	200 mM/ml	

Comments:
1. Addition of TdR (0.03 mM) inhibited TdR kinase. Inhib. relieved by CdR as long as TdR did not exceed 0.5 mM.
2. Amethopterin stimulated TdR kinase as did 5-fluorodeoxyuridine. Both reagents prevented growth (mitosis).
3. Puromycin prevented amethopterin stimulation.

Ref.	TdR Kinase Source	pH	Mg^{++}	ATP	ATP/Mg^{+2}
32	Regenerating rat liver	8.0	3 μM	3 μM	1
			0.012 mM/ml	0.012 mM/ml	

Comments:
1. Preincubation effectively (37°) destroys TMP kinase while TdR kinase was unaffected.

TdR	Incub. Time and Temp.	Buffer	Total Assay Vol. ml	Other Reagents	Stop Reaction With	Separation of Product
$TdR\text{-}C^{14}$ (1) 16.7 mμM= 0.5 μC (2) 200 mμM= 0.5 μC	37° 60 min	Tris-HCl 0.2 M (to bring up to final volume)	0.25	3-P-glycerate 6 mM	Ice	"Disc" elution
$TdR\text{-}C^{14}$ 2 mM/ml	5 min preincub. 0–30 min at 37°	0.2 M Tris-HCl 200 μl	0.5	3-P-glycerate 120 mM	Cool and TCA ppt	Paper chrom.
15 mμM $TdR\text{-}H^{3}$ 0.225 μC 0.00006 mM/ml	37° (preincubation also)	Tris-HCl 20 μM	0.25	3-P-glycerate 2.5 μM	Cool and TCA ppt	Paper chrom.

Appendix 2. Some Characteristics of Assay Systems for TMP Kinase

Ref.	Enzyme Source	Homogenizing Medium	Buffer in Assay/ml	pH	Mg^{++}/ml	ATP/ml	Mg^{++}/ATP	dTMP Concentration in Assay	KF or NaF/ml	Other Compounds in Assay/ml
53	Landschutz ascites tumour cells	0.1 M phosphate buffer pH 8.1	Tris (100 μM)	7.6	5 μM	5 μM	1	(P^{32}-TMP) 20 mμM	–	–
74	*E. coli* cells infected by T_2 phage		Tris (80 μM)	7.5	8 μM	4.8 μM	1.667	(P^{32}-TMP) 400 mμM	–	–
33	Mouse C3H liver (normal) and ascites hepatoma 129(P)	0.05 M pH 7.4 glycylglycine + 0.01 M K_2HPO_4	Tris (40 μM)	7.4	16 μM	8 μM	2	(TMP-2-C^{14}) 64 mμM	–	–
32	Regenerating rat liver	Buffered sucrose	Tris 80 μM	8.0	12 μM	12 μM	1	(TMP-H^3) 66 mμM	4 μM	10 μM 3-phosphoglycerate
218	Leukemic myeloblasts (avian) induced by BAL strain A virus		Tris 2.0×10^5 μM	7.75	3.6×10^4 μM	3.6×10^4 μM	1	(TMP-2-C^{14}) 6500 mμM	–	7.5×10^3 μM 3-phosphoglycerate + KCl
212	Human erythrocytes	Sucrose-KCl pH 8.0	–	8.0	8.5 μM	8.5 μM	1	(TMP-2-C^{14}) 6.5 mμM	–	8.5 μM 3-phosphoglycerate
7	Oak silkworm *Antheraea pernyi*	0.25 M (100) sucrose 0.2 tris (6) pH 7.7 0.15 M KCl (4)	Tris 1.0×10^5 μM	7.7	5×10^3 μM	5×10^3 μM	1	(TMP-2-C^{14}) 1.65×10^4 mμM	–	–
42	*Rana pipiens*	0.05 M pH 7.8 glycylglycine + 0.01 M K_2HPO_4	Same as homogenizing medium	7.8	40 μM	40 μM	1	(TMP-2-C^{14}) 50 mμM	–	–

References

1. Okazaki, R., and Kornberg, A., *J. Biol. Chem.*, **239**, 269 (1964).
2. Okazaki, R., and Kornberg, A., *J. Biol. Chem.*, **239**, 275 (1964).
3. Shoup, G. D., Prescott, D. M., and Wyckes, J. R., *J. Cell Biol.*, **31**, 295 (1966).
4. Johnson, R. A., and Schmidt, R. R., *Biochim. Biophys. Acta*, **129**, 140 (1966).
5. Sachsenmaier, W., and Ives, D. H., *Biochem. Zeit.*, **343**, 399 (1965).
6. Brookes, V. J., *Biochim. Biophys. Acta*, **119**, 268 (1966).
7. Brookes, V. J., and Williams, C. M., *Proc. Natl. Acad. Sci. U. S.*, **53**, 770 (1965).
8. Bresnick, E., and Thompson, U. B., *J. Biol. Chem.*, **240**, 3967 (1965).
9. Ives, D. H., Morse, P. A., and Potter, V. R., *J. Biol. Chem.*, **238**, 1467 (1963).
10. Bianchi, P. A., Butler, J. A. V., Crathorn, A. R., and Shooter, K. V., *Biochim. Biophys. Acta*, **53**, 123 (1961).
11. Weissman, S. M., Smellie, R. M. S., and Paul, J., *Biochim. Biophys. Acta*, **45**, 101 (1960).
12. Bukovsky, J., and Roth, J. S., *Advances in Enzyme Regulation*, **2**, 371 (1964).
13. Bukovsky, J., and Roth, J. S., *Cancer Res.*, **25**, 358 (1965).
14. Bresnick, E., Thompson, U. B., Morris, H. P., and Liebelt, A. G., *Biochem. Biophys. Res. Commun.*, **16**, 278 (1964).
15. Roth, J. S., *Life Sci.*, **3**, 1145 (1964).
16. Bresnick, E., and Karjala, R. J., *Cancer Res.*, **24**, 841 (1964).
17. Morse, P. A., Jr., and Potter, V. R., *Cancer Res.*, **25**, 499 (1965).
18. Brent, T. P., Butler, J. A. V., and Crathorn, A. R., *Nature*, **207**, 176 (1965).
19. Littlefield, J. W., *Biochim. Biophys. Acta*, **95**, 14 (1965).
20. Stubblefield, E., and Mueller, G. C., *Biochem. Biophys. Res. Commun.*, **20**, 535 (1965).
21. Lieberman, I., Abrams, R., Hunt, N., and Ove, P., *J. Biol. Chem.*, **238**, 3955 (1963).
22. Dubbs, D. R., and Kit, S., *Exptl. Cell Res.*, **33**, 19 (1964).
23. Bresnick, E., Williams, S. S., and Mosse, H., *Cancer Res.*, **27**, 469 (1967).
24. Beltz, R. E., *Biochem. Biophys. Res. Commun.*, **9**, 78 (1962).
25. Bollum, F. J., Anderegg, J. W., McElya, A. B., and Potter, V. R., *Cancer Res.*, **20**, 138 (1960).
26. Kim, J. H., Gelbard, A. S., and Perez, A. G., *Cancer Res.*, **27**, 1301 (1967).
27. Gray, E. D., Weissman, S. M., Richards, J., Bell, D., Keir, H. M., Smellie, R. M. S., and Davidson, J. N., *Biochim. Biophys. Acta*, **45**, 111 (1960).
28. Fausto, N., and Van Lancker, J. L., *J. Biol. Chem.*, **240**, 1247 (1965).
29. Pitot, H. C., Peraino, C., Lamar, C., Jr., and Kennan, A. L., *Proc. Natl. Acad. Sci. U. S.*, **54**, 845 (1965).
30. Canellakis, E. S., and Mantsavinos, R., *Biochim. Biophys. Acta*, **27**, 643 (1958).
31. Beltz, R. E., *Arch. Biochem. and Biophys.*, **99**, 304 (1962).
32. Behki, R. M., and Morgan, W. S., *Arch. Biochem. and Biophys.*, **107**, 427 (1964).
33. Kielley, R. K., *Cancer Res.*, **23**, 801 (1963).
34. Bianchi, P. A., *Biochim. Biophys. Acta*, **55**, 547 (1964).

35. Potter, R. L., and Nygaard, O. F., *J. Biol. Chem.*, **238,** 2150 (1963).
36. Bresnick, E., *J. Biol. Chem.*, **240,** 2550 (1965).
37. Canellakis, E. S., Jaffee, J. J., Mantsavinos, R., and Krakow, J. S., *J. Biol. Chem.*, **234,** 2096 (1959).
38. Roth, J. S., and Askew, J., *Biochim. Biophys. Acta,* **130,** 28 (1966).
39. Maley, F., and Maley, G. F., *Biochemistry,* **1,** 847 (1962).
40. Weinstock, I. M., and Dju, M. Y., *Life Sci.*, **6,** 797 (1967).
41. Scholl, A., Herrmann, H., and Roth, J. S., *Life Sci.*, I, 9 (1968).
42. Atkinson, B., and Roth, J. S., unpublished results.
43. Wanka, F., *Z. Naturforsch.*, **22,** 91 (1967).
44. Wanka, F., and Walboomers, J. M., *Z. Pflanzenphysiol.*, **55,** 458 (1966).
45. Wanka, F., Vasil, I. K., and Stern, H., *Biochim. Biophys. Acta,* **85,** 50 (1964).
46. Hotta, Y., and Stern, H., *Proc. Natl. Acad. Sci. U. S.*, **49,** 648 (1963).
47. Hotta, Y., and Stern, H., *J. Biophys. Biochem. Cytol.*, **11,** 311 (1961).
48. Roth, J. S., and Koths, M., unpublished results.
49. Stone, G. E., Miller, O. L., Jr., and Prescott, D. M., *J. Cell Biol.*, **25,** 171 (1965).
50. Branton, D., and Jacobson, L., *Stain Technol.*, **37,** 239 (1962).
51. Feinendegen, L. E., and Bond, V. P., *Exp. Cell Res.*, **27,** 474 (1962).
52. Hacker, B., Ph.D. Thesis, Univ. of New York at Buffalo, 1967.
53. Grav, H. J., and Smellie, R. M. S., *Biochem. J.*, **94,** 518 (1965).
54. Bianchi, P. A., Crathorn, A. R., and Shooter, K. V., *Biochim. Biophys. Acta,* **61,** 728 (1962).
55. Bresnick, E., Thompson, U. B., and Lyman, K., *Arch. Biochem. and Biophys.*, **114,** 352 (1966).
56. Baker, B. R., Schwan, T. J., and Santi, D. V., *J. Med. Chem.*, **9,** 66 (1966).
57. Bresnick, E., and Williams, S. E., *Biochem. Pharmacol.*, **16,** 503 (1967).
58. Sartorelli, A. C., and Booth, A. B., *Mol. Pharmacol.*, **3,** 71 (1967).
59. Grav, H. J., and Smellie, R. M. S., *Biochem. J.*, **89,** 486 (1963).
60. McAuslan, B. R., and Joklik, W. K., *Biochem. Biophys. Res. Commun.*, **8,** 486 (1962).
61. Ives, D. H., *J. Biol. Chem.*, **240,** 819 (1965).
62. Bianchi, P. A., Farina, M. V., and Polli, E., *Biochim. Biophys. Acta,* **91,** 323 (1964).
63. Kit, S., and Dubbs, D. R., *Virology,* **26,** 16 (1965).
64. Breitman, T. R., *Biochim. Biophys. Acta,* **67,** 153 (1963).
65. Morris, N. R., Reichard, P., and Fischer, G. A., *Biochim. Biophys. Acta,* **68,** 93 (1963).
66. Delamore, I. W., and Prusoff, W. H., *Biochim. Biophys. Acta,* **80,** 220 (1964).
67. Moore, E. C., and Hurlbert, R. B., *J. Biol. Chem.*, **241,** 4802 (1966).
68. Bojarski, T. B., and Hiatt, H. H., *Nature,* **188,** 1112 (1960).
69. Pitot, H. C., Peraino, C., Pries, N., and Kennan, A. L., *Advances in Enzyme Regulation,* **3,** 359 (1965).
70. Hiatt, H. H., and Bojarski, T. B., *Biochem. Biophys. Res. Commun.*, **2,** 35 (1960).
71. Kielley, R. K., *Biochem. Biophys. Res. Commun.*, **10,** 249 (1963).
72. Kielley, R. K., *Biochim. Biophys. Acta,* **53,** 150 (1961).

73. Nelson, D. J., and Carter, C. E., *Mol. Pharmacol.,* **3,** 341 (1967).
74. Bello, L. J., van Bibber, M. J., and Bessman, M. J., *Biochim. Biophys. Acta,* **53,** 194 (1961).
75. Kokolis, P., Weiss, F., and Weiss, A. J., *Proc. Am. Assoc. Cancer Res.,* **7,** 37 (1966).
76. Kit, S., Dubbs, D. R., and Piekarski, L. J., *Biochem. Biophys. Res. Commun.,* **8,** 72 (1962).
77. Stern, H., in "Developing Cell Systems and Their Control," 18th Growth Symposium 1960, Ronald Press, New York, N. Y., p. 135.
78. Stern, H., *J. Biophys. Biochem. Cytol.,* **9,** 271 (1961).
79. Hotta, Y., and Stern, H., *Proc. Natl. Acad. Sci. U. S.,* **49,** 861 (1963).
80. Hotta, Y., and Stern, H., *J. Cell Biol.,* **25,** 99 (1965).
81. Stern, H., and Hotta, Y., *Brookhaven Symp. Biol.,* **16,** 59 (1963).
82. Gelbard, A. S., Kim, H. S., and Eidinoff, M. L., *Cancer Res.,* **26,** 748 (1966).
83. Baserga, R., Estensen, R. D., Peterson, R. O., and Layde, J. P., *Proc. Natl. Acad. Sci. U. S.,* **54,** 745 (1965).
84. Mayfield, E. D., Jr., Liebelt, R. A., and Bresnick, E., *Cancer Res.,* **27,** 1652 (1967).
85. Shen, S. R. C., and Schmidt, R. R., *Arch. Biochem. and Biophys.,* **115,** 13 (1966).
86. Bessman, N. J., *J. Biol. Chem.,* **234,** 2735 (1959).
87. Hecht, L. I., Potter, V. R., and Hebert, E., *Biochim. Biophys. Acta,* **15,** 134 (1954).
88. Canellakis, E. S., in Proc. IVth Int. Biochem. Congr. Vienna, Pergamon Press, London, 1958, p. 75.
89. Cannon, W. D., Jr., and Breitman, T. R., *Biochem.,* **6,** 810 (1967).
90. Bolton, E. T., *Proc. Natl. Acad. Sci. U. S.,* **40,** 764 (1954).
91. Lesley, S. M., and Graham, A. F., *Canad. J. Microbiol.,* **2,** 17 (1956).
92. Lichtenstein, J., Barner, H. D., and Cohen, S. S., *J. Biol. Chem.,* **235,** 457 (1960).
93. Weiss, A., Zagerman, A., and Kokolis, P., *Fed. Proc.,* **24,** 330 (1965).
94. Main, R. K., Cole, L. J., and Walwick, E. R., *Int. J. Radiobiol.,* **7,** 45 (1963).
95. Fausto, N., Uchiyama, T., and van Lancker, J. L., *Arch. Biochem. and Biophys.,* **106,** 447 (1964).
96. Beltz, R. E., and Applegate, R. L., *Biochem. Biophys. Res. Commun.,* **1,** 298 (1959).
97. Hiatt, H. H., and Bojarski, T. B., *Cold Spring Harbor Symp. Quant. Biol.,* **26,** 367 (1961).
98. Pitot, H. C., Peraino, C., Lamar, C., Jr., and Lesher, S., *Science,* **150,** 901 (1965).
99. Pitot, H. C., and Peraino, C., *J. Biol. Chem.,* **239,** 1783 (1964).
100. Greengard, O., Spanier, B. B., and Baker, G. T., *Fed. Proc.,* **25,** 285 (1965).
101. Greengard, O., Baker, G. T., Horowitz, M. L., and Knox, W. E., *Proc. Natl. Acad. Sci. U. S.,* **56,** 1303 (1966).
102. Stevens, L., and Stocken, L. A., *Biochem. Biophys. Res. Commun.,* **7,** 315 (1962).

103. Meyers, D. K., Hemphill, C. A., and Townsend, C. M., *Canad. J. Biochem. and Physiol.*, **39,** 1043 (1961).
104. Meyers, D. K., *Canad. J. Biochem. and Physiol.*, **40,** 619 (1962).
105. Maley, G. F., Lorensen, M. G., and Maley, F., *Biochem. Biophys. Res. Commun.*, **18,** 364 (1965).
106. Lieberman, I., and Ove, P., *J. Biol. Chem.*, **237,** 1634 (1962).
107. Lieberman, I., Abrams, R., and Ove, P., *J. Biol. Chem.*, **238,** 2141 (1963).
108. Gebert, R. A., and Potter, V. R., *Fed. Proc.*, **23,** 268 (1964).
109. Duckworth, D. H., and Bessman, M. J., *J. Biol. Chem.*, **242,** 2877 (1967).
110. Nohara, H., and Kaplan, A. S., *Biochem. Biophys. Res. Commun.*, **12,** 189 (1963).
111. Bello, L. J., van Bibber, M. J., and Bessman, M. J., *J. Biol. Chem.*, **236,** 1467 (1961).
112. Bessman, M. J., and Bello, L. J., *J. Biol. Chem.*, **236,** PC 72 (1961).
113. Kit, S., and Dubbs, D. R., *Biochem. Biophys. Res. Commun.*, **11,** 55 (1963).
114. Kit, S., Dubbs, D. R., and Piekarski, L. J., *Biochem. Biophys. Res. Commun.*, **11,** 176 (1963).
115. Kit, S., Piekarski, L. J., and Dubbs, D. R., *J. Mol. Biol.*, **6,** 22 (1963).
116. Dubbs, D. R., and Kit, S., *Virology*, **22,** 214 (1964).
117. Kit, S., and Dubbs, D. R., *Biochem. Biophys. Res. Commun.*, **13,** 500 (1963).
118. Centifanto, Y., and Kaufman, H. E., *Proc. Soc. Exptl. Biol. and Med.*, **120,** 23 (1965).
119. Kit, S., Dubbs, D. R., de Torres, R. A., and Melnick, J. L., *Virology*, **27,** 453 (1965).
120. Sheinin, R., *Virology*, **28,** 47 (1966).
121. Littlefield, J. W., and Basilico, C., *Nature*, **211,** 250 (1966).
122. Kit, S., Dubbs, D. R., and Frearson, P. M., *Cancer Res.*, **26,** 638 (1966).
123. Ohno, S., *J. Biochem.*, **61,** 277 (1967).
124. Takahashi, M., Ueda, S., and Ogino, T., *Virology*, **30,** 742 (1966).
125. Kaplan, A. S., Ben-Porat, T., and Coto, C., in "Molecular Biology of Viruses," J. S. Colter and W. Paranchych, Eds., Academic Press, New York, N. Y., 1967, 527 pp.
126. Fiala, S., Fiala, A., Tobar, G., and McQuilla, H., *J. Natl. Cancer Inst.*, **28,** 1269 (1962).
127. Stein, I. D., and Rubini, J. R., *Proc. Soc. Exp. Biol. and Med.*, **120,** 374 (1965).
128. Trubowitz, S., Feldman, D., Morgenstern, S. W., and Hunt, V. W., *Biochem. J.*, **80,** 369 (1961).
129. Eker, P., *J. Biol. Chem.*, **240,** 419 (1965).
130. Butler, J. A. V., Bianchi, P., Crathorn, A. R., and Shooter, K. V., *J. Chim. Phys.*, 972 (1961).
131. Bianchi, P. A., Butler, J. A. V., Crathorn, A. R., and Shooter, K. V., *Biochim. Biophys. Acta*, **48,** 213 (1961).
132. Rubini, J. R., Keller, S., and McCall, M. S., *Cancer Res.*, **24,** 655 (1964).
133. Chang, L. O., and Looney, W. B., *Cancer Res.*, **25,** 1817 (1965).
134. Fritzon, P., *European J. Biochem.*, **1,** 12 (1967).
135. Aposhian, H. V., *Biochem. Biophys. Res. Commun.*, **18,** 230 (1965).

136. Kahan, F., Kahan, E., and Riddle, B., *Fed. Proc.,* **23,** 318 (1964).
137. Kahan, F. M., *Fed. Proc.,* **22,** 406 (1963).
138. Schneider, W. C., *J. Biol. Chem.,* **216,** 287 (1955).
139. Schneider, W. C., *J. Natl. Cancer Inst.,* **18,** 569 (1957).
140. Schneider, W. C., and Brownell, L. W., *J. Natl. Cancer Inst.,* **18,** 579 (1957).
141. Rotherham, J., and Schneider, W. C., *J. Biol. Chem.,* **232,** 853 (1958).
142. Potter, R. L., Schlesinger, S., Buettner-Janusch, V., and Thompson, L., *J. Biol. Chem.,* **226,** 381 (1957).
143. Stewart, P. A., Quastler, H., Skougaard, M. R., Wimber, D. R., Wolfsberg, M. F., Perrotta, C. A., Ferber, B., and Carlough, M., *Radiation Res.,* **24,** 521 (1965).
144. Logsdon, D. F., and Kundel, H. L., *SAM TR*-65-77 (1965).
145. Huntley, G. H., and Lajtha, I. G., *Int. J. Rad. Biol.,* **5,** 447 (1962).
146. Nygaard, O., and Potter, R. L., *Radiation Res.,* **16,** 243 (1962).
147. Xeros, N., *Nature,* **194,** 682 (1962).
148. Bootsma, D., Budke, L., and Vos, O., *Exp. Cell Res.,* **33,** 301 (1964).
149. Peterson, D. F., and Anderson, E. C., *Nature,* **203,** 642 (1964).
150. Puck, T. T., *Science,* **144,** 565 (1964).
151. Kim, J. H., Kim, S. H., and Eidinoff, M. L., *Biochem. Pharmacol.,* **14,** 1821 (1965).
152. Morris, N. R., and Fischer, G. A., *Biochim. Biophys. Acta,* **68,** 84 (1963).
153. Goulian, M., and Beck, W. S., *Biochim. Biophys. Acta,* **129,** 336 (1966).
154. Biswas, C., Hardy, J., and Beck, W. S., *J. Biol. Chem.,* **240,** 3631 (1965).
155. Soska, J., and Soskova, L., *Folia Biol.,* **5,** 425 (1959).
156. Ord, M. G., and Stocken, L. A., *Biochim. Biophys. Acta,* **29,** 201 (1958).
157. Bettendorf, G., Künkel, H. A., and Maass, H., *Z. ges. exptl. Med.,* **133,** 87 (1960).
158. Gentry, G. A., Morse, P. A., Ives, D. H., Gebert, R., and Potter, V. R., *Cancer Res.,* **25,** 509 (1965).
159. Greulich, R. C., Cameron, I. L., and Thrasher, J. D., *Proc. Natl. Acad. Sci. U. S.,* **47,** 743 (1961).
160. Adlestein, S. J., and Kohn, H. I., *Biochim. Biophys. Acta,* **138,** 163 (1967).
161. O'Brien, J. S., *Cancer Res.,* **22,** 267 (1962).
162. Friedkin, M., in "The Kinetics of Cellular Proliferation," F. Stohlman, Jr., Ed., Grune and Stratton, New York, N. Y., 1959, p. 97.
163. Cleaver, J. E., and Holford, R. M., *Biochim. Biophys. Acta,* **103,** 654 (1965).
164. Eker, P., *J. Biol. Chem.,* **240,** 2607 (1965).
165. Cooper, R. A., Perry, S., and Breitman, T. R., *Cancer Res.,* **26,** 2267 (1966).
166. Cooper, R. A., Perry, S., and Breitman, T. R., *Cancer Res.,* **26,** 2276 (1966).
167. Cameron, I. L., *Nature,* **209,** 630 (1966).
168. Friedkin, M., and Roberts, D., *J. Biol. Chem.,* **207,** 245 (1954).
169. Friedkin, M., and Roberts, D., *J. Biol. Chem.,* **207,** 257 (1954).
170. Friedkin, M., and Kalckar, H., in "The Enzymes," 2nd ed., P. Boyer, H. Lardy, and K. Myrbäck, Eds., Academic Press, New York, N. Y., 1961, p. 237.

171. Razzell, W. E., and Casshyap, P., *J. Biol. Chem.,* **239,** 1789 (1964).
172. Baker, B. R., *J. Med. Chem.,* **10,** 297 (1967).
173. Baker, B. R., and Kawazu, M., *J. Med. Chem.,* **10,** 302 (1967).
174. Baker, B. R., and Kawazu, M., *J. Med. Chem.,* **10,** 313 (1967).
175. Baker, B. R., and Kawazu, M., *J. Med. Chem.,* **10,** 311 (1967).
176. Baker, B. R., Kawazu, M., Santi, D. V., and Schwan, T. J., *J. Med. Chem.,* **10,** 304 (1967).
177. Ledemann, E., and Wacker, A., *Z. Naturforsch.,* **22,** 42 (1967).
178. Zimmerman, M., and Seidenberg, J., *J. Biol. Chem.,* **239,** 2618 (1964).
179. Zimmerman, M., *J. Biol. Chem.,* **239,** 2622 (1964).
180. Jacquez, J. A., *Biochim. Biophys. Acta,* **61,** 265 (1962).
181. Breitman, T. R., and Bradford, R. M., *Biochim. Biophys. Acta,* **138,** 217 (1967).
182. Zimmerman, M., *Biochem. Biophys. Res. Commun.,* **8,** 169 (1962).
183. Nakamura, H., and Sugino, Y., *J. Biol. Chem.,* **241,** 4917 (1966).
184. Mourad, N., and Parks, R. E., Jr., *Biochem. Biophys. Res. Commun.,* **19,** 312 (1965).
185. Norman, A. W., Wedding, R. T., and Kay Block, M., *Biochem. Biophys. Res. Commun.,* **20,** 703 (1965).
186. Ratliff, R. L., Weaver, R. H., Lardy, H. A., and Kuby, S. A., *J. Biol. Chem.,* **239,** 301 (1964).
187. Coutsogeorgopoulos, C., Hacker, B., and Mantsavinos, R., *Biochem. Biophys. Res. Commun.,* **20,** 129 (1965).
188. Canellakis, E. S., Karmen, H. O., and Morales, D. R., *Proc. Natl. Acad. Sci. U. S.,* **53,** 184 (1965).
189. Reichard, P., and Estborn, B., *J. Biol. Chem.,* **188,** 839 (1951).
190. Plentl, A. A., and Schonheimer, R., *J. Biol. Chem.,* **153,** 203 (1944).
191. Brown, G. B., Roll, P. M., and Weinfield, H., in "Phosphorus Metabolism," Vol. 2, W. D. McElroy, and B. Glass, Eds. Johns Hopkins Univ. Press, Baltimore, Md., 1952, p. 388.
192. Holmes, W. L., Prusoff, W. H., and Welch, A. D., *J. Biol. Chem.,* **209,** 503 (1954).
193. Gaito, R. A., and Prusoff, W. H., *Biochem. Pharmacol.,* **11,** 323 (1962).
194. Rueckert, R. R., and Mueller, G. C., *Cancer Res.,* **20,** 1584 (1960).
195. Stubblefield, E., and Mueller, G. C., *Cancer Res.,* **22,** 1091 (1962).
196. Mueller, G. C., Kajiwara, K., Stubblefield, E., and Rueckert, R. R., *Cancer Res.,* **22,** 1084 (1962).
197. Holz, G. G., Jr., *J. Protozool.,* **13,** 2 (1966).
198. Klemperer, H. G., and Haynes, G. R., *Biochem. J.,* **108,** 541 (1968).
199. Ol'shanetskaya, A. D., and Novikova, M. A., *Vop. Med. Khim.,* **13,** 432 (1967).
200. Powell, W. F., *Biochim. Biophys. Acta,* **55,** 979 (1962).
201. Sakuma, K., and Terayama, H., *J. Biochem.,* **61,** 504 (1967).
202. Scholtissek, C., *Biochim. Biophys. Acta,* **145,** 228 (1967).
203. Scholtissek, C., *Biochim. Biophys. Acta,* **145,** 238 (1967).
204. Scholtissek, C., *Biochim. Biophys. Acta,* **155,** 23 (1968).
205. Seifert, J., and Riman, J., *Neoplasma,* **14,** 237 (1967).

206. Sachsenmaier, W., Fournier, D. V., and Gürtler, K. F., *Biochem. Biophys. Res. Commun.*, **27,** 655 (1967).
207. Baugnet-Mahieu, L., Goutier, R., and Semal, M., *Radiation Res.*, **31,** 808 (1967).
208. Duspiva, F., and Hansen-Delkeskamp, E., *Z. Naturforschg.*, **206,** 582 (1965).
209. Cleaver, J. E., "Thymidine Metabolism and Cell Kinetics," in "Frontiers of Biology," Vol. 6, John Wiley, New York, N. Y., 1967.
210. Eker, P., *J. Biol. Chem.*, **240,** 2607 (1965).
211. Shimabayashi, Y., and Iwamoto, K., *Agr. Biol. Chem.* (Japan), **30,** 947 (1966).
212. Farina, M. V., Polli, E. E., and Bianchi, P. A., *Biochim. Biophys. Acta,* **103,** 512 (1965).
213. Sugino, Y., Ann. Report, *Inst. Virus Res. Kyoto Univ.*, **9,** 1 (1966).
214. Kit, S., de Torres, R. A., and Dubbs, D. R., *Cancer Res.*, **26,** 1859 (1966).
215. Eker, P., *J. Biol. Chem.*, **241,** 659 (1966).
216. Green, M., Pina, M., and Chaboga, V., *J. Biol. Chem.*, **239,** 1188 (1964).
217. Russell, W. C., *Nature,* **212,** 1537 (1966).
218. Seifert, J., and Riman, J., *Folia Biologica* (Praha), **11,** 156 (1956).

Implications of the Kinetic Behavior of a Soluble Cytochrome Oxidase Preparation *

A. J. DAVISON † *and W. W. WAINIO*
Department of Biochemistry and the Bureau of Biological Research
Rutgers—The State University
New Brunswick, New Jersey

The importance of cytochrome oxidase ‡ (or cytochrome $a + a_3$) derives from the fact that it is the terminal enzyme in the mitochondrial respiratory chain of all aerobic organisms. As its name implies it oxidizes ferrocytochrome *c*, and as its terminal position in the respiratory chain indicates, it is itself oxidized by atmospheric oxygen. Thus the reaction sequence is:

4 ferrocytochrome *c* X cytochrome oxidase $2O^{-2} + 4H^{+} \rightarrow 2H_2O$
4 ferricytochrome *c* X cytochrome oxidase$^{(-4)}$ O_2

Much of the early kinetic information was obtained with the insoluble enzyme, *i.e.,* with the enzyme still attached to the inner mitochondrial membrane. With the advent of the use of surface active agents to prepare the enzyme, it became possible to study the reaction mechanism under conditions where both the substrate and the enzyme

* Supported by grants from the U. S. Public Health Service (AM-06828), the National Science Foundation (G24032), and the South African C.S.I.R.

† From a thesis submitted by A. J. Davison to the Graduate Faculty, Rutgers—The State University, May 1964. Present address: Department of Physiology and Pharmacology, Medical School, University of Cape Town, Observatory, Cape Town, South Africa.

‡ Enzyme Commission number: 1.9.3.1; Systematic name: Cytochrome *c*: oxygen oxidoreductase.

were in a soluble state. It was of interest to know whether the kinetic behavior of a relatively pure soluble preparation of cytochrome oxidase was the same as that of an insoluble preparation.

In all preparations studied by us and by others, the reaction was first order for any given concentration of cytochrome *c*, although the apparent first order rate constant decreased as the initial substrate concentration was increased. Frajola and Vestling[3] interpreted this behavior as a combination of first and zero order kinetics. Smith and Conrad[15] and Slater[14] concluded that the enzyme was inhibited by both substrate and product, either of which they assumed could form an inactive complex with the enzyme. McGuinness and Wainio[11] interpreted the results as due to product inhibition, but not substrate inhibition. They showed that the rate dependence on initial substrate concentration was close to zero order and suggested that the decrease in rate as a given reaction proceeded was due, not to the decrease in substrate concentration, but rather to the accumulation of product.

Minnaert[12] expressed this behavior in terms of the rate equation:

$$v = \frac{\text{Constant} \times [S]}{\text{Constant} \times [P + S] + \text{Constant}}$$

We have derived an identical equation from the kinetics of the soluble cytochrome oxidase "Preparation 2-0.8-10-1.5" of Wainio.[19] Yonetani and Ray[21] have since published results on their soluble cytochrome oxidase preparation in which they obtain the algebraically equivalent rate equation:

$$v = \frac{V[S]}{K_m + [P_o + S_o]\dfrac{K_m}{K_i} + \left(1 - \dfrac{K_m}{K_i}\right)[S]}$$

where V = maximum velocity, K_m = the dissociation constant for the ferrocytochrome *c*-cytochrome oxidase complex, K_i = the dissociation constant for the ferricytochrome *c*-cytochrome oxidase complex, $[P_o]$ = initial concentration of product, and $[S_o]$ = initial concentration of substrate.

The establishment of a mathematical formulation which describes the results of all workers should permit a rigorous interpretation of the results. If speculation is not carried beyond the limits of the available evidence, current controversies should now be resolved.

Details of our kinetic findings are presented here, together with a plausible mechanistic interpretation.

Methods and Materials

The cytochrome oxidase used was "Preparation 2-0.8-10-1.5" prepared by the method of Wainio.[19] It had a heme to protein ratio of 0.88 and an activity (in the presence of added lecithin) of 5.7 sec^{-1} (mg protein/3 ml)$^{-1}$ when measured as described by Wainio.[18] Beef heart cytochrome *c* (type IV, Sigma Chemical Company, St. Louis, Missouri) was used as substrate. It was reduced with dithionite and aerated. Sodium phosphate buffer was used to give a pH equal to 6.0 in the cuvette at a final concentration of 0.1 M.

The progress of the reaction was followed at 25° by recording absorbance at 550 mμ for 60 sec, using a Carey model 11 recording spectrophotometer. The apparent first order rate constants given are $\frac{\Delta \ln \text{Absorbance}}{\Delta t}$. The theoretical initial velocity was calculated by extrapolating to zero product concentration. The difference in the molar absorptivity of cytochrome *c* at 550 mμ was taken to be 1.85×10^4 M^{-1} cm^{-1} after Margoliash.[10]

Results

The effects of the initial concentrations of substrate $[S_o]$ and enzyme $[E]$ on the apparent first order rate constant (k) are shown in Figure 1. Clearly $1/k = k'[S_o] + k''$ and since $v = k[S]$,

$$v = \frac{[S]}{k'[S_o] + k''}.$$

When the reaction is studied in the presence of added product it becomes evident that it is not $[S]$ which determines k but the total concentration of cytochrome *c*, ($[S] + [P]$), as in Figure 2. Normally, this distinction is unimportant since at the start of the reactions initial product concentration is small and $[S_o] \simeq [S] + [P]$. This observation, however, excludes the suggestion that the activity of the enzyme is determined by its initial exposure to substrate. Clearly then, the rate equation is better formulated as:

$$-\frac{d[S]}{dt} = \frac{k_1[S][E]}{k_2([S] + [P]) + k_3}$$

since this expression contains only concentrations of species actually present in the system. This expression is preferable to the mathematically equivalent equation of Yonetani and Ray[21] which contains the

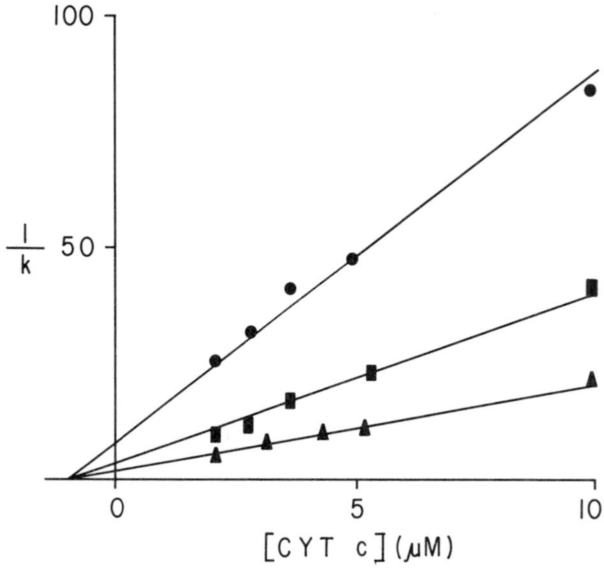

Figure 1. Effect of cytochrome c on the reciprocal of the apparent first order rate constant at three cytochrome oxidase concentrations: ● 0.05 ml of enzyme solution; ■ 0.10 ml of enzyme solution; ▲ 0.20 ml of enzyme solution.

quantity $[S_o]$, since $[S_o]$ is mechanistically unintelligible. The equation is of the form obtained when steady-state rate equations are solved by the method of King and Altman,[6] and thus is suitable for testing hypothetical mechanisms. The fact that the first order rate constant remains constant throughout any given reaction results from the fact that $([S]+[P])$ does not change. For each molecule of substrate consumed, a molecule of product is formed.

The effects of enzyme concentration on the coefficients of the rate equation are shown in Figures 3 and 4. From the linear relationships obtained the complete rate equation can be written:

$$-\frac{d[S]}{dt}=\frac{k_1[S][E]}{k_2[S]+k_2[P]+k_3}.$$

No term involving oxygen concentration appears in the equation, since the reaction is not appreciably slowed when the oxygen concentration is decreased by as much as 90% (Wharton and Griffiths [20]).

In a manner consistent with this equation, the nature of the inhibition by product was found to be strictly competitive (Figure 5). $K_m(k_3/k_2)$

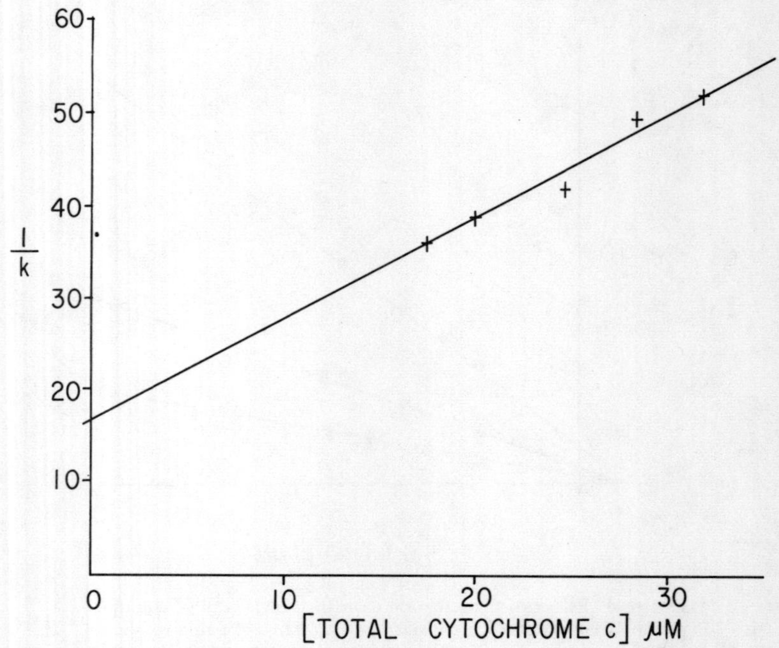

Figure 2. Dependence of the reciprocal of the apparent first order rate constant on the total cytochrome *c* concentration.

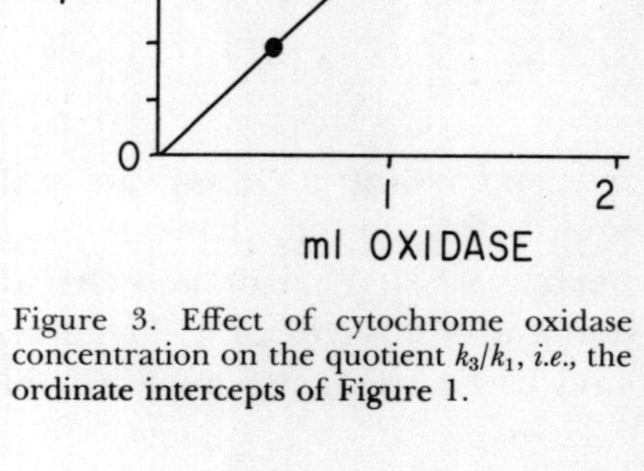

Figure 3. Effect of cytochrome oxidase concentration on the quotient k_3/k_1, *i.e.*, the ordinate intercepts of Figure 1.

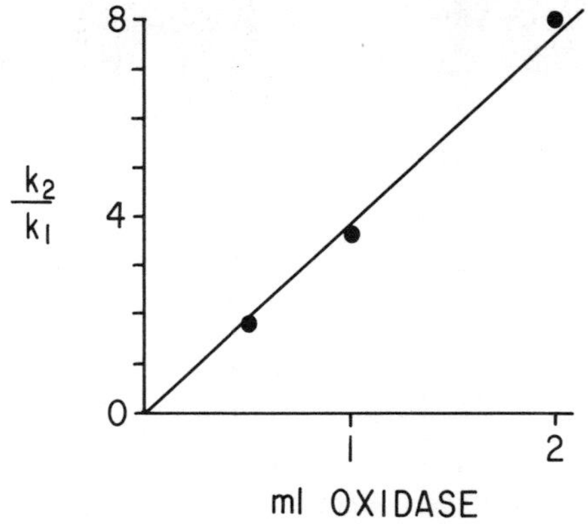

Figure 4. Effect of cytochrome oxidase concentration on the quotient k_2/k_1, *i.e.*, the slopes of Figure 1.

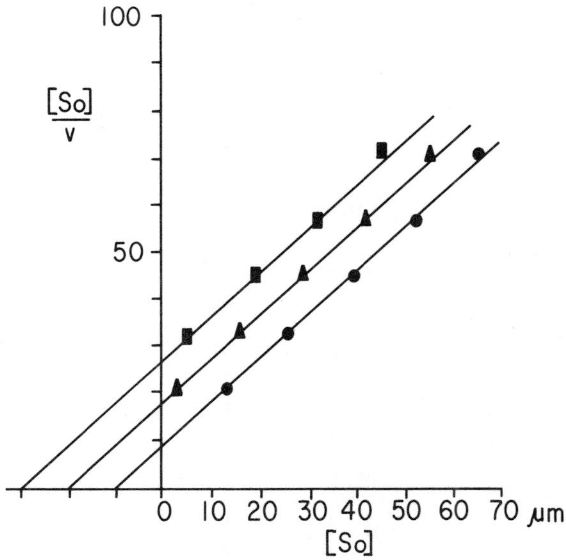

Figure 5. Dependence of $[S_o]/v$ on $[S_o]$ at two concentrations of added product: ● without added ferricytochrome *c*; ▲ with 10 μM ferricytochrome *c*; ■ with 20 μM ferricytochrome *c*.

was calculated to be 8.50×10^{-6} M. For product concentrations of 10 μM and 20 μM, K_i was found to be 8.49×10^{-6} M and 8.47×10^{-6} M, respectively. Thus K_m equalled K_i within experimental error. The values obtained by Yonetani and Ray[21] were 8.5×10^{-6} M for K_m and 8.3×10^{-6} M for K_i.

Discussion

In view of the diversity of conclusions drawn from identical kinetic data, it is worthwhile considering what conclusions necessarily arise from the available evidence, and what speculations are the simplest required to account for the observed facts. For this purpose the formulation:

$$v = \frac{k_1[S][E]}{k_2[S] + k_2[P] + k_3}$$

is preferred to that of Yonetani, because it contains all the factors on which the rate depends, because it eliminates initial concentrations which are mechanistically uninterpretable since the substrate can never be added in the fully reduced state, and because it can be compared with rate equations derived from hypothetical mechanisms since it has the same mathematical form.[2]

Two conclusions can be drawn with some certainty. First, the absence of a negative term in the numerator implies that the reaction is irreversible. Second, the fact that the coefficients of $[S]$ and $[P]$ in the denominator are identical implies that the enzyme system has the same affinity toward product as toward substrate.[1] The irreversibility is not unexpected in view of the large negative ΔF for the reaction between ferrocytochrome *c* and oxygen. The equal affinity of the enzyme for substrate and for product is also predictable in view of their nearly identical structures.[16,17] These two conclusions are reflected in the facts that product inhibition is found to be strictly competitive and that K_i is identical to K_m within the limits of experimental error.

Let us now consider what is the simplest mechanism consistent with the established facts. The mechanism:

$$E \xrightarrow{\overset{S}{\downarrow}} ES \xrightarrow{\overset{P}{\uparrow}} E$$

is the simplest which yields the required rate equation, but cannot be consistent with both irreversibility and equal affinity for substrate and product, since if the reaction is irreversible it would follow that the reaction of the enzyme with product is less favored than its reaction with substrate. The symbol E represents the enzyme, and in accordance with the physical and chemical evidence indicating its homogeneity it is treated here as a single enzyme with an unspecified number of functional units. The simplest mechanism consistent with the kinetic findings is:

$$E \xrightarrow[k_{-1}]{S \downarrow k_{+1}} ES \xrightarrow{O_2 \; k_{+2}} EP \xrightarrow[k_{-3}]{k_{+3} \uparrow P} E$$

which, given the assumption that $k_{+1}=k_{-3}$ and $k_{-1}=k_{+3}$, yields the steady-state rate equation:

$$-\frac{d[S]}{dt}=\frac{k_{+1}k_{-1}k_{+2}[S]}{(k_{+1}k_{-1}+k_{+1}k_{+2})[S]+(k_{+1}k_{-1}+k_{+1}k_{+2})[P]+k_{+1}k_{-1}+k_{-1}k_{+2}}.$$

(Equation 1)

This is one of the two mechanisms considered as feasible by Minnaert. Both this and his other more complex mechanisms are, however, unsatisfactory, since in the reaction with oxygen only a single electron could be transferred from ES to the oxygen molecule, whereas the oxygen must acquire four electrons before the second product, water, can be formed. The simultaneous reaction of four ES molecules with a single oxygen molecule is not feasible sterically. Dissolved, partially-reduced oxygen could not be an intermediate in the formation of water as suggested by King and Lee,[7] since these free radicals are so reactive that the possibility of their finding the way to the binding sites of four successive molecules of ES can be neglected. Therefore, assuming that the partially reduced oxygen remains bound to the enzyme until completely reduced to water, the mechanism necessarily becomes:

$$E \xrightarrow{S\downarrow} SE \xrightarrow{B\downarrow} SEB \longrightarrow PEB' \xrightarrow{P\uparrow} EB' \xrightarrow{S\downarrow} SEB' \longrightarrow PEB''$$

$$PEB'' \xrightarrow{P\uparrow} EB'' \xrightarrow{S\downarrow} SEB'' \longrightarrow PEB''' \xrightarrow{P\uparrow} EB''' \xrightarrow{S\downarrow} SEB''' \longrightarrow PEB''''$$

$$PEB'''' \xrightarrow{Q\uparrow} PE \xrightarrow{P\uparrow} E$$

where B = oxygen, Q = water, and EB', EB'', EB''', EB'''' are enzyme-oxygen complexes in increasing stages of reduction, protonated to an unspecified degree.

This mechanism is analogous to that proposed by LuValle and Goddard.[9] It requires simplification, however, before the kinetic consequences can be profitably determined. In view of the absence of [B] from the empirical rate equation and the known rapidity of oxygen binding[4] this step can be assumed to be kinetically inaccessible. The reaction is pseudo-zero order with respect to water and therefore the step containing Q can also be neglected.

The overall reaction thus reduces to a series of four analogous cycles:

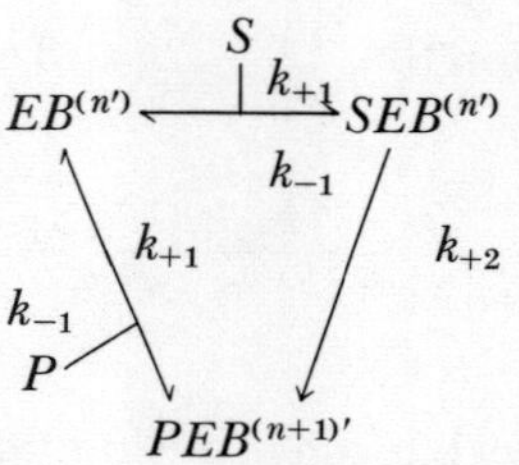

If k_{+2} is identical for four cycles or if k_{+2} is very large, the overall kinetic behavior is that of any individual cycle. If not, the overall kinetic behavior may be considered to be that of the slowest cycle.

The rate equation predicted by such a mechanism is as given in Equation 1, but the mechanism, unlike the simpler mechanism of Minnaert, adequately accounts for the stoichiometry of the reaction.

The above is the simplest mechanism which accounts for the observed kinetics and yet is consistent with the stoichiometry of the reaction. Mechanisms such as that of Smith and Conrad[15] and Slater[14] are consistent with both, but involve the additional and unnecessary assumption of an extra inhibitory binding site for substrate on the enzyme. The above treatment accounts fully for the assumed "inhibition" of cytochrome oxidase by substrate and product in terms of simple mass action effects. It also provides a simple explanation of the existence of an "oxygenated" form of the oxidase reported by various workers (Okunuki and Sekuzu,[13] Davison,[1] Lemberg and Mansley[8]) since in the absence of a continued supply of electrons from ferrocytochrome *c*, reduction of the enzyme-oxygen complex would be arrested in an intermediate state with properties identical to those described.

A further simplification can be made if k_{+2} is assumed to be large in comparison with k_{+1} and k_{-1}, *i.e.*, if electron transfer within the ternary complex is fast in comparison with substrate binding. In this case, coefficients not containing k_{+2} may be neglected. Thus:

$$-\frac{d[S]}{dt} = \frac{k_{+1}k_{-1}[S][E]}{k_{+1}[S] + k_{+1}[P] + k_{-1}}.$$

An estimate of the elementary rate constants, k_{+1} and k_{-1}, for the binding and desorption of substrate and product, respectively, may then be obtained from the slopes and intercepts of figures, since

$$k_{+1} = \frac{1}{\text{intercept } [E]} \qquad \text{and} \qquad k_{-1} = \frac{1}{\text{slope } [E]}.$$

Approximate values for k_{-1} and k_{+1}, obtained from Figure 1, were found to be 1.5×10^7 M^{-1} sec^{-1} and 1.6×10^7 M^{-1} sec^{-1}, respectively, at 25°C. The latter value is reasonably consistent with those obtained directly by Gibson *et al.*[5] using stopped-flow techniques for the Yonetani preparation, 4×10^6 M^{-1} sec^{-1}, and for the Griffiths and Wharton preparation, 4×10^7 M^{-1} sec^{-1}, at 21° and 20°, respectively. However, since Gibson has evidence that electron transfer within the complex is not as fast as assumed, this simplification may be of doubtful validity.

Summary

The empirical rate equation for a partially purified, soluble, cytochrome oxidase preparation agrees with that found by Minnaert for the oxidase activity of fragmented mitochondria. It is suggested that the simplest explanation of the observed kinetics is a mechanism in which molecules of ferrocytochrome *c* successively transfer electrons to a cytochrome oxidase-oxygen complex, such as that of "oxygenated" cytochrome oxidase. On this basis it is possible to account for the presumed "inhibition" of cytochrome oxidase by substrate and by product on the basis of mass action effects alone.

References

1. Davison, A. J., *Dissertation Abst.,* **25,** 6192 (1965).
2. Davison, A. J., *Nature,* **210,** 1161 (1966).
3. Frajola, W. J., and Vestling, C. S., *J. Biol. Chem.,* **209,** 677 (1954).
4. Gibson, Q. H., and Greenwood, C., *Biochem. J.,* **86,** 541 (1963).
5. Gibson, Q. H., Greenwood, C., Wharton, D. C., and Palmer, G., *J. Biol. Chem.,* **240,** 888 (1965).
6. King, E. L., and Altman, C., *J. Phys. Chem.,* **60,** 1375 (1956).
7. King, T. E., and Lee, C. P., *Biochim. et Biophys. Acta,* **37,** 342 (1960).
8. Lemberg, R., and Mansley, G. E., *Biochim. et Biophys. Acta,* **118,** 19 (1966).
9. LuValle, J. E., and Goddard, D. R., *Quart. Rev. Biol.,* **23,** 197 (1948).

10. Margoliash, E., *Biochem. J.*, **56,** 535 (1964).
11. McGuinness, E. T., and Wainio, W. W., *J. Biol. Chem.*, **237,** 3273 (1962).
12. Minnaert, K., *Biochim. et Biophys. Acta,* **50,** 23 (1961).
13. Okunuki, T., and Sekuzu, I., *Seitai no Kagaku,* **5,** 279 (1954).
14. Slater, E. C., *Biochem. Soc. Symp.*, **15,** 76 (1958).
15. Smith, L., and Conrad, H., *Arch. Biochem. Biophys.*, **63,** 403 (1956).
16. Ulmer, D. D., *Biochemistry,* **4,** 902 (1965).
17. Urry, D. W., and Doty, P., *J. Am. Chem. Soc.*, **87,** 2756 (1965).
18. Wainio, W. W., *J. Biol. Chem.*, **216,** 593 (1955).
19. Wainio, W. W., *J. Biol. Chem.*, **239,** 1402 (1964).
20. Wharton, D. C., and Griffiths, D. E., *Arch. Biochem. Biophys.*, **96,** 103 (1962).
21. Yonetani, T., and Ray, G. S., *J. Biol. Chem.*, **240,** 3392 (1965).

VI. Viral and Neoplastic Diseases

A Model for the Study of the Effect of Viral Stress on Protein Metabolism

ROBERT L. SQUIBB
Laboratories of Disease and Environmental Stress
Rutgers—The State University
New Brunswick, New Jersey

Introduction

We deem it a privilege and honor to participate in this tribute to a long-time friend and colleague, Dr. James B. Allison, and to acknowledge his part in helping to create our Laboratories of Disease and Environmental Stress.

Rather than relate the entire scope of our research program, a summation of the development and standardization of an experimental model for research in disease and nutrition interrelationships is presented. This we believe best reflects Dr. Allison's major interest. No attempt has been made to give an extensive literature review in this important field but pertinent references have been inserted where necessary.

General Methodology

Studies related to the influence of diet and disease on protein metabolism require an exacting model. This is particularly essential where ultramicro biochemical measurements or electron microscopic studies are to be carried out.

The chick

The chick has a number of advantages over other species. Aside from the ready availability of good breeding stock at reasonable cost, there is a considerable reference background on the chicken's nutritional requirements,[1] and on its physiology and pathology.[2] We have found that working with avian diseases is far less hazardous to technical personnel yet they have practical application to diseases affecting man. For example, the Newcastle disease virus which will be discussed here is a myxovirus in the same group as measles, mumps, and parainfluenza.

For our research we have been using newly hatched White Leghorn cockerel chicks of known breeding. They are maintained in community or single all-wire cages in air-conditioned rooms with controlled lighting; water and feeding are *ad libitum* or restricted according to the design of a particular experiment. Individual body weights are recorded at appropriate intervals. Animals designated for infection are transferred to similar but isolated rooms at 4 weeks of age where each chick is inoculated with Newcastle disease virus (NDV). This 4-week preliminary period allows the chick's metabolism to pass from a poikilothermic to a homeothermic state and for parental immunity to NDV to be depleted.

Growth rates of control chicks in all experiments are monitored with standard reference curves developed for a particular diet. Failure of an experimental control group to meet a reference curve is considered indicative of the presence of an unforeseen stress which invalidates the trials.

Biochemical methods

The parameters observed were selected to be representative of protein metabolism and be sensitive to the changes that may occur in this dynamic process. With this in mind, DNA, RNA, tissue protein, and the free amino acids are determined simultaneously on one or more tissues.

Procedures for chemical analyses are the same for all experiments. Work is not begun until samples for an entire experiment are collected. Due to large numbers and to equalize all levels of experimental error, samples of an experiment are analyzed in lots each of which contain an equal representation of the different treatments and sampling periods.

The tissues are analyzed for DNA and RNA using modifications of the Schmidt-Thannhauser method suggested by Hutchison and Munro[3] and Wannemacher *et al.;*[4] protein by the biuret technique which is standardized by the Kjeldahl procedure; and free amino acids by the method of Squibb.[5]

The Newcastle disease virus

A Rutgers-Grun strain of NDV is used for all experiments and is injected intramuscularly at 28 days of age. The inoculum may be diluted to obtain from 0 to 70% mortality with better than 99% confirmed infection. At LD_{50} the course of the infection follows a well-defined cycle with the first 72 hours post inoculation termed the incubation period (Figure 1). During this stage there are no effects on body weight or feed and water intake. The next 72 hours are the period of active involvement during which the greatest proliferation of the virus and rise in antibody titer occurs, feed and water consumption drop sharply, definite clinical symptoms are noted, and nearly all the mortality is recorded. Initiation of recovery begins approximately 144 hours post inoculation. Depending on the level of inoculum used, clinical symptoms of the birds at this time range from none to acute paralysis in approximately 35% of the survivors. Liver weight increases significantly during the incubation period, a phenomenon which we have found to correlate with an increased nitrogen retention; [6] during active involvement liver weight is significantly reduced and then gradually increases in the recovery period.

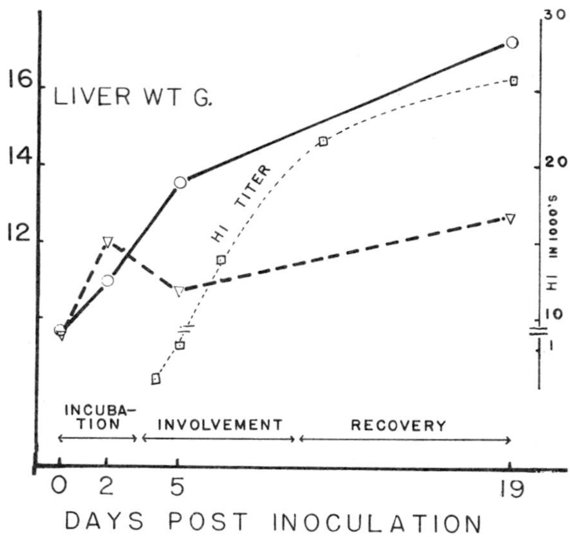

Figure 1. Liver weights and antibody titers of chicks infected with a 10^{-1} concentration of Newcastle disease virus. Solid line noninfected chicks; dotted line NDV infected.

Body temperatures of the chicks during the disease cycle are presented in Figure 2. Temperatures that do not rise above 108°F are associated with birds that recover completely or show only mild respiratory symptoms at the end of the 14-day disease period. Abnormally high temperatures (>108°F) lasting more than 48 hours, followed by a significant drop below normal levels (106°F) are associated with acute neurotropic involvement, which includes paralysis and death.

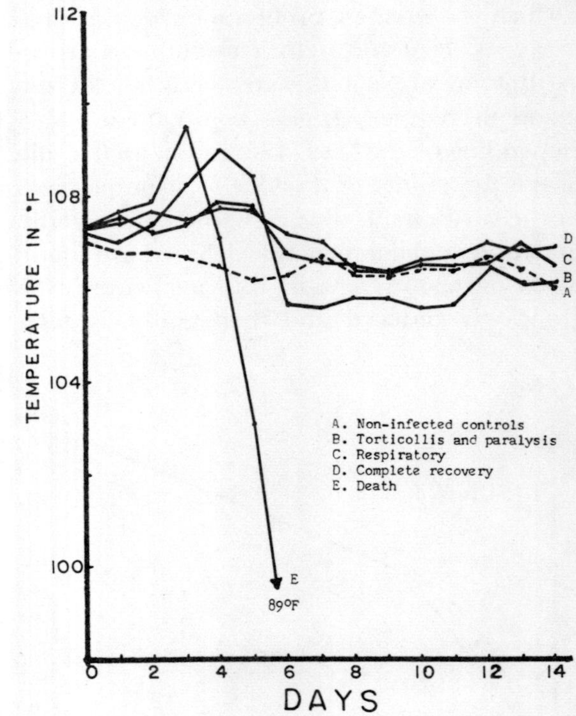

Figure 2. Relation of body temperatures to clinical symptoms of NDV infected chicks.

Nitrogen Retention During Virus Infection

Allison and Fitzpatrick[7] stressed the value of nitrogen balance studies. They pointed out that nitrogen and specific amino acid requirements of the liver are influenced by the synthesis of the proteins of blood and tissue for maintenance and repair, etc. Obviously, in the disease state there may be so much stress on the metabolic processes that these requirements cannot be fulfilled.

Our initial studies concerned the effects of diet and disease on nitrogen retention in young rapidly growing chicks wherein an NDV inoculum capable of producing 50% mortality was employed. *Ad libitum,* feed-restricted, and force-fed groups were used in order to evaluate the effects of feed intake on nitrogen retention. Food restriction was accomplished by limiting the intake of noninfected controls to that of infected chicks. Force-feeding into the crop was initiated as soon as the first depression of food intake was noted in the virus-infected birds, with the amount equated to the *ad libitum* intake of the noninfected controls.

Figures 3 and 4 show that NDV significantly increased nitrogen retention during the NDV incubation period; as previously noted (Figure 1), this increased nitrogen retention was found to correlate with a rise in viral titer and antibody production. During the initial stages of our work the rise in nitrogen retention was interpreted as being part of the de-

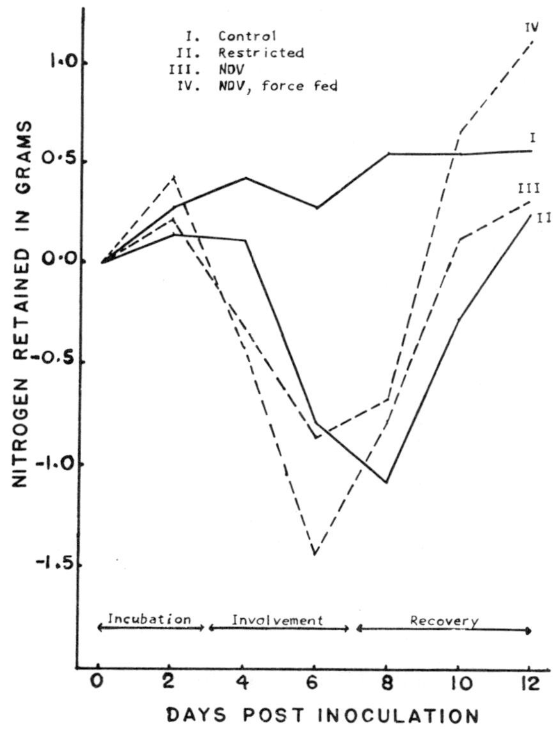

Figure 3. Nitrogen retention of control and surviving NDV infected chicks during a disease cycle.

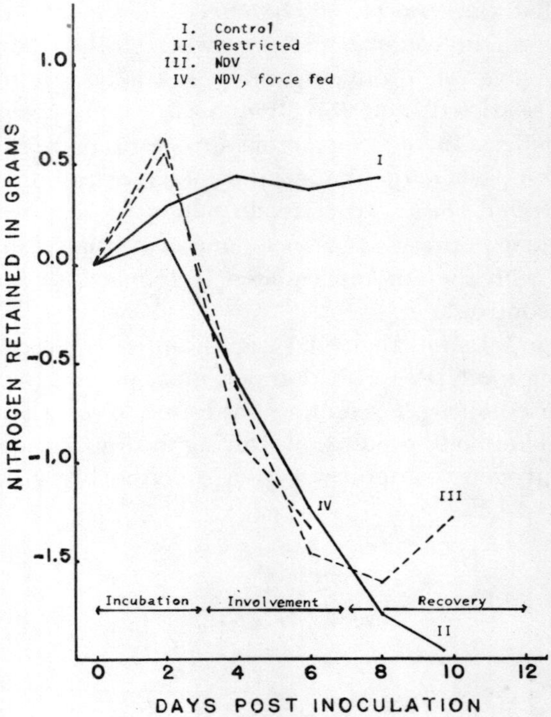

Figure 4. Nitrogen retention of control and NDV infected chicks that failed to survive during a disease cycle.

fense mechanism. This interpretation has lost its validity, however, as the standardization of the model progressed.

In the active involvement period, or time of peak mortality, a sharp depression in nitrogen retention was observed which correlated with a drop in food intake. It is of interest to note that the depletion curves (Figure 3) for the survivors were nearly identical to those of the chicks that succumbed to the disease (Figure 4). The data indicate that the disease effect was so strong that even force-feeding did not reverse the drop in nitrogen retention. This finding suggests that nutrient passage from the crop and absorption from the gut was inhibited by the infection. Following the active involvement stage of the NDV, food intake slowly returned to normal and the rate of increase in nitrogen retention in the survivors surpassed that of the noninfected controls.

While force-feeding was ineffective during the active involvement stage of the disease, this procedure increased nitrogen retention during recovery, suggesting that the technique reactivated the digestive processes at an earlier time in the recovery period.

In another experiment when chicks were given an immunizing dose of NDV (one that would not affect livability or produce severe clinical symptoms) nitrogen retention again was increased (Figure 5); however, this time it remained above normal throughout the first 8 days post inoculation. But regardless of the continuing higher levels of nitrogen retention there were no apparent effects on HI titer levels.

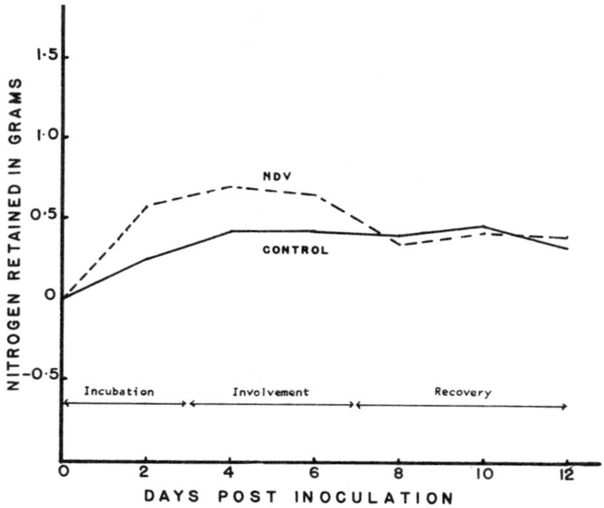

Figure 5. Nitrogen retention of control and NDV immunized chicks during a disease cycle.

While nitrogen is utilized in defense mechanisms, the data here indicate that the initial rise is a reactive process. Evidence for this supposition lies in the observation that nitrogen retention was maintained during immunization wherein dietary intake was not affected, and that animals which died had a higher nitrogen retention than those that survived.

An Index of Disease Involvement

Even though the same amount and strength of inoculum are injected intramuscularly into each chick, the degree of reaction to the infection always varies widely in any group; that is, some may die while others may be partially or totally paralyzed or show no signs of infection whatsoever. Many susceptible chicks often take longer to develop symptoms than others. Since these variables are the true reflection of a normal, artificially infected population they naturally must be considered in order to properly interpret the biochemical data resulting from experimental treatments. Therefore, to further standardize the model and develop an index of disease involvement, a trial was designed to study the effect of severity of infection on the biochemical parameters selected. Chicks were injected with NDV at an LD_{50} level and tissue samples were obtained at regular intervals over the entire disease cycle. When the biochemical analyses were completed the data were grouped and arrayed according to increasing severity of clinical symptoms.

Table I shows that in totally susceptible birds as the intensity of disease involvement rose, *e.g.*, from no apparent symptoms to imminent death, body weights and temperatures were linearly depressed while liver as percent of body weight increased. In the liver, a principal target organ of NDV involvement, DNA increased linearly with increasing disease involvement. This is believed to result from tissue contraction resulting from the starvation interaction. In other words, greater numbers of cells are present per gram of fresh tissue and thus a greater amount of DNA. RNA synthesis and tissue protein were not greatly affected save at the time of death.

Table I. Effect of Degree of Newcastle Disease Virus Involvement on Nucleic and Free Amino Acids in Avian Liver

Newcastle Disease Symptoms *	Body Wt. (g)	Body Temp. (°F)	Liver/Body (%)	DNA	RNA	Protein	Lys	His	Arg	Asp Acid	Ala	Val	Leu †
				(mg g wet liver)									
Noninfected	434	105.2	2.3	1.65	14.9	206	.91	.53	.80	1.50	.91	.68	.95
NDV-infected													
None	414	106.1	3.1	1.88	14.8	211	.58	.44	.54	1.11	.66	.47	.71
Respiratory	401	106.5	3.4	2.02	14.8	211	.51	.48	.49	1.23	.67	.41	.59
Partial paralysis	358	106.5	3.4	1.95	14.5	207	.53	.39	.53	1.06	.67	.42	.59
Full paralysis	323	100.5	3.5	2.18	14.4	212	.46	.47	.46	1.15	.62	.35	.50
Dying	303	90.6	3.9	2.44	14.1	208	.49	.35	.43	.77	.56	.39	.52
Dead	310	81.0	4.2	2.49	13.4	224	.28	.37	.30	.55	.62	.30	.31

* In order of increasing severity.

† Leucines combined.

The RNA values were of interest because they did not indicate any viral contribution, yet NDV is an RNA virus. It was also evident that with an increasing clinical index of involvement the total effect of the disease, which included diet intake confounding, linearly depressed all the individual free amino acids.

Disease versus Starvation Effects

Since dietary intake declined as disease involvement increased, it became evident that the decrease in body weight was a summation of the disease effect plus the interaction of a depressed dietary intake. Therefore, the next step in the standardization of the model was to separate the effects of reduced food intake during infection from the effects of the disease *per se.* In this phase food was withheld from both noninfected and infected chicks during the active involvement period of the NDV; water was continued *ad libitum.*

Table II shows that starvation *per se* was the principal factor in depressing body weights and increasing DNA per gram of fresh liver tissue. While starvation reduced RNA slightly, as would be expected, it did not significantly affect tissue protein and RNA within the 3-day active involvement period.

Table II. Interaction of Starvation and Newcastle Disease Virus on the Nucleic and Free Amino Acids in Avian Liver

Treatment	Body Wt.	DNA	RNA	Protein	Lys	His	Arg	Asp Acid	Ala	Val	Leu *
	(g)				(mg/g wet liver)						
Control, *ad lib.*	382	1.79	16.0	210	1.08	.46	.91	1.23	1.36	.82	1.23
Control, starved	299	2.23	15.5	249	1.18	.41	.89	1.17	1.37	.80	1.16
NDV, starved	275	2.46	16.9	223	.85	.39	.72	1.09	1.08	.73	1.02
NDV, repleted (9)	357	1.70	15.3	225	.87	.45	.64	1.05	.93	.54	.83
NDV, repleted (19)	603	1.92	17.4	270	.62	.34	.48	.80	.71	.43	.57
					(mg/mg DNA)						
Control, *ad lib.*			8.9	117	.60	.26	.51	.69	.76	.46	.69
Control, starved			7.0	112	.53	.18	.40	.52	.61	.36	.52
NDV, starved			6.9	91	.35	.16	.29	.44	.44	.30	.41
NDV, repleted (9)			9.0	132	.51	.26	.38	.62	.55	.32	.49
Ndv, repleted (19)			9.1	141	.32	.18	.25	.42	.37	.22	.30

* Leucines combined.

The highly significant reduction of the free amino acid pool could not be attributed solely to dietary and/or disease factors when the values observed during the recovery or repletion stage of the NDV cycle were considered. Body weight increases observed at 9 and 19 days post inoculation indicated a high state of protein anabolism during the repletion period. This phenomenon accelerated the requirements for the *materia prima* of synthesis, as indicated by the uniformly high depletion rate of each of the free amino acids. It is interesting to note that this rapid growth reduced the pool in the liver to a greater extent than that observed during the involvement stage of the infection.

Diurnal Rhythms

One little researched area is that of normal fluctuations or diurnal oscillations of tissue constituents. These normal phenomena are the reflections of the ebb and flow of the many body processes as affected by diet, environment, growth, various stresses, etc. In the continuing standardization of our model, these oscillations were deemed to have

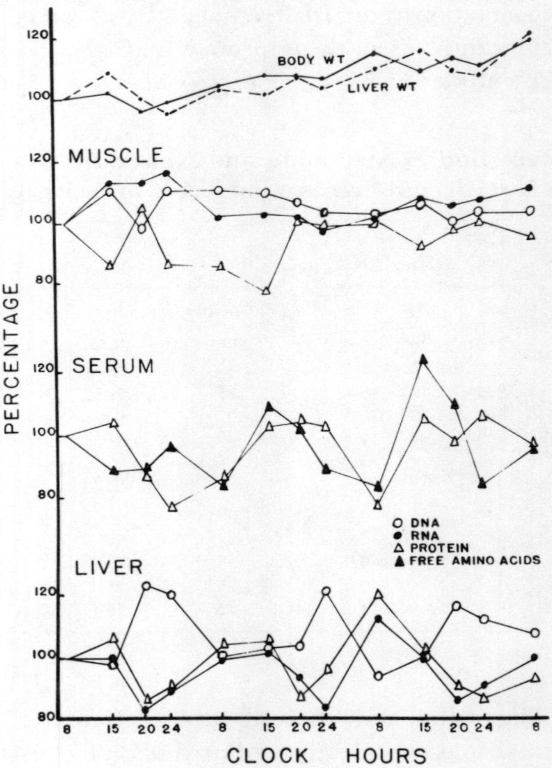

Figure 6. Body and liver weights of control chicks and diurnal oscillations of nucleic and total free amino acids in muscle, serum and liver.

an important bearing on the evaluation and interpretation of research findings.

In the first series of experiments[8] liver samples were obtained over the first 24 hours of the NDV active involvement period. Distinct diurnal fluctuations were noted in the nucleic and free amino acids of control and infected chicks. Based on these preliminary studies, animal management practices were modified so that in addition to controlled temperature and lighting the caretaker was only permitted to enter the rooms once a day and always at the same hour. Instead of a single daily sampling, tissues were obtained from time of NDV inoculation (0800 hours) and again at 1500, 2000 and 2400 hours for one or more consecutive days from both noninfected and infected chicks.

Figures 6 and 7 show the results obtained in one such trial where constant light was used. Again, significant diurnal fluctuations were found

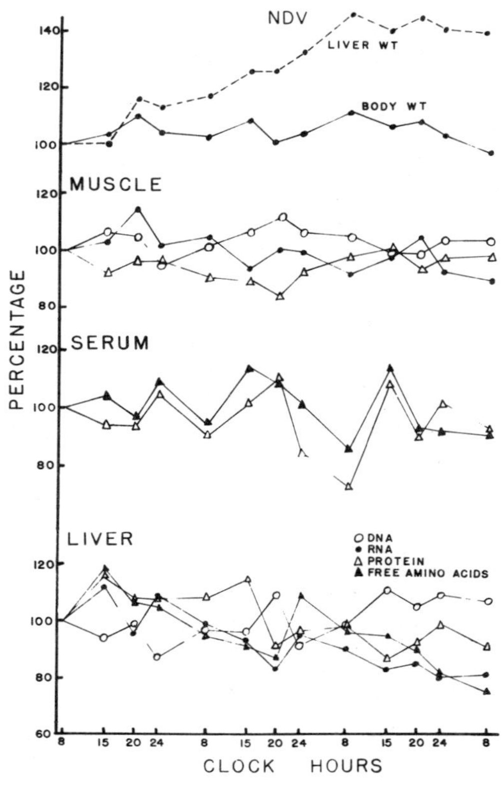

Figure 7. Body and liver weights and diurnal oscillations of nucleic and free amino acids in muscle, serum and liver of NDV infected chicks.

in liver, muscle, and serum. Dietary intake could not be implicated since food consumption was normal for both treatment groups. Moreover, earlier work[8] showed that rhythms occurred during the active involvement stage of NDV when inanition is acute. Nor could the oscillations be attributed to changes in moisture content of the various tissues studied.[9]

Of the particular biochemical parameters observed, especially interesting were the diurnal changes in liver nucleic acids and protein. Highest levels of DNA in the controls occurred in the evening and lowest in the daytime. RNA and protein synthesis in the liver, on the other hand, were found to be highest during daylight hours.

Using the amplitudes of the oscillations as indicators, it is apparent that DNA, RNA and protein synthesis were more dynamic in liver than muscle; the free amino acids, however, were equally dynamic in all three tissues.

When values for control and NDV chicks are compared with respect to clock hours it can be seen that the NDV caused a desynchronization of all parameters in the liver within 12 hours after inoculation. Moreover, by the end of the 72-hour incubation period the free amino acid levels also were linearly depressed. Desynchronization did not reach the same degree in the muscle as in the liver; but the rhythms in the serum of control and infected chicks remained in synchronization, with the NDV curve slightly higher than the controls.

Based on these data, it was necessary to question what effect a normal light-dark schedule would have on the diurnal rhythms. Figures 8 and 9 compare the results of two trials, one with constant light (LL) and the other with animals maintained on a 12/12 light-dark regimen (LD) with the light on from 0800 to 2000 hours daily. Control and NDV chicks were under test and tissue samples were taken from both groups from 0 to 72 hours post inoculation.

The LD regimen entrained patterns of liver weight, DNA, RNA and free amino acids to definite clock hours. A comparison of the curves of the noninfected controls conditioned to LL with those of the LD controls shows obvious changes in the patterns. Of particular interest is the phase shift in DNA wherein peaks observed at 2000 to 2400 hours were shifted to 0800 hours. This reversal of DNA peaks also occurs in other species and tissue constituents when light schedules are inverted.[10,11] The LL regimen appeared to dampen oscillations of liver weights and RNA and to desynchronize the free amino acids.

When the interactions between exogenous (light) and endogenous (disease) factors are examined it can be seen that there were other

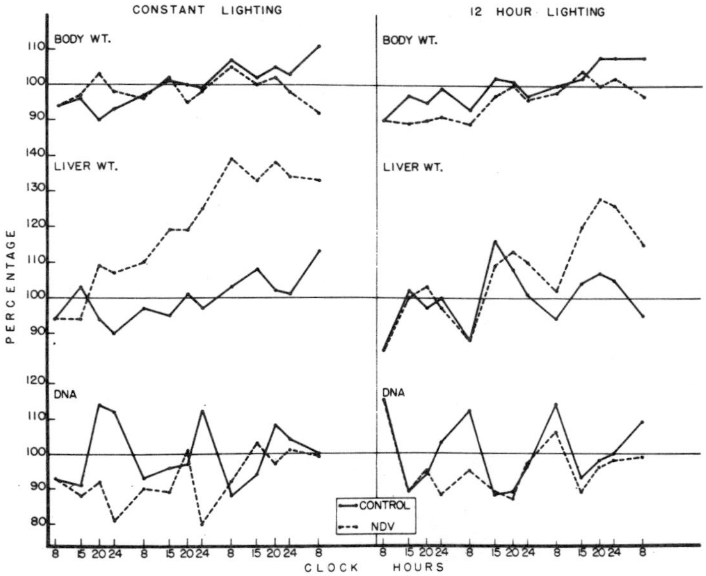

Figure 8. Body and liver weights and liver DNA of control and NDV infected chicks on constant and 12-hour lighting regimens, in terms of percentage change from control average for a 72-hour period.

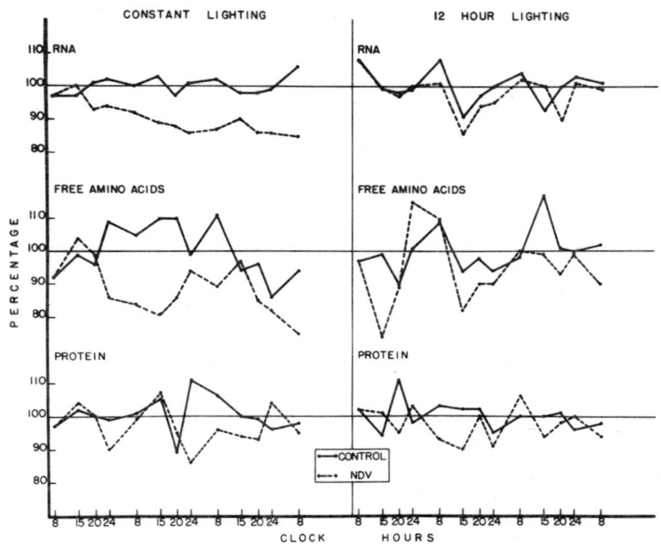

Figure 9. Liver RNA, tissue protein and free amino acid pool of control and NDV infected chicks on constant and 12-hour lighting regimens, in terms of percentage change from control average for a 72-hour period.

changes in the diurnal rhythm patterns. These observed interactions could not be attributed to differences in the intensity of disease involvement as final mortality in both NDV groups was approximately 50%.

A comparison of the patterns of the various parameters in the NDV tissues under the two lighting regimens shows that the LD schedule was a powerful synchronizer during the 72 hour incubation period of the virus. This was evident from the fact that the LD schedule held liver weights, DNA, RNA and the total free amino acids in reasonable synchronization to the curves of the controls. On the other hand, under LL infected livers showed a linear increase in weight, DNA values became desynchronized from controls, and RNA and total free amino acid levels were depressed.

The Nature of the Free Amino Acid Pool

The fact that the diurnal oscillations of the free amino acid pool in control chicks were of considerable magnitude suggested the need to know more about the relationship of the individual free amino acids to one another. In other words, how constant is the make-up of a free amino acid pool, and is it affected by the observed diurnal fluctuations of the pool itself? It is recognized that in the course of determining the free amino acids of a tissue the results are representative of many different proteins. Fortunately, in the case of liver, parenchymal cells comprise approximately 85% of the organ. Free amino acid extracts of this tissue, therefore, are fairly representative of a single type of cell. It can be assumed that in a normal rapidly growing individual, supplied with food and water *ad libitum,* diet intake and tissue metabolism are organized to maintain the *materia prima* in proper ratio for the code requirements of parenchymal tissue.

Data supporting this hypothesis are presented in Figures 10 and 11. Here it can be seen that in spite of the wide daily fluctuations in the free amino acid pool in the liver and serum of both control and infected groups there is a remarkable constancy between the ratios of individual free amino acids. The patterns of the pool oscillations in the liver, however, were desynchronized by the infection, that is, the peaks and troughs in the diseased animals were shifted so that they did not correlate with control values or clock hours.

Carrying this study further, free amino acid levels in the livers of control groups and those subjected to various stresses in 11 experiments over a 3-year period were re-examined for their "within pool" ratios.

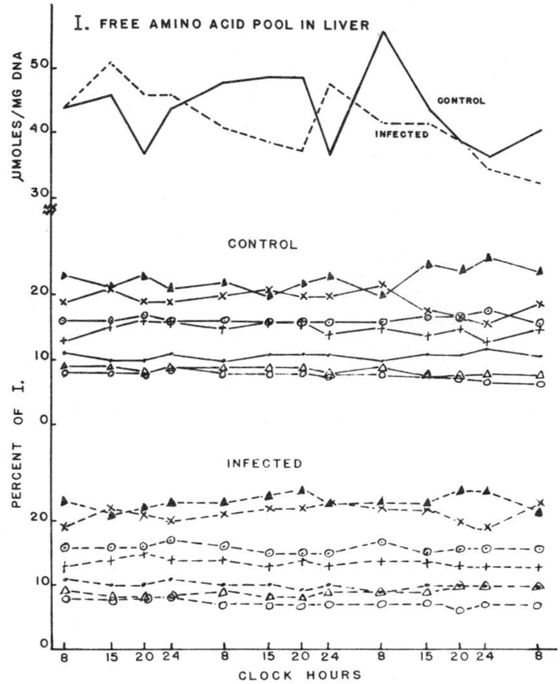

Figure 10. Daily oscillations of the liver free amino acid pool and its components during an NDV infection. ▲ Alanine; △ arginine; × aspartic acid; ○ histidine; ⊙ leucines; ● lysine; + valine.

As shown in Table III, when compared to control values for more than 200 individuals, extremes of dietary imbalance and environmental temperatures, recorded for at least 24 hours and sometimes as long as 96 hours, failed to change the within pool ratios in the liver. However, this was not considered surprising since these particular stresses did not affect body weights.

On the other hand, in the presence of a tuberculosis or Newcastle disease virus infection where body weights were depressed, the within pool ratios were changed significantly. While the table indicates there were ratio differences between the bacterial and viral diseases, in both infections the amino acid ratio most affected was lysine. Specificity may be indicated by the fact that in tuberculosis the ratio of aspartic acid to the other amino acids increased while that of the leucines decreased. Little or no effect was apparent on the alanine and valine ratios. With

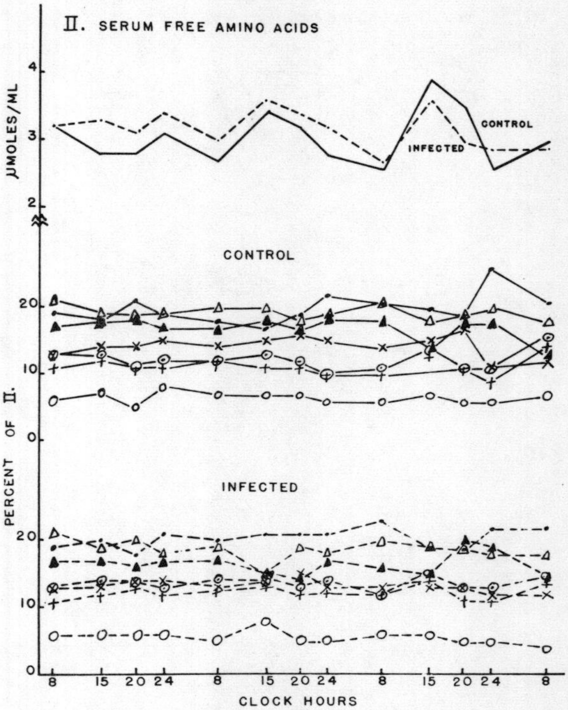

Figure 11. Daily oscillations of total quantities and individual serum free amino acids during an NDV infection. ▲ Alanine; △ arginine; × aspartic acid; ○ histidine; ⊙ leucines; ● lysine; + valine.

Table III. Effect of Stress on Free Amino Acid Ratios in Avian Liver

Free Amino Acid	Control * (%)	4x Dietary Lysine Requirement (%)	High Environmental Temperature (%)	Avian Tuberculosis (%)	Newcastle Disease Virus (%)
Lys	11 ± .11 †	12	11	4	8
His	6 ± .41	7	6	8	8
Arg	8 ± .18	8	7	5	7
Asp A	20 ± .83	18	20	36	30
Ala	25 ± .89	25	26	26	24
Val	13 ± .42	14	13	12	10
Leu ‡	17 ± .43	16	17	9	13
Total	100	100	100	100	100

* Values taken from 11 experiments comprising approximately 200 individuals.
† Standard error of the mean.
‡ Leucines combined.

NDV, on the other hand, histidine and aspartic acid increased while the leucines and valine decreased. The ratio of valine remained constant.

Summary

We believe the greatest difficulty and challenge facing the biologist is to gain an understanding of the diurnal fluctuations in the components of the various tissues being studied. These oscillations are present in all biological systems and reflect the dynamism of the many metabolic processes that occur. Many of them have defined, highly reproducible patterns. We have demonstrated here that an endogenous (disease) and an exogenous (light) agent can affect these fluctuations. Space does not permit our describing other factors that can affect them such as adverse temperatures, dietary imbalances, etc.

There is increasing evidence that the patterns of these oscillations in in a number of tissue constituents may be specific and consequently may prove to be useful diagnostic tools. Fluctuations in treatment groups, plotted against control values and clock hours, have revealed an almost immediate desynchronization due to experimental treatment.

We would venture to predict that the most meaningful research in the future, and the greatest advance in knowledge about the intact system will result from experiments carefully designed to evaluate the patterns of fluctuations in tissue components of normal vs. treated groups.

Acknowledgment

This is to acknowledge the valuable assistance of Marie Utzinger in the preparation of this manuscript.

References

1. Ewing, W. R., "Poultry Nutrition," 5th ed., Ray Ewing Co., Pasadena, Calif., 1963.
2. Biester, H. E., and Schwarte, L. H., "Diseases of Poultry," 4th ed., Iowa State Press, Ames, Iowa, 1959.
3. Hutchison, W. C., and Munro, H. N., *The Analyst,* **86,** 768 (1961).
4. Wannemacher, R. W., Banks, W. L., and Wunner, W. H., *Anal. Biochem.,* **11,** 320 (1965).
5. Squibb, R. L., *Nature,* **199,** 1216 (1963).
6. Sanslone, W. R., and Squibb, R. L., *J. Nutr.,* **76,** 86 (1962).

7. Allison, J. B., and Fitzpatrick, W. H., "Dietary Proteins in Health and Disease," C. C. Thomas, Springfield, Ill., 1960.
8. Squibb, R. L., *Nature,* **202,** 1138 (1964).
9. Lyons, M. M., Squibb, R. L., and Siegel, H., *Nature,* **216,** 1113 (1967).
10. Halberg, F., Galicich, J. H., Ungar, F., and French, L. A., *Proc. Soc. Exp. Biol. & Med.,* **118,** 414 (1965).
11. Dzickanowski, D., and Nowak, A., *Act. Physiol. Polon.,* **13,** 700 (1962).

Investigations of Experimental Mammary Tumors *

RUSSELL HILF †
Squibb Institute for Medical Research
New Brunswick, New Jersey

Cancer of the breast will be found in one of every 18 to 19 women in the United States during her lifetime. It is anticipated that at least 60,000 new cases of mammary cancer will occur in the United States per year and the annual mortality rate from this disease is approximately 25,000. While there is no doubt that early diagnosis and surgical extirpation can be successful in many cases, surgical removal of this neoplasm, either by simple or radical mastectomy, does not usually lead to a prolonged disease-free state and recurrence is common within a 5 year period. Detection of a tumor by the usual palpation techniques may be quite late relative to the time that a neoplasm has been present. MacDonald[1] has estimated that the time between the occurrence of a neoplastic cell and the discovery of a palpable mass may be 8 years or even longer. If this is a valid estimate, and, considering that the median survival time of the patient with untreated mammary cancer is 27–30 months, then the preclinical phase of this disease may occupy 3/4 of the total life time of this cancer.

There are several procedures currently employed for the treatment of women with recurrent breast cancer. Surgical oophorectomy in the pre-menopausal patient or oophorectomy-adrenalectomy in the post-

* The studies performed in the author's laboratory were supported by contracts S-43-ph-2395 and PH43-65-1050 from the Endocrine Evaluation Branch, General Laboratories and Clinics, National Cancer Institute, National Institutes of Health, Bethesda, Maryland.

† Present address: Biochemistry Department, University of Rochester School of Medicine and Dentistry, Rochester, N. Y.

menopausal patient has produced temporary regression of tumors in approximately 35–45% of the cases.[2,3] Hypophysectomy, performed subsequent to the recurrence of disease in the patients having shown responses to hormonal ablation, may induce a temporary remission in approximately 50% of these patients. If a patient has shown a response to any of the ablative procedures, this patient may also respond to hormonal therapy such as the administration of either androgen or estrogen. Objective response to estrogen therapy approaches 35% and response to androgen treatment about 20% in women with recurrent, disseminated advanced breast cancer. Data have been obtained indicating that the responses to hormonal treatment are accompanied by a genuine improvement of the patient and prolongation of life.[4,5] The clinician can also use chemotherapeutic agents, primarily in patients demonstrating no response to hormonal manipulation.

A classical approach to the study of a disease involves the development of a laboratory model closely simulating the clinical syndrome. Having such a model, intensive study of its characteristics, including sensitivity to chemical agents, can lead to the uncovering of therapeutic compounds. It is, therefore, imperative that laboratory models be developed if chemotherapeutic control of breast cancer is to be achieved. It is particularly pertinent to breast cancer that a spectrum of models be found since the clinical disease may be hormone-dependent, hormone-responsive, or independent of hormonal control. If models can be developed, it should be possible to discern reasons for observable similarities and differences in their responsiveness to a variety of agents and to obtain an understanding of the means by which their growth can be controlled.

Studies in our laboratory were initiated to ascertain and characterize the biochemical properties of experimental mammary carcinomas. It was deemed essential that gross changes in the weight and growth of these neoplasms be correlated with histologic and biochemical alterations. This approach to the study of mammary tumors seemed logical in view of the large body of data that is now being acquired by the study of hepatomas of varying growth rates. Weber and his colleagues have shown that the rate of glycolysis of these liver tumors correlated with the rate of their growth.[6,7] This finding has been confirmed by Burk *et al.*[8] In a recent review of these data, Weber has developed a "molecular concept of neoplasia" and demonstrated a relationship between the loss of gluconeogenesis in the rapidly growing hepatomas with a decrease in certain catabolic pathways of protein and nucleic acid metabolism.[9] It would appear from these data that the enzyme machinery of the neoplasm has been altered in such a manner as to result in a continual growth of cells and a loss of normal function. Examination of patterns of enzyme activity in these hepatomas has revealed that each line of tumor is a distinct biochemical entity, despite some morphological similarity.[10]

Other relevant studies, those of Pitot and his colleagues,[11] have demonstrated that some of the hepatomas are capable of responding to the same stimuli that cause the induction of certain enzymes in normal liver.[12] This is of prime importance in the control of neoplasms of endocrine target organs, since these tissues are regulated in both growth and function by the hormonal milieu.[13]

Dunning reported that the R3230AB tumor responded to hormonal treatment with characteristic changes in morphology.[14] During our studies with this neoplasm, a slow-growing variant line arose in one of the maintenance animals. This new tumor line, which has been called R3230AC, was found to consist primarily of epithelial cell elements with very little connective tissue or stroma. Considerable connective tissue elements are present in the R3230AB carcinoma. Since we were in the process of examining the biochemical characteristics of the parent R3230AB tumor and its responses to hormonal manipulations, the availability of this new neoplastic line afforded the opportunity of comparing two morphologically different but related mammary adenocarcinomas in the same strain.

A summary of some of our findings in the R3230AB and R3230AC carcinomas is presented in Table I. Both neoplasms responded to treatment with estrogen by a decrease in growth and a concomitant appearance of secretory activity, as reflected by vacuolization of the epithelial cells. This secretory response was very marked in the R-3230AC carcinoma and continuous treatment with estrogen resulted in the accumulation of whitish, milk-like fluid resembling rat milk in regard to its free fatty acid content and the fact that the electrophoretic properties of the proteins present were similar to those of the casein and whey proteins of rat milk, but differing from it by its much lower lactose content ($^1/_{50}$ of that in rat milk).[15] The estrogen-induced secretory response was accompanied by elevations in the R3230AB tumor, in glucose-6-phosphate dehydrogenase and malate dehydrogenase (decarboxylating) activities, enzymes associated with the production of NADPH, an essential cofactor for protein synthesis and lipogenesis.[16–18] Administration of estrogen also produced an increase

Table I. Comparison of Two Mammary Tumors and Mammary Glands and Their Responses to Estrogen Treatment

Measurement	R3230AB		R3230AC		Mammary Glands	
	Control	Estrogen	Control	Estrogen	Control	Estrogen
Weight at 3 weeks (gms)	20.5	13.0	4.0	2.0	–	Increased
Morphology	Fibroadenoma	Secretory changes induced	Adenocarcinoma	Marked secretory activity – milk like fluid accumulates	Adipose tissue – few epithelial cells	Stimulation of epithelial cell no.
RNA (μg/mg)	6.6	6.6	6.0	7.0	0.7	2.5
DNA (μg/mg)	4.9	4.2	6.6	5.0	1.0	3.1
Anaerobic lactic acid production *in vitro* (μg)						
glucose substrate	31	45	0	0	0	0
Malic acid substrate	9	36	0.8	1.2	1.2	4.3
Enzyme activity per Mg/DNA						
G6PD	0.32	0.82	1.37	3.84	0.29	1.03
ME	0.23	0.59	0.95	1.66	0.56	0.74
ICD	0.84	0.89	0.90	0.81	0.59	0.72
GPI			12.1	11.5	8.14	5.02
PGM			0.12	0.18	0.039	0.058
αGPD			0.15	0.07	4.60	2.74
Lipids (μg/mg/DNA)						
Cholesterol			0.33	0.36	0.70	0.64
Free fatty acids			1.96	7.19	12.80	3.89
Triglycerides			5.15	30.38	280	67.7

Enzymes assayed were: G6PD, glucose-6-phosphate dehydrogenase; ICD, isocitrate dehydrogenase, decarboxylating; ME, malate dehydrogenase, decarboxylating; GPI, glucosephosphate isomerase; PGM, phosphoglucomutase; αGPD, α-glycerolphosphate dehydrogenase; GDH, glutamate dehydrogenase; CL, citrate lyase; and AAT, aspartate aminotransferase.

in the levels of free fatty acids and triglycerides, an increase in the activity of phosphogl ucomutase, no increase in glucosephosphate isomerase, and a decrease in DNA levels. Little or no *in vitro* conversion of glucose to lactic acid was observed in the R3230AC tumor from control or estrogen-treated tumor-bearing animals whereas the R3230AB neoplasm exhibited anaerobic glycolysis under these conditions. Thus, administration of estrogen did not produce an increase in the rate of glycolysis of the R3230AC neoplasm but did appear to stimulate metabolism toward the production of a secretory product.

To examine the mechanism of these biochemical responses to estrogen, several experiments were performed employing inhibitors of protein and nucleic acid synthesis. First, it was necessary to determine the time course of events following a single injection of estrogen. For the R3230AC neoplasm, it was found that a significant elevation in both glucose-6-phosphate dehydrogenase and malate dehydrogenase (decarboxylating) activities occurred within 2–4 days after administration of estradiol valerate. Incorporation of radioactive leucine into protein and uridine into RNA was markedly increased after administration of estrogen. Concomitant administration of either actidione or actinomycin D with estrogen prevented the estrogen-induced increases in enzyme activities and protein synthesis, suggesting that the estrogen-induced enzyme changes resulted from a stimulation of protein synthesis *de novo* in this adenocarcinoma.[19,20] The decrease in tumor growth caused by treatment with estrogen was, therefore, not correlated with an inhibition of the metabolism of this neoplasm but rather, with a stimulation of both protein synthesis and lipogenesis. The hormone-induced elevation in metabolic activity resulted in the production of secretory products, diverting these metabolic precursors from use for cell multiplication. It would seem that the R3230AC neoplasm has retained the capacity to respond to the stimulus provided by estrogen, an example of a control mechanism still operative in a tumor. This apparent paradox of heightened metabolic activity and decreased growth rate may be somewhat analogous to that described in the clinical report by Huseby and Thomas.[21] These investigators reported that in several women with breast cancer, estrogen treatment caused a stimulation of secretory activity of normal cells of the breast occurring simultaneously with a regression of the mammary neoplasm. Assuming a finite or limiting quantity of available substrate, it appears that the hormone was able to direct metabolic substrates away from pathways used for growth of the cancer.

The study of neoplasia is really one of comparisons, either of tumors with other tumors or of tumors with normal tissue.[22] The normal mammary gland is an organ that displays striking differences in metabolic activity depending on the physiological state of the host. In our studies, we used the mammary glands of the tumor-bearing animal for biochemical assays in order to compare them with the responses of the neoplastic tissue to hormonal agents. The very thorough recent studies of Baldwin,[23,24] who determined several enzyme activities in mammary glands of rats at various stages prior to and during pregnancy, as well as during post-partum lactation, along with the extensive studies of Abraham and his colleagues on lipogenesis of the mammary gland,[25,26] are most helpful in making comparisons between mammary neoplasms and the mammary gland. The results obtained after analyses of mammary glands of the host bearing the R3230AC tumor are presented in Table I. Estrogen administration caused an increase in both RNA and DNA levels of the mammary gland. There are several similarities between the type of response observed in mammary glands and the R3230AC mammary adenocarcinoma after treatment with estrogen such as the hormone-induced elevation in glucose-6-phosphate dehydrogenase, malate dehydrogenase (decarboxylating), and phosphoglucomutase activities as well as the decreased activities of glucosephosphate isomerase and α-glycerolphosphate dehydrogenase. There are, however, striking differences in the absolute levels of enzyme activity/mg DNA of the normal and neoplastic mammary tissues. Nevertheless, these results demonstrate that the R3230AC neoplasm has retained some of the characteristics of the mammary gland response to estrogen. These data suggest that the R3230AC neoplasm is an estrogen-sensitive tumor and perhaps it is representative of some of the neoplasms in women who respond to estrogen therapy. Since this experimental neoplasm is not hormone-dependent but is autonomous, it may represent a metabolic state more akin to that which is found in women with recurrent breast disease. This experimental tumor displays certain responses that are similar to those reported in the clinical literature. We should continue to characterize our experimental tumor models in the hope that they may be relevant to the disease in women; this approach will verify the utility of the model.

As mentioned earlier, approximately 50% of the clinical cases of breast cancer do not show a beneficial response to hormonal manipulation. What criteria should be used to establish the usefulness of an experimental mammary tumor that does not respond to hormonal treatment? One such neoplasm is being studied in our laboratory. This tumor, HMC, was not altered in growth rate or RNA and DNA levels after treatment with estrogen. Table II lists the effects of estrogen on various enzyme activities in the HMC tumor. These data clearly show that the enzyme activities measured are not significantly altered by

Table II. Influence of Hormonal Treatment on the Biochemistry of the HMC Mammary Tumor

Measurement	Control (diluent)	Treatment Estradiol Valerate (10 mg/kg/wk)
Weight (gm)	28.53 ± 2.10	32.29 ± 3.31
RNA (μg/mg)	7.36 ± 0.31	7.17 ± 0.40
DNA (μg/mg)	6.80 ± 0.22	7.26 ± 0.25
RNA/DNA Ratio	1.06 ± 0.05	0.99 ± 0.03
Enzyme activity (μmoles/min/mg DNA)		
G6PD	1.297 ± 0.054	1.145 ± 0.046
ICD	0.217 ± 0.008	0.187 ± 0.005
ME	0.077 ± 0.005	0.067 ± 0.004
GPI	3.740 ± 0.146	3.336 ± 0.103
PGM	0.096 ± 0.008	0.081 ± 0.012
αGPD	0.016 ± 0.005	0.007 ± 0.001
GDH	0.010 ± 0.002	0.008 ± 0.001
CL	0.004 ± 0.001	0.005 ± 0.001
AAT	0.467 ± 0.025	0.344 ± 0.012

Treatment was initiated on the day after tumor transplantation. Animals were sacrificed 21 days after tumor transplantation and tumors weighed, portions taken and frozen. Samples were kept frozen until assayed. Values are Mean ± Standard Error.

Enzymes assayed were: G6PD, glucose-6-phosphate dehydrogenase; ICD, isocitrate dehydrogenase, decarboxylating; ME, malate dehydrogenase, decarboxylating; GPI, glucosephosphate isomerase; PGM, phosphoglucomutase; αGPD, α-glycerolphosphate dehydrogenase; GDH, glutamate dehydrogenase; CL, citrate lyase; and AAT, aspartate aminotransferase.

hormonal treatment. In an earlier study, acid and alkaline phosphatase activities responded to treatment with estrogen in a similar manner in the HMC and R3230AB tumors.[27] Since the HMC tumor has been shown to be inhibited significantly by nonhormonal agents such as 5-fluorouracil, cytoxan, thioTEPA, and hydroxyurea, it may well represent a useful experimental tumor model for the study of its biochemical characteristics as compared to the hormone-responsive neoplasms.

Can an exhaustive study of experimental mammary tumors supply enough knowledge to improve the results obtained by chemotherapy? It seems logical that a study of experimental breast neoplasms would lead to acquisition of information more pertinent to management of breast cancer than examining thousands of compounds in systems which may or may not have any clinical counterpart. A complicating factor in the establishment of a laboratory model for breast cancer is that the disease recurring after ablative procedures may no longer be amenable to the same therapy. It would appear that the therapeutic regimen only affected a portion of the neoplastic cell population, the recurrent neoplastic cells having been temporarily affected in their growth and multiplication. This indicates the need for developing multiple models, each model having at least one parallel counterpart in the clinical disease. Such an approach was suggested by Vollmer, who suggested the "phase model" concept for the investigation of new agents for chemotherapy.[28] Certainly, the approaches to date have not resulted in, or led to, any striking advances in the treatment of breast cancer. On the basis of survival figures, Shimkin concluded that there has been no improvement in the overall survival of women with breast cancer in the last 30 years.[29,30] Shimkin went on to state that these conclusions do not indicate that therapy of breast disease is of no avail but rather, that newer approaches to prognosis and treatment are required.

One such approach is the work of Bulbrook and his associates in England.[31] These investigators have found that the ratio of urinary excretion of androgens and corticoids is of value in determining the prognosis of the patient, at least in terms of a response to adrenalectomy or hypophysectomy. Other studies of human breast cancer are underway to arrive at a better understanding of the clinical disease. Several studies of human breast cancer tissue have yielded interesting biochemical data, suggestive of the existence of differences in enzyme activity related to the histological grading of the tumor.[32–34] We have initiated an investigation of human breast neoplasms, as well as of benign lesions, examining several biochemical endpoints as we have done in the experimental tumors. Results to date, although preliminary, indicate the existence of a rather homogeneous pattern of enzyme activities in infiltrating ductal carcinomas and statistical analyses of these data have revealed the occurrence of several highly significant correlations among these enzyme activities.

However, it should be emphasized that any approach to the study of cancer utilizing human neoplastic tissue may not elucidate the events involving the fundamental alteration from normal. Studies of the pre-neoplastic state are required and to this end, the investigations of DeOme and his colleagues on the hyperplastic alveolar nodules of mice need to be extended to other species.[35,36] Recently, studies along these lines in the rat following carcinogen feeding have been published, although these studies have not yet definitely proven that the hyperplastic changes observed are truly pre-neoplastic.[37,38] Farber *et al.* have also examined pre-neoplastic changes in the liver following ethionine feeding and it will be most interesting to learn of their findings in relation to the ethionine-induced hepatomas.[39,40] Brennan and co-workers have reported that a 20 minute exposure of normal mammary glands to dimethylbenz(a)anthracene *in vitro* resulted in mammary carcinomas on subsequent transplantation of the carcinogen-exposed tissue.[41] It would be most important to learn of the biochemical changes that occur during this very brief exposure to carcinogen. Extensive studies have been performed utilizing methylcholanthrene and dimethylbenz (a) anthracene, carcinogens capable of inducing mammary tumors in a high percentage of animals fed these agents. Biochemical studies of these carcinogen-induced neoplasms have been reported[42–44] and a systematic examination of enzyme activities, nucleic acids and lipids in these neoplasms, as influenced by hormonal treatment of the tumor-bearing animal, is currently in progress in our laboratory.[45]

It remains to be proven that any experimental mammary tumor is similar to clinical breast cancer. The studies being conducted to examine the biochemical and morphological characteristics of both the experimental neoplasms and human tumors could provide the data needed for establishing this relationship between the laboratory tumors and the clinical disease. If such a correlation is found, the results obtained from experiments with such a laboratory system should provide useful data for improving the chemotherapy of breast cancer.

References

1. MacDonald, I., *Am. J. Surg.,* **111,** 435 (1966).
2. Segaloff, A., *Cancer Chemother. Rpt.,* **41,** Suppl. No. 1 (1964).
3. Tagnon, H. J., Coune, A., Heuson, J. C., and Van Rymenent, M., in "New Trends in the Treatment of Cancer," L. Manuila, S. Moles, and P. Rentchnick, Eds., Springer-Verlag, New York, 1967, p. 126.
4. Cooperative Breast Cancer Group, *J. Am. Med. Assoc.,* **188,** 1069 (1964).

5. Segaloff, A., in "Recent Progress in Hormone Research," Vol. 22, G. Pincus, Ed., Academic Press, New York, 1966, p. 351.
6. Weber, G., and Morris, H. P., *Cancer Res.,* **23,** 987 (1963).
7. Sweeney, M. J., Ashmore, J., Morris, H. P., and Weber, G., *Cancer Res.,* **23,** 995 (1963).
8. Burk, D., Woods, M., and Hunter, J., *J. Natl. Cancer Inst.,* **38,** 889 (1967).
9. Weber, G., and Lea, M. A., in "Advances in Enzyme Regulation," Vol. 4, G. Weber, Ed., Pergamon Press, New York, 1966, p. 115.
10. Potter, V. R., *Cancer Res.,* **24,** 1085 (1964).
11. Pitot, H. C., and Cho, Y. S., *Progr. Exptl. Tumor Res.,* **7,** 158 (1965).
12. Pitot, H. C., *Ann. Rev. Biochem.,* **35,** 335 (1966).
13. Wenner, C. E., in "Advances in Enzyme Research," Vol. 29, F. F. Nord, Ed., Interscience Publishers, New York, 1967, p. 321.
14. Dunning, W. F., in "Biological Activities in Relation to Cancer," G. Pincus and E. P. Vollmer, Eds., Academic Press, New York, 1960, p. 225.
15. Hilf, R., *Science,* **155,** 826 (1967).
16. Hilf, R., Michel, I., Bell, C., Freeman, J. J., and Borman, A., *Cancer Res.,* **25,** 286 (1965).
17. Hilf, R., Michel, I., and Bell, C., *Cancer Res.,* **26,** 865 (1966).
18. Hilf, R., Michel, I., and Bell, C., in "Recent Progress in Hormone Research," Vol. 23, G. Pincus, Ed., Academic Press, New York, 1967, p. 229.
19. Hilf, R., Michel, I., Silverstein, G., and Bell, C., *Cancer Res.,* **25,** 1854 (1965).
20. Hilf, R., Goldenberg, H., and Bell, C., *Cancer Res.,* **27,** 1485 (1967).
21. Huseby, R. A., and Thomas, L. B., *Cancer,* **7,** 54 (1954).
22. Knox, W. E., in "Advances in Cancer Research," Vol. 10, A. Haddow and S. Weinhouse, Eds., Academic Press, New York, 1967, p. 117.
23. Baldwin, R. L., and Milligan, L. P., *J. Biol. Chem.,* **241,** 2058 (1966).
24. Baldwin, R. L., *J. Dairy Sci.,* **49,** 1533 (1967).
25. Abraham, S., and Chaikoff, I. L., *J. Biol. Chem.,* **234,** 2249 (1959).
26. Abraham, S., Matthes, K. J., and Chaikoff, I. L., *Biochem. Biophys. Acta,* **49,** 268 (1961).
27. Hilf, R., Johnson, M. M., Breuer, C., Freeman, J. J., and Borman, A., *J. Natl. Cancer Inst.,* **31,** 541 (1963).
28. Vollmer, E. P., in "Hormonal Steroids," Vol. 2, L. Martini and A. Pecile, Eds., Academic Press, New York, 1965, p. 351.
29. Shimkin, M. B., *Cancer,* **20,** 1039 (1967).
30. Hickey, R. C., *Cancer,* **20,** 1060 (1967).
31. Bulbrook, R. D., and Hayward, J. L., in "Current Concepts in Breast Cancer," A. Segaloff, K. K. Meyer, and S. DeBakey, Eds., The Williams and Wilkins Co., Baltimore, 1967, p. 115.
32. Smith, J. A., King, R. J. B., Meggitt, B. F., and Allen, L. N., *Brit. J. Cancer,* **20,** 335 (1966).
33. Muir, G. G., and Fawcett, A. N., *Brit. J. Cancer,* **20,** 274 (1966).
34. Goldberg, D. M., Pitts, J. F., and Ayre, H. A., *Brit. J. Cancer,* **21,** 312 (1967).
35. DeOme, K. B., Faulkin, L. J., Jr., Bern, H. A., and Blair, P. B., *Cancer Res.,* **19,** 515 (1959).
36. DeOme, K. B., Nandi, S., Bern, H. A., Blair, P. B., and Pitelka, D., in "The

Morphological Precursors of Cancer," L. Severi, Ed., Division of Cancer Research, Perugia, 1962, p. 349.
37. Beuving, L. J., Faulkin, L. J., Jr., DeOme, K. B., and Bergs, V. V., *J. Natl. Cancer Inst.,* **39,** 423 (1967).
38. Beuving, L. J., Bern, H. A., and DeOme, K. B., *J. Natl. Cancer Inst.,* **39,** 431 (1967).
39. Epstein, S., Ito, N., Merkow, L., and Farber, E., *Cancer Res.,* **27,** 1702 (1967).
40. Merkow, L. P., Epstein, S. M., Caito, B. J., and Bartus, B., *Cancer Res.,* **27,** 1712 (1967).
41. Brennan, M. J., Grau, W. H., and Singley, J. A., *Proc. Amer. Assoc. Cancer Res.,* **I,** 9 (1966).
42. Rees, E. D., and Huggins, C., *Cancer Res.,* **20,** 963 (1960).
43. Hershey, F. B., Johnston, G. Murphy, S. M., and Schmitt, M., *Cancer Res.,* **26,** 265 (1966).
44. Rees, E. D., Shuck, A. E., and Ackermann, H., *J. Lipid Res.,* **7,** 396 (1966).
45. Hilf, R., Goldenberg, H., Bell, C., Gruenstein, M., Meranze, D. M., and Shimkin, M. B., *Cancer Res.,* **29,** 977 (1969).

Research on Tumor-Host Relationships and Chemotherapy of Canine Malignancies

JOHN R. MC COY, V.M.D.
School of Medicine
Rutgers University
New Brunswick, New Jersey

The dog develops cancer with what is presently judged to be about one-half the frequency of man and most frequently of all the domesticated animals. It is one of the few companions of man that is permitted to live out its lifespan in essentially the same environment as that of its master.

The position of the dog is unique; as a rule this animal is present within a family group as a pet, sharing a dietary regime similar to that of his human owner which mirrors to some degree the economic and nationalistic factors that govern the selection of foods by a given family.

Similarly, the development of this relationship has caused a rise of a field of veterinary medical practice devoted to control of the diseases of the dog. Enlightened use of these professional facilities has led to the study of this animal at various ages during its lifespan and, mainly at veterinary colleges, to its clinical examination, and to necropsy for the purpose of discovering what disease has terminated its lifespan. Research on combatting such diseases as distemper which caused large losses in young dogs and on nutritional problems has resulted in the attainment by greater numbers of these animals of their "cancer age." This has revealed a definite tendency for the canine to develop neoplastic diseases beginning in its sixth year.

Neoplasia in the dog possibly occurs in about 10% of the dog population. The most frequent cases are those of the mammary gland, skin and adnexae, such as sweat glands and sebaceous glands and basal cell, squamous cell and epidermoid carcinoma, mast cell sarcoma, fibrosarcoma, neurofibrosarcoma, osteosarcoma, hemangiosarcoma, and malignant melanoma are common. Sustentacular cell tumor and seminoma of the testes are frequently seen, but teratomas are rare. Malignant lymphoma is frequently seen.

Neoplasms noted by their paucity among dogs, in contrast with the man, are chorionepithelioma, mesothelioma and tumors of the heart, spleen, gall bladder, and the genito-urinary, gastrointestinal, respiratory and central nervous systems, as well as myeloid leukemia.

The preponderance of the dog's neoplasms are, therefore, in surface regions readily available for direct measurement of gross size and for repeated biopsies.[1–44, 46–49]

It is appropriate to note here the increasing frequency with which epidemiological data are being published by schools of veterinary medicine, some large noncollegiate veterinary institutions, and interested practitioners, through data retrieval techniques, which are fairly new upon the scene. The basic deficiency, in terms of valid statistical comparison with the occurrence of neoplastic disease in man, is a means of determining similarities, or lack thereof, in the biological behavior of neoplasms of man and the dog with the viewpoint to evaluating the efficacy of results to be anticipated from the chemotherapeutic, surgical, and radiotherapy of neoplastic diseases. A masterful compilation of existing information on this subject is to be found in the work of Mulligan.[39] It cannot be denied that there will always be a marked limitation in terms of epidemiology in canine cancer, in that, unless dogs are of purebred origin they are not issued birth certificates, death certificates, insurance policies, or various types of licenses other than that concerned with community registration. Census figures taken at the municipal level by people of various interests and degrees of responsibility are known to be quite inaccurate and subject to rather abrupt revisions. While dogs, unlike the species destined to provide man with food and fiber, live a full natural life, their owners are frequently unable to support financially extended periods of therapy or study necessary to have many cases become statistically valid sources of information. To date, we lack an international nomenclature and statistical code for the tabulation of neoplastic disease. This whole area needs to be organized on both national and international levels, in order to facilitate correspondence and data exchange between investigating agencies and individuals.

A step in this direction has been taken in the formation of the Epidemiology Branch, National Cancer Institute, U. S. Public Health Service, whose mission it is to further the interest and feasibility of comparative cancer research by facilitating and coordinating national and, if indicated, international programs in comparative cancer research. It may be said that the feasibility of future participation of veterinary medicine in all fields of comparative degenerative disease research is being enhanced by the excellent work being done in the many veterinary medical agencies that have undertaken parts of the whole problem that fall within their spheres of competency.[43]

In 1948, the author initiated a pilot investigation study in conjunction with the Bureau of Biological Research of Rutgers University, Dr. James B. Allison, Director, and Dr. Moses L. Crossley, then a Director of Research for American Cyanamid Company, under the auspices of the Damon Runyon Memorial Cancer Fund, for the purposes of determining whether or not the procurement of dogs with spontaneous neoplastic disease and their treatment with chemotherapeutic agents of interest were feasible projects to undertake. The experience was encouraging and in 1950 the National Cancer Institute of the National Institutes of Health of the U. S. Public Health Service granted funds to place the study on a larger, but still accurately controlled, study of chemotherapy in spontaneous canine cancer. The early agents studied were those in the class known as alkylating agents whose main pharmacologic action appeared to be the hydrolysis of deoxyribonucleic acid (DNA), and were found to have some slight proportional affinity for the increased amounts of this nucleic acid known to be present in malignant cells. These radiomimetic agents duplicated responses gained with X-ray therapy or radioisotopes, with the advantage of systemic distribution of the compound at effective levels and relatively wide tolerance of these levels by the host cells which were mature and not concerned with replication, as opposed to cells of the neoplasm which were concerned only with reproduction of their numbers leading finally to the death of the host. The canine patients and control dogs under study were housed in metabolism cages and fed semipurified diets of known composition and biological value. This permitted the study of the tumor-free dog for comparative measures of the effects of a therapeutic agent on the host, through induction of protein depletion, for instance, in order to study the latter's influence on the dosage (safety) levels for a given chemotherapeutic agent (Figures 1, 2).[61]

It was found that these measurements could be conducted over protracted periods of time in the same animal depending on the nature of the neoplastic disease presented. Thus, cases of malignant lymphoma might live for a few weeks untreated, or for more than fourteen months when treated with agents such as 2,4,6-triethylenemino-s-triazine (TEM), or N-(3 oxapentamethylene)-N′,N″-diethylenephosphoramide (MEPA). Dogs bearing other neoplasms lent themselves to more prolonged study, such as those with mammary cancer, seminoma, sustentacular cell tumor, fibrosarcoma, malignant melanoma, and mast cell sarcoma. These ran courses measured in years. This permitted not only observations of the effect of a chemotherapeutic agent on the visible symptom of neoplastic disease represented by the tumor, but measurements of other factors as well, such as exploring the means of administration of

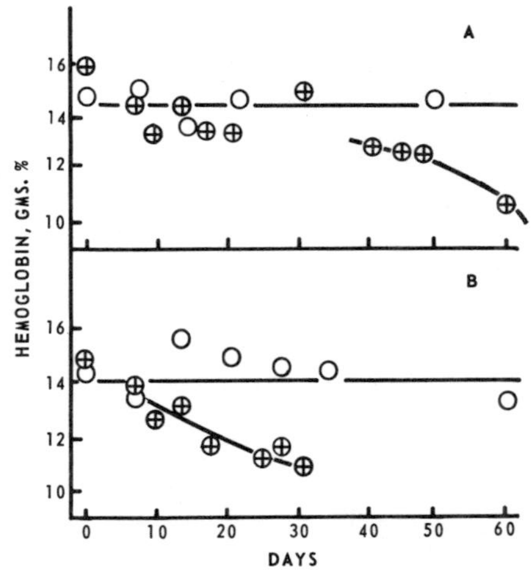

Figure 1. The average hemoglobin response to MEPA in the food of dogs maintained on a low casein diet (A) and a protein-free diet (B).

A—○ = 3 controls fed 0.2 gm of nitrogen per day per kg of body weight; ⊕ = 3 received 5.0 mg orally of MEPA per kg of body weight.

B—○ = 3 fed protein-free diet; ⊕ = 3 received 5.0 mg orally of MEPA per kg of body weight per day while fed a protein-free diet.

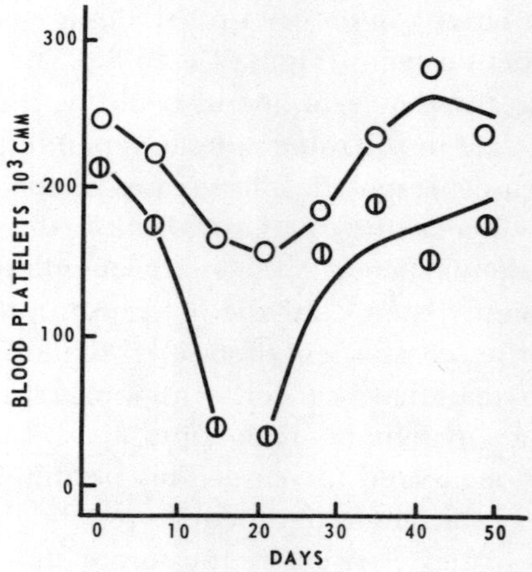

Figure 2. Blood platelet counts in each of 2 normal dogs fed a low casein diet and given three intravenous injections (arrows) of 1.0 mg of N-(3-oxapentamethylene)-N′,-N″-diethylenephosphoramide (MEPA) per kg of body weight. ○ = dog 1; ⦶ = dog 2.

the agent under study in order to determine the level of maximum therapeutic effect and the maximum total dose tolerated. For instance, the dividing line between the therapeutic and toxic doses of MEPA per kilogram of body weight given intravenously for three successive days was 1.5 milligrams per kilogram of body weight. Orally, on the same dosage regime the average patient tolerated 3 milligrams of MEPA per kilogram. The total amount given to some patients was over four grams in the case of MEPA. It was possible to study the degree of therapeutic effectiveness against certain groups of neoplasms for as long as seven years and, concurrently, to study the effect of the agent upon the physiological status of the host in terms of the latter's resistance to its toxic effects coupled with the depleting effects of his disease.[45,50,51,60,61] As a sequel to this, efforts calculated to enhance the efficiency of host resistance to the toxic effects of the agent under study were made through nutritional supplementation. It was found that, if dogs that were being depleted by a tumor or the effect of poor appetite or the toxicity of an agent under study, when their diet was supplemented by sulphur-bearing amino acids the treatment period was lengthened for at least three times over that which it took for toxicity to develop in control dogs (Figure 3).[61]

It was possible to obtain ample material for histological and biochemical analysis from the tumor tissue when neoplasms of appropriate type and size were under study and, of course, various body fluids and tissues were sampled for similar determinations. The reduction of the albumin-globulin ratio in tumor-bearing dogs has been shown and, as well, the duplicating effect of this type of depletion by the administration of toxic chemotherapeutic agents has been reported.[43] Experiments were designed with regard to measuring the effect of the neoplastic disease, with and without treatment, upon the host's so-called

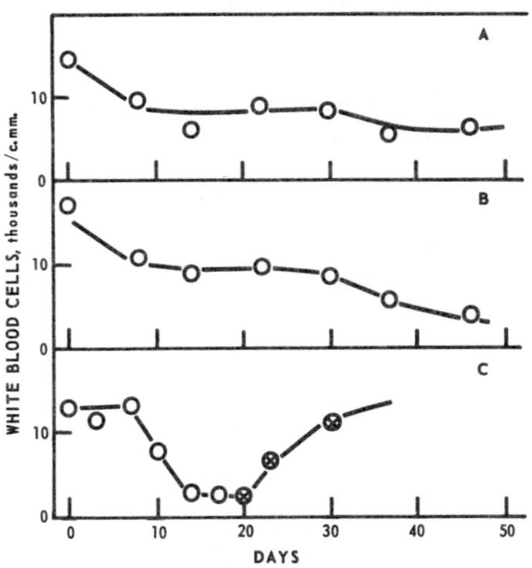

Figure 3. The average white blood cell response to N-(3-oxapentamethylene)-N′,N″-diethylenephosphoramide (MEPA) given intravenously to dogs maintained on a low casein diet.

A—○ = 3 were fed 0.2 gm of nitrogen plus 0.67% of methionine and 0.67% of glycocyamine per day per kg of body weight, plus 1.0 mg of MEPA per kg of body weight intravenously thrice weekly throughout the experimental period of 45 days.

B—○ = 3 were fed 0.6 gm of nitrogen per day per kg of body weight, plus 1.0 mg of MEPA per kg of body weight intravenously thrice weekly throughout the experimental period.

C—○ = 3 were fed 0.2 gm of nitrogen per day per kg of body weight, plus 1.0 mg of MEPA per kg of body weight intravenously thrice weekly for two weeks. ⊗ = MEPA discontinued.

protein reserves by elucidating electrophoretic patterns of serum proteins (Table I), the uptake of S35, l-methionine by plasma proteins and alterations in serum lipid metabolism, especially with regard to the migration of serum lipoproteins electrophoretically, mainly in the beta-2 group.[45] Serum glycoproteins were measured similarly and all of these demonstrated the depletion of the host by its disease, depletion of a type similar to that induced by nutritional inadequacy in protein levels and/or amino acid composition of certain experimental diets in normal adult dogs.[51]

Table I. Average Serum Values
(grams per 100 cc)

	Control	Malignant Lymphoma
No. of dogs	14	17
Total Protein	6.03	5.80
Albumin	3.48	2.54
Alpha 1	.12	.12
Alpha 2	.17	.23
Alpha 3	.39	.70
Beta 1	.59	.59
Beta 2	.64	.61
Gamma	.64	1.09
Total Globulins	2.55	3.33
A/G Ratio	1.63	.85

The average serum values in untreated control dogs and dogs with malignant lymphoma prior to chemotherapy, on commercial diets.

Neoplastic disease was shown to deplete its host in a fashion that was predictable and could actually be surpassed in rate of depletion by depleting diets fed to control dogs, the difference being that, in the latter case, prolonged administration of these diets did not result in the death of the hosts. Concurrently, clinical pathological methods were used to detect variations owing to extrinsic factors such as infections of local or general nature and included determination of red and white blood cell counts, hemoglobin determinations, and sedimentation rates as well as Schilling index calculations, activities of serum glutamic oxalacetic and pyruvic transaminases and lactic dehydrogenase. Blood urea nitrogen values and routine urinalyses were done, with occasional electrophoretic examination of the urine for possible exotic protein excretion. Of primary interest was the activity of serum ribonuclease at pH 8, in terms of its correlation with tumor growth. The enzyme activity in most of the groups other than malignant lymphoma (Figure 4) gave indication of possible value in determining the initial susceptibility of the tumor cells to the action of the chemotherapeutic agent under study, for a lowering of the initially high enzyme activity (above 4 O.D. units) was a positive indication of therapeutic response and was commonly confirmed by a following reduction in the size of the neoplasm. It also showed promise of efficacy in use as a prognosticatory tool, in that adaptation of the tumor cells to the presence of the agent was usually signalled by a rise in enzyme activity, thus calling for a change in the therapeutic regimen being administered (Figure 5).

Such was not the case, however, with regard to the tumor growth rate in highly and moderately virulent malignant lymphomas, which did not present a high enzyme activity initially in approximately 50% of the cases. But, even here, therapeutic response was represented by a drop in enzyme activity in the serum as a therapeutic response was gained, even though the initial baseline determinations of enzyme activity were within normal range (Figure 6).

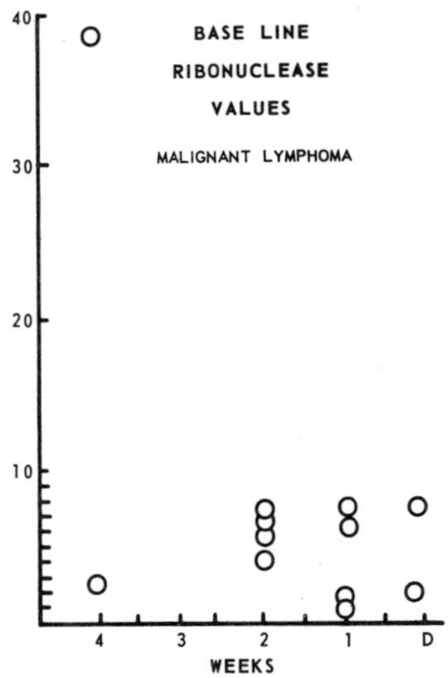

Figure 4. The initial serum ribonuclease activity in 12 dogs with malignant lymphoma, prior to chemotherapy.

Utilizing normal dog tissues the ribonucleic acid (RNA), deoxyribonucleic acid (DNA) and total protein content were determined in milligrams per gram of tissue and reported for brain, pancreas, kidney, skeletal muscle, heart muscle, liver, spleen, and lymph node (Table II).

Of these, the spleen and lymph node values determined have since been compared with their leukotic counterparts from dogs with malignant lymphoma. Concurrently, the ribonuclease activity of the normal tissues at each half pH level from 3.5 through 9 was investigated. From these studies, the commonest "peaks" for spleen, lymph node,

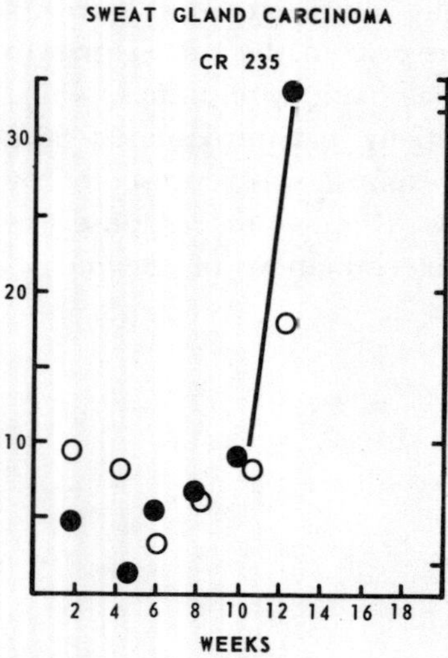

Figure 5. Sweat gland carcinoma. Case treated with Dihydro-E-73, 0.5 mg/kg body weight, single dose intravenously 3 days, 4 days withdrawal between weekly series. Terminated by euthanatasia.

● = tumor volume measured in cubic centimeters read on right ordinate of graph. Open circles, ○, represent triplicate sample averages of serum ribonuclease activity measured in optical density units read on left ordinate of graph.

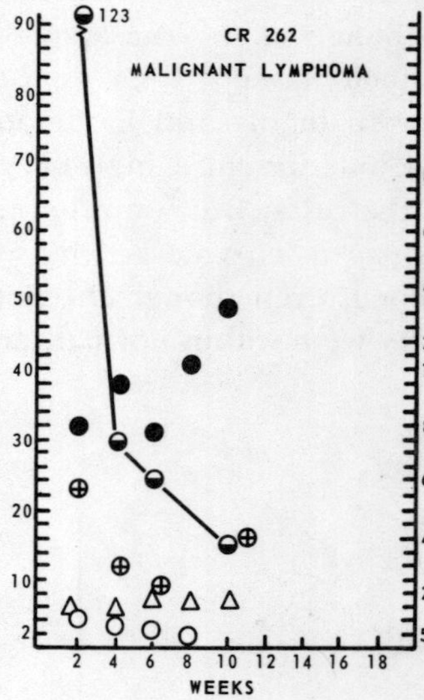

Figure 6. A dog with malignant lymphoma treated with Dihydro-E-73, 0.5 mg/kg body weight, single dose intravenously 3 days, 4 days withdrawal between weekly series. Terminated by euthanatasia.

○ = serum ribonuclease activity; △ = total serum proteins; ⊕ = serum glutamic oxalacetic transaminase activity; ◒ = serum glutamic pyruvic transaminase activity; ● = serum lactic dehydrogenase activity. All enzyme activities are read on left ordinate of graph as optical density units, as are the mg % of total serum proteins. LDH activity is read on right ordinate.

Table II. Average Tissue Values of Five Control Dogs in mg per gm of Tissue (Triplicate Runs)

	Protein	RNA	DNA	RNA/ DNA	Protein/ DNA
Kidney	113.30	3.89	1.46	2.77	80.75
Brain	87.58	1.61	.57	1.55	53.36
Pancreas	121.19	20.98	.34	62.04	353.24
Skeletal Muscle	191.58	1.17	.51	2.28	375.56
Cardiac Muscle	172.31	1.58	.59	2.68	292.95
Liver	156.93	5.33	1.20	4.46	106.59
Spleen	160.79	4.68	3.01	1.55	53.36
Lymph Node	107.15	11.21	3.28	3.44	33.83

Tissue values from five clinically normal mixed breed dogs fed commercial diets. Triplicate samples were run. All values are expressed as milligrams per gram of tissue.

and serum occurred at pH 5.8 and 7.8 and were used as the comparison points for similar determinations of spleen, lymph node, and serum activity in untreated dogs with malignant lymphoma. From the values obtained, the ratios of RNA/DNA and protein/DNA were calculated in both groups of dogs and it may be stated that (Figure 7), compared to the control samples, the protein values of the lymphomatous spleen, and lymph node cells was lower and the DNA content was lower as was the ribonuclease activity at pH 5.8 and 7.8 The RNA content was significantly higher. The RNA/DNA ratio was, therefore, significantly higher in the leukotic cells. This did not support the premise of higher leukotic cell protein synthesis for the protein/DNA ratios were similar for normal and lymphomatous cells and the protein content of the lymphomatous cells was actually lower per gram of tissue. It was reasoned that this might reflect increased cell fluid content which was supported by the slightly lower DNA value in the lymphomatous cell. When the high content of RNA was considered, especially in the presence of the lower ribonuclease activity at most pH levels in the lymphomatous cells the indication was that the RNA was complexed in such fashion as to obviate hydrolysis by the enzyme ribonuclease which may indicate its coupling with ribosomal material or binding with protein, although it is not active in protein synthesis. The position of this RNA in the cell is unknown by reason of our methods of extraction and it may, therefore, be of nuclear, cytoplasmic or possibly of exogenous

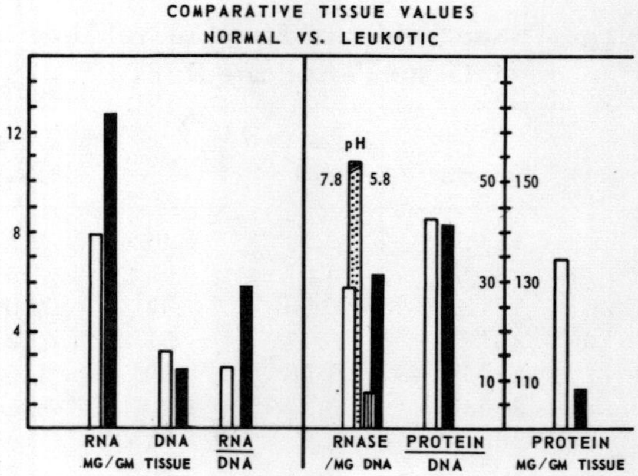

Figure 7. Comparative lymphoma tissue values: normal vs. leukotic.

Solid bars represent tissue values from five untreated canine cases of malignant lymphoma, clinically depleted, not advanced, and free of secondary complications.

Open bars represent control dog tissue values.

A dotted bar represents the pH 5.8 activity of leukotic tissue ribonuclease (RNASE).

A striped bar represents the pH 7.8 activity of ribonuclease (RNASE) in normal tissue.

Necropsies were performed on all immediately following euthanatasia.

origin.[51] This has led us to our latest undertaking in the area of cancer research at Rutgers, for we are presently designing experiments in which major portions of the tumors are removed surgically and the proteins altered chemically for reinjection into the host, in order to serve as antigens which might stimulate recognition of and, hopefully, the rejection of the tumor tissue by the host. Simultaneously, serum samples are removed prior to surgery for antibody screening in an effort to determine the presence or absence of antitumor antibody, which has been shown to exist in other species—in recognizable amounts if not in sufficient quantity to permit the host to rid himself of his disease. Hypothetically, the excitation of recognition and the resulting formation of antibodies which would be adsorbed to the surface of the tumor cell in amounts of sufficient quantity and quality to lead to its lysis are the goals sought. At the same time, purification of antigen to rid it of all proteinaceous reference material residing in the normal tissues of the host as well as his tumor will be accomplished. The use of such purified antigen to gain further antibody production from the host against particulate matter contained in his neoplastic cells will, it is hoped, lead to the formation of specific antibodies that, in conjunction with the complement of the host serum, will permit recognition and adsorption of antibody that will result in phagocytosis of the tumor debris by leucocytes. The question will be, then, as to whether or not resistant neoplastic cells will be produced during the early phases of treatment which will represent a new generation of neoplastic cells refractory to further use of passive immunity technics aimed at their elimination.

Concurrently, over the past eight years, a study has been made in cooperation with the New Jersey State Department of Public Health under the administrative direction of Dr. Oscar Sussman to determine the occurrence of leukemia in the families of owners of dogs suffering from malignant lymphoma. This was based upon an original experimental design whose object it was to determine the public health significance of this human-canine relationship.[34] Coincidental evidence of malignant lymphoma occurring in a pet dog following the development of lymphoma in a boy in the family eight months prior to the appearance of the disease in the pet dog has been reported.[38] The comprehensive survey of the epizootiologic characteristics of feline and canine leukemia and lymphoma undertaken by Dorn, *et ux,* in two counties in California for a two and a half year period warrants reporting here. The analysis of 93 canine and 106 feline leukemia and lymphoma cases through a registry was maintained at a constant rate of accession and without seasonal variation. Adjusting for the portion of the animal population not given veterinary attention, these investigators found the average annual lymphoma incidence was 24.0 per one hundred thousand dogs and 41.6 per one hundred thousand cats. The canine rates reported were thus higher than any previously ascertained and both the canine and feline incidence rates were higher than the combined leukemia-lymphoma annual incidence rate for man, the feline rates being about two and a half times higher than the incidence rate in man subject to adjustment for the differences in years of life expectancy or lifetime risk. In man the leukemia incidence rate according to national figures is 7.4 per one hundred thousand and the lymphoma incidence rate is 9 per one hundred thousand, totaling 16.4 per one hundred thousand human population. Sex differences were not significant in the dog. The incidence of approximately half visceral forms and half peripheral forms of lymphatic involvement in terms of site of origin supports existing data.[28,33,36,37]

Current research in canine cancer is expanding into many areas of investigation beyond that of empirical chemotherapy, and meaningful statistical data are accumulating, though at a rate hardly comparable to man's experience with neoplastic diseases. This is, however, stated in the light of our knowledge that the dog, in the majority of forms, malignant lymphoma being a notable exception, does not develop cancer as frequently as does man. Statistically, significant findings are not obviated by this, however, and definite contributions to the field of comparative medicine will accrue through critically evaluated therapeutic experience in the dog and cat.

Schools of Veterinary Medicine, internationally organized, with the cooperation of private veterinary agencies and practices, will continue to gather important data.

The chemotherapy of canine tumors is a field in relative infancy for veterinarians and, to date, serves the practitioners more in a palliative than curative sense in terms of end result. Random cases undergo periods of remission when treated with today's armamentarium of anticancer drugs, but the final therapeutic moiety eludes us as those neoplastic cells unaffected by the drug used, in variable percentages, adapt themselves to the presence of the therapeutic agent and reproduce as a chemically selected tumor, now resistant to, but not dependent upon, the particular agent. In our experiences, the primary malignant neoplasm has even regressed totally, never to return, but telemetastases have implanted, have resisted the cytotoxic effect of the agent and have appeared clinically as a new mutant form of the neoplasm, refractory to the drug.

Where possible to apply it properly, surgical removal of a neoplasm remains the method of choice. Successful reversion of marked infiltrative tendencies in mammary tumor surgical recurrences may be anticipated from the use of alkylating agents prior to surgery, especially in neoplasms of the mammary glands.[60]

Currently, categories of chemotherapeutic agents available to the practitioner include:

1. Antimetabolites
   a. Folic acid antagonists
      1.′ Aminopterin
      2.′ Amethopterin (Methotrexate)
      3.′ Pyrimethamine (Daraprim)
   b. Purine and pyrimidine analogs
      1.′ 6-mercaptopurine
      2.′ 6-thioguanine
      3.′ 8-azuguanine

4.′ 5-fluorouracil
5.′ 6-azauracil; 6 azauridine
6.′ 1-aminocyclopentanecarboxylic acid
c. Specific amino-acid antagonists
1.′ L-asparaginase
2. Cytotoxins
a. Anti-tumor antibiotics
1.′ Actinomycins (-D)
2.′ Mitomycin C
3.′ Azaserine
4.′ Cycloheximide (Dihydro-E-73)
5.′ Streptonigrin
b. Alkylating agents
1.′ Nitrogen mustard
2.′ Melphalan
3.′ Sarcolysin
4.′ Triethylene melamine TEM
5.′ Phosphoramides
a.′ TEPA
b.′ Thio-TEPA
c.′ MEPA
6.′ Cyclophosphamide
c. Others
1.′ Urethane
2.′ Colchicine
3.′ Vincristine
4.′ Vinblastine
5.′ Radioactive isotopes
6.′ Auxins
3. Endocrine Agents
a. Cortisone
b. Prednisone
c. Prednisolone
d. Triamcinalone
e. Testosterone propionate
f. 8 delta-testalactone
g. Diethylstilbestrol
h. ECP, cyclopentylpropionate ester
4. Others
a. Dimethylacetamide
b. N-desacetylthiocolchicine

c. Bayer E39, bis-ethylene-iminobenzoquinone
d. Y^{90} DTPA
e. Human placental extract, Livingstone
f. Coley's toxins

Most of those listed have given rise to negative data in terms of therapeutic success; none has proved successful in a significant number of cases. The hope lies in the thousands of possible anti-tumor compounds being screened by various agencies each year. No organic moiety of our world-wide ecology will escape investigation and new compounds are being synthesized in laboratories by the thousands.

Recently, a milestone was passed in the reports of the effectiveness of anti-tumor therapy with an enzyme, L-asparaginase, extracted from cultures of *E. coli.* Therapeutic response was disappointing, for while regressions of tumors were gained in a high percentage of the cases receiving 500,000 units twice a day for five days, the remission periods were of variable duration, resumption of tumor growth being common to each case.[57,58] This may be a matter of dosage inadequacy. The salient feature of the drug's use is to be found in it being the first instance in which a known fundamental biochemical difference exists between the normal and the neoplastic cell, the latter's inability to synthesize L-asparagine when its host-based source is removed by the circulating catabolic enzyme, L-asparaginase, and the resultant death of the tumor cell from what is, basically, an induced nutritional deficiency of an essential amino acid. The normal cell synthesizes asparagine and is not affected by the induced deficiency.

Owing to current highly expensive culture and extraction methods, however, when coupled with the temporary effects gained by its therapeutic use, L-asparaginase appears to be more notable as a unique event than attainable as a therapeutic drug.

Experience is being accumulated in the use of various agents. Rather than dwell in length on their indicated uses, which might prejudice the veterinary practitioner in his choice of drug, only those uses are catalogued in which regressions of the symptom represented by the tumor led to variable remission periods. Otherwise, a reported dosage level is listed that at least conveys the comfort to the practitioner of knowing that survival of the patient was an appreciative feature of the therapeutic course, especially in the cytotoxic groups, wherein exceedingly powerful pharmacological actions are concurrent with anti-tumor effects.

As to general factors governing use of such agents successfully, certain steps should be undertaken in sequence prior to instituting therapy:

1. Establish the baseline values in terms of the status of the individual patient. Most frequently the hematogram is concerned with establishing safe dose levels, for proliferative centers of cell reproduction are most frequently responsive to the systemic action of an anti-cancer drug, and this is usually manifested by a general or selective decrease in the circulatory leucocyte population. It has been our experience that the erythropoietic system of the dog is generally resistant to toxic effects.[61] This is a species difference from the human patient, in whom no such advantage appears to be the rule.

White blood cell counts should be determined at least twice, a few days apart, prior to therapy, as should enzyme activity measurements or other biochemical evaluations of the presenting state of the case.

2. Initiation of therapy in the presence of leucocytosis is warranted. It may represent either the effect of the tumor upon the host or secondary infectious processes concerned with the presence of the neoplasm, such as ulceration of the tumor. In any case, the outlook is for an anticipated drop toward or below normal. In our hands, a white blood cell count of lower than 5,000 cells per cubic millimeter in the dog is cause for concern, owing to the possibility of an otherwise slight infection becoming a source of an overwhelming infectious episode owing to the lack of proper host defenses.

3. Initiation of therapy in the face of a leucopenia is worthy of great caution. It may represent onset of a secondary viral infection in the host, or, it may represent an advanced depletion of somatic reverses by the neoplastic disease. In such a case the clinician may rightly ponder the question: Will institution of chemotherapy at such an advanced stage merely duplicate the depletion already presented in this case? Will it not serve to hasten, rather than retard, an unsuccessful termination?

In our experience, the decision is usually to proceed. First, such depletion favors the chemotherapeutic response. A markedly depleted host may well contain a depleted tumor, one in its most susceptible state. Some regressions are dramatic under these circumstances, while a patient presented in good to fair clinical condition may make a less favorable response in the presence of similar neoplastic disease. This is especially true in malignant lymphoma, but usually not following the occurrence of ocular hemorrhages in this disease. Determination of a complicating viral disease, such as distemper occurring in an older depleted cancer patient, is possible. Treatment should not be undertaken in such a case.

4. Reverse, when possible, the catabolic effects of neoplasia on the

host by the institution of nutritional repletion in whatever form, from amino acid parenteral technics to feeding of therapeutic diets or commercial formulae supplemented with 1% methionine (dry weight basis). Do not use vitamins indiscriminately if the patient is not eating well, for these will serve the depleting mechanism that is operative, not the repletion of the host, who will be unable to accomplish repletion without extrinsic supplies of needed nutrients to rebuild his body pools of protein, especially.

Frequently, however, the clinician is faced with a dog that has no desire to eat, so typical of advanced nutritional depletion in the canine, and whose separation from its owner further complicates the situation. Under the circumstances we ask the owner to attend the case daily, and to ply the patient with favorite viands, in an effort to reverse this otherwise prejudicial situation.

5. There are still circumstances under which euthanatasia is the ultimate expression of affection and compassion for intractable disease states.

During the course of chemotherapy the investigator must watch for signs of toxicity peculiar to the individual case, *i.e.*, sudden hemorrhage, evidence of arthritis, dyspnea from lysis of pulmonary metastases and accompanying edema and congestion, vomiting, hematuria, convulsions, melena, tenesmus, deafness, blindness, alopecia, etc. Withdrawal of the agent is logical, as is the initiation of corrective therapy.

In the leucopenic phase, if such is attained, close watch for flaring infections of local and focal origin must be maintained and corrective treatment undertaken, for such a patient is in an extremely susceptible state, and may be overwhelmed by seemingly minor episodes of infectious challenge, *i.e.*, otitis externa, corneal ulcer, decubita, etc.

Renal function is of prime concern for in chemotherapy, unlike the situation in surgical ablation, removal of the lysed portions of tumor debris must proceed, as in any situation regarding protein catabolism, through the kidneys via the liver. A depleted host has difficulty handling this function at best. A dehydrated, depleted host has little chance to avoid final endogenous toxemia, for these end-products of tumor lysis accumulating in his body fluids are certainly lethally toxic. Fluid therapy is a requisite frequently during the course of therapy. It is mandatory during periods of active tumor regression when a state of diuresis is highly desirable, at times in conjunction with drastic antihistamine therapy.[61]

Remission periods, whatever level of tumor regression attained, are usually signalled by return of appetite, often ravenous, and a well-balanced diet, fortified with vitamins, especially of the B-group, is indicated and well-received. Dramatic repletion and accompanying psychosomatic return to normalcy are rewarding experiences to the successful clinician and owner.

Cases terminating fatally should be necropsied and, at least, a gross pathology report should be recorded with the institution or the drug's manufacturer in the interest of accumulating data on therapeutic and/or toxic manifestations of the drug.

The armamentarium presently available to the veterinary practitioner includes the following. Some are investigational and suitable records must be kept for federal and company analysis; others are available commercially:

1. Cyclophosphamide—used in doses of 2.5 mg/kg body weight/day, orally, in mammary neoplasms and others. Or, in regimens of 7–10 mg/*pound* body weight a day for 1 week, or 5 mg/kg a day for 8 days, then 2 mg/kg for 5 days, or 3 mg/kg for 9 days, all orally administered. In malignant lymphoma the dose may be necessarily lowered to 1 mg/*pound* per day if massive tumor regression occurs. Osteosarcoma did not respond. Transmissible venereal tumor did not respond.

2. Methotrexate—used in doses of 0.20–0.70 mg/kg, has been used to good effect in acute leukemia of children, choriocarcinoma and embryonal carcinoma of the testis, combined with actinomycin and chlorambucil in the adult human. Citrovorum factor administered concurrently aids in restriction of toxic effects.

3. Vinblastine—care is necessary to insure against leakage into the perivascular tissues because of its highly irritating property. In malignant lymphoma 0.10–0.4 mg/kg once a week, intravenously.

4. TEM—triethylene melamine in doses of 0.3 mg/kg/day intravenously for 3 days with a 4 day interval between series. Useful in malignant lymphoma for one or two remissions, then falls drastically in therapeutic effectiveness, requiring higher and toxic doses. Useful in melanoma, squamous cell carcinoma, temporarily.

5. Thio TEPA—useful at doses of 0.5 mg/kg intravenously for 3 days with a 4 day interval between series. Useful in various neoplastic situations, may be combined with testosterone propionate (4–15 mg/day for 5 days) in treating mammary cancer. Useful in perianal adenoma in conjunction with ECP (1–2 mg i.m.monthly).

6. 6-mercaptopurine—used in malignant lymphoma in doses of 1 mg/*pound* per day, orally, in one dose to effect.

7. Prednisone—effective in malignant lymphoma in doses of 1–2 mg/*pound* per day orally and in combination with a protective broad spectrum antibiotic. May be used up to 50 mg/day orally in treating mastocytoma. Is more desirable than cortisone, since it produces no electrolyte imbalance. May be used for protracted periods of time.

Agents of interest alone, combined, or in sequential combinations, in addition to those mentioned above are: 5 fluorouracil, vincristine, chlorambucil, magnesium sulfate, sodium thiosulfate, vitamin E, methylene blue (2%), Myleran, actinomycin D, Bayer E39 and Y^{90} DTPA. MEPA should be evaluated again.

References

CANINE TUMOR SURVEYS

1. Barron, C. N., Saunders, L. A., Seibold, H. R., and Heath, M. K., *Am. J. Vet. Research,* **24,** 1263–1269 (1963).
2. Bloom, F., *Arch. Path.,* **33,** 661–676 (1942).
3. Brodey, Robert S., *Am. J. Vet. Research,* **21,** 787–812 (1960).
4. Brodey, R. S., Sauer, R. M., and Medway, W., *J.A.V.M.A.,* **143,** 471–495 (1963).
5. Brodey, R. S., and Craig, P. H., *J.A.V.M.A.,* **147,** 1628–1643 (1965).
6. Brodey, R. S., Reid, C. F., and Sauer, R. M., *J.A.V.M.A.,* **148,** 29–43 (1966).
7. Cotchin, E., *Brit. Vet. J.,* **109,** 248–257 (1953).
8. Cotchin, E., *Vet. Res.,* **66,** 879–885 (1954).
9. Cotchin, E., *J. Comp. Path. & Therap.,* **65,** 115–129 (1955).
10. Fidler, I. J., and Brodey, R. S., *J.A.V.M.A.,* **151,** 710–715 (1967).
11. Fowler, E. H., Wilson, G. P., III, Roenigk, W. J., and Koestner, A., *J.A.V.M.A.,* **149,** 281–285 (1966).
12. Krook, L., *Acta Microbiol. Scand.,* **35,** 407–422 (1954).
13. McGrath, J. T., "Neurologic Examination of the Dog," Lea & Febiger, Philadelphia, Pa., 1960.
14. Moulton, J. E., "Tumors in Domestic Animals," Univ. of California Press, Berkeley and Los Angeles, Calif., 1961.
15. Mulligan, R. M., "Neoplasms of the Dog," Williams and Wilkins Co., Baltimore, Md., 1949.
16. Mulligan, R. M., *Am. J. Path.,* **25,** 339–355 (1949).
17. Mulligan, R. M., *Am. J. Vet. Research,* **22,** 345–351 (1961).
18. Nielsen, S. W., Schroder, J. D., and Smith, D. L. T., *J.A.V.M.A.,* **124,** 28–35 (1954).
19. Nielsen, S. W., and Cole, C. R., *Am. J. Vet. Research,* **19,** 417–432 (1958).
20. Nielsen, S. W., and Hovara, A., *Am. J. Vet. Research,* **21,** 813–830 (1960).
21. Nielsen, S. W., and Cole, C. R., *Am. J. Vet. Research,* **21,** 931–948 (1960).
22. Nielsen, S. W., and Cole, C. R., *Am. J. Vet. Research,* **22,** 663–672 (1961).
23. Owen, L. N., and Stevenson, D. E., *Res. Vet. Sci.,* **2,** 117–129 (1961).

24. Ribelin, W. E., and Bailey, W. S., *Cancer,* **11,** 1242–1246 (1958).
25. Saunders, L. S., and Barron, C. N., *Cancer Res.,* **18,** 234–245 (1958).
26. Schlotthauer, C. F., and Millar, J. A. S., *J.A.V.M.A.,* **119,** 442 (1951).
27. Schlotthauer, C. F., and Millar, J. A. S., "Tumors in Dogs," Symposium on Newer Knowledge About Dogs, Gaines Dog Research Center, Kankakee, Ill., 1955.
28. Smith, H. A., and Jones, T. C., "Veterinary Pathology," Lea & Febiger, Philadelphia, Pa., 1961.
29. Stubbs, E. L., and Furth, J., *Am. J. Path.,* **10,** 275–286 (1934).
30. Stunzi, H., *Arch. Path.,* **54,** 423–429 (1952).
31. Thrasher, J. P., Ichinose, H., and Pitot, H. C., *Am. J. Vet. Research,* **24,** 808–817 (1963).
32. Willis, R. A., "The Pathology of Tumors," Butterworth, Inc., Washington, D. C., 1960.

EPIDEMIOLOGY

33. Backgren, W., *Acta Vet. Scand.,* **6,** Suppl. 1, 80 (1965).
34. Cohen, D., Booth, S., and Sussman, O., *Am. J. Vet. Research,* **20,** 1026–1031 (1959).
35. Cohen, D., Brodey, R. S., and Chen, S. M., *Am. J. Vet. Research,* **25,** 1776–1779 (1964).
36. Dorn, C. R., Taylor, D. O. N., Chaulk, L. E., and Hibbard, H. H., *Am. J. Pub. Health,* **56,** 254–265 (1965).
37. Dorn, C. R., Taylor, D. O. N., and Hibbard, H. H., *Am. J. Vet. Research,* **28,** 993–1001 (1967).
38. Drusin, L. M., Finkbeiner, J. A., McCoy, J. R., and Miller, D. G., *J.A.M.A.,* **196,** 219–222 (1966).
39. Mulligan, R. M., *Ann. N. Y. Acad. Sci.,* **108,** Art. 3, 642–690 (1963).
40. Priester, W. A., *J.A.V.M.A.,* **146,** 971–973 (1965).
41. Taylor, D. O. N., *J.A.V.M.A.,* **147,** 1492–1497 (1965).
42. Wagstaff, D. J., Goyings, L. S., and Langhans, A. F., *Am. J. Vet. Research,* **28,** 1479–1482 (1967).
43. Whipple, H. E., Ed., *Ann. N. Y. Acad. Sci.,* **108,** Art. 3, 617–1326 (1963).

EXPERIMENTAL

44. Adams, E. W., Moore, E. G., and Carter, L. P., *Am. J. Vet. Res.,* **29,** 1241–1244, 1968.
45. Allison, J. B., Wannemacher, R. W., Jr., Russell, T., and McCoy, J. R., *Cancer Research,* **18,** 394–397 (1958).
46. Bloom, F., and Meyer, L., *Am. J. Path.,* **21,** 683–715 (1945).
47. Herman, L. H., Slaughter, L. J., and Martin, D. P., *J.A.V.M.A.,* **151,** 1308–1310 (1967).
48. Kasza, L., *Am. J. Vet. Research,* **25,** 1178–1185 (1964).
49. McBride, N. L., "A Canine Lymphoma: Possible Etiologic and Immunologic Significance to Burkett's Lvmphoma of Man," *Proc. A.A.H.A.,* 147–150 (1967).
50. McCoy, J. R., *Ann. N. Y. Acad. Sci.,* **76,** Art. 3, 850–854 (1958).

51. McCoy, J. R., "Progress Report: Studies on the Host-Tumor Relationship in Canine Leucosis, Malignant Lymphoma. Comparative Leukemia Research," Pergamon Press, Oxford and New York, 1965.
52. Moier, H., *Zentralbl. Vet.-Med.,* **4,** 633–688 (1957).
53. Prier, J. E., Schaffer, B., and Skelley, J. F., *Nature,* **202,** 611–613 (1964).
54. Prier, J. E., Boszel, J., Craig, P. M., and Cain, D. F., *Am. J. Vet. Research,* **28,** 183–189 (1967).
55. Rosanoff, E. I., *Proc. Soc. Exp. Biol. and Med.,* **100,** 384–387 (1959).
56. Squire, R. A., *Am. J. Vet. Research,* **26,** 97–107 (1965).
57. Wheeler, G. P., *Cancer Res.,* **22,** 651–688 (1962).

CLINICAL CHEMOTHERAPY

58. Brodey, R. S., "Treatment of Canine Lymphoma with L-Asparaginase," *Proc. A.V.M.A.,* **151** (Dallas) (1967).
59. Kirk, R. W., "Current Veterinary Therapy: 1966–1967," W. Saunders Co., Philadelphia, Pa., 1966.
60. McCoy, J. R., Crossley, M. L., and Allison, J. B., *Proc. Am. Assoc. Cancer Research,* **1,** 37 (1953).
61. McCoy, J. R., Allison, J. B., Crossley, M. L., and Wannemacher, R. W., *Am. J. Vet. Research,* **17,** 90–97 (1956).
62. McCoy, J. R., *Federation Proc.,* **19,** 396 (1960).
63. Owen, L. N., *J. Small Animal Practice,* **1,** 116–129 (1960).
64. Owen, L. N., *Brit. J. Cancer,* **18,** 407–418 (1964).
65. "Progress in Canine Practice, Infections, Infestations and Neoplasms," Pt. 2, 1, American Veterinary Publications, Santa Barbara, Calif., 1967.
66. Sodikoff, C. H., *J.A.V.M.A.,* **149,** 1735–1740 (1966).

James Boyd Allison—An Appreciation

John R. McCoy, V.M.D.

ALLISON, James Boyd, educator; b. Punxsutawney, Pa., Jan. 28, 1901; s. James Boyd and Elizabeth (Robinson) A.; B.S., Pa. State U., 1923; Ph.D., Ia. State Coll., 1927; m. Dorothy Lewis, Dec. 31, 1926; children—Marjorie (wife of Dr. Arthur Marshall Smith), Dorothy (Mrs. D. Jeffrey Bostock). Asst. Ia. State Coll., 1923-26, indsl. fellow, 1926-27; instr. biochemistry Rutgers U., 1927-29, asst. prof., 1929-41, asso. prof., 1941-45, prof. physiology and biochemistry, 1945-—, dir. Bur. Biol. Research, 1944-59; director of the research council, 1959-—; lecturer physiology College Phys. and Surg., Columbia, 1943-45. Liaison, sci. adv. bd. Q.M. Food and Container Inst., 1948, chmn. subcom. nutrition Q.M. Inst., 1954-—; chmn. com. amino acids Food and Nutrition Bd., 1954-—; com. protein requirements FAO, 1955; mem. food nutrition bd. NRC, 1956; mem.-at-large Council of Gordon Research Conf., 1956; tech. adv. com. Instituto de Nutricion de Centro America y Panama, 1956. Mem. bd. health, Highland Park, N.J., 1955. Trustee Middlesex Gen. Hosp., New Brunswick, N.J. Mem. Am. Assn. Cancer Research, Am. Chem. Soc., A.A.A.S., Soc. Exptl. Biol. and Medicine, N.Y. Acad. Sci. (pres. 1962), Am. Inst. Chemists (chmn. N.J., 1953-54), Am. Society Biological Chemists, American Institute Nutrition (treasurer 1959-62), Sigma Xi, Phi Beta Kappa, Phi Lambda Upsilon, Alpha Zeta, Phi Kappa Phi, Gamma Sigma Delta. Presbyn. (elder, trustee). Rotarian (pres. New Brunswick 1953-54). Mem. editorial bd. Jour. Nutrition, 1952-56. Home: 409 Grant Av., Highland Park, N.J. Office: Bureau of Biological Research, Rutgers U., New Brunswick, N.J.

—*Who's Who in America* 1964–1965

Well, there you have them, the credentials of a career marked by perseverance, integrity, deep religious commitment, profound philosophical contemplation, and technical genius; of scholarship, honors and citations; of missions of humaneness, telescopically compiled as a measure of one human life, one career, spanning close to four decades of ever-increasing contributions to the welfare of man; the chronicle of achievements so many and varied that, in reflection, you might envision a mighty and lofty genius of thunderous voice and intensive, almost militaristic drive bent upon a stern course of scientific production—damn the torpedoes, full speed ahead!

Not so!

He was a simple man. He was a quiet man. He was a child of God. His commitment was deep and direct. He was free of hypocritical religiosity. He tithed to the Presbyterian Church. He served as an Elder. He was a confidant of the minister. He was a major force in the construction of its magnificent church in New Brunswick.

You might prognosticate that such attributes, compounded, would lead to self-righteousness, to self-satisfaction, to condescension, and to pontifications long of text and warm of wind.

Not so!

He was a family man, beloved. The devotion of his wonderful wife, Dorothy, and the happy, affectionate warmth that pervaded the lives of their daughters, Marge and Dottie, well prepared these young women for the happy homes that they in turn have established; both wonderful mothers, having been so deeply loved and so wisely reared by Dot and Jim.

You might opine, then, that such a group so self-sufficient, so protected, so deeply united in family love would be headed by a man of introspection, of introversion to the point of being innately spartan, provincial.

Not so!

It was a hot sultry July day many decades ago. Punxsutawney, in a shallow valley west of Pittsburgh, was a lovely, typical small Pennsylvania town initially agricultural; in later years discovery of subsurface coal pockets had led to small family mining operations, their proceeds supplementing family incomes, enough of a source of affluence to make Punxsutawney inviting to traveling entertainers who eked out a precarious living during the summer season.

One of them was heralded as "Hugo, the Human Fly" and a fascinated Punxsutawney populace was held breathless as Hugo flitted up steeples, bank buildings, and City Hall, flirting with death as he climbed barehanded up sheer brick and wood walls.

On the ground a certain youngster, that Allison boy Jim, was taking in Hugo's act. Garsh! If that wasn't the best way to see the town? Or to look out over the valley whenever a fellow wanted to? Late that night, a neighbor coming home from a meeting stopped in her tracks. Could it be that someone was robbing the Allisons? Young Jim's father owned the local hardware store and his grandfather was a physician, still practicing in his seventies, so a relative affluence blessed their lot in life.

Yessirree, there *was* someone climbing slowly right up the side of the Allison house. But who robs a house in a night shirt? Small figure, too. Or, was that a girl robber in a loose dress, bundled up around her middle that way? The neighbor stole closer for a better look up at the intruder in the dim gas lit night.

Not so! Garsh!!!!

There was the time, too, when that wickedly hot month of July was casting its summery spell of the Fourth over Jim, inducing an underlying adventurousness which, coupled with his Scotch-Irish ancestry, had persuaded him and another lad to sneak through a special board in the fairgrounds fence that gave entry to the track a special flavor and saved a quarter, that is if you weren't seen by the constable. They were. In fact, he appeared to have been waiting for that particular board to creak open, that's how close he was. Well, sir, J. B. sized the situation up fast, because the constable had waited for them to get *in* a little before he made *his* rush. Only one way to go, across the race track! Away they went, right into the middle of a motorcycle race! Jim's neighbors watched from the stands with horror as the two boys ran directly onto the track just as the cycles rounded the far turn and roared down the straightaway at a fierce thirty. They shrieked in terror as the bikes veered off course in a desperate but futile effort to avoid striking the youngsters. Both boys, unconscious, were taken to a Punxsutawney funeral parlor. Jim's physician grandfather was called. Concussion and a broken shoulder for Jim, other boy about the same—off to the hospital with them both!

The shoulder and collar bone plagued Jim a good deal from then on, especially during the cold, wet winters in the East. You'd think that experience might have caused some restriction on future activities, especially athletics.

Not so!

Cross country runner, cyclist, and wrestler were the accolades gained at Penn State during his student years.

As a matter of fact, it wasn't supposed to *be* his institution, Penn State. Jim's mother was a graduate of a small college in Grove City and he was slated to matriculate there that fall and become a mathematician. An all-round student with the highest standing in his class, he had passed up the chance to be valedictorian because a girl who was second had cried so hard at not gaining the spotlight that Jim insisted she give the graduation address. He might well have been expected to breeze along a smooth and disciplined course, passively geometrizing and trigonometrizing around classes at Grove City, just following in the family footsteps.

Not so!

Some of Jim's cycling friends had been accepted at Penn State. So had Dorothy Lewis, the pretty girl Jim had dated during high school days. It was 1919, the war was over, life seemed carefree and the future for an enlightened mankind seemed secure.

So Jim went off to Penn State with his buddies while *they* matriculated, just for the ride, and to see Dotty in her new role as a first-year home economics student. In fact, he liked it so much that he attended classes with them, a utopian situation that persisted until Thanksgiving. At that time a keenly observant registrar, who had counted one head too many amongst the freshman, called him on the carpet for an interview that must have gone pretty well, for Jim was told to return after the holiday with his high school transcript, his family's permission, and money. So he did, and graduated Phi Kappa Phi, *cum laude* in 1923. Phi Beta Kappa was to follow at Rutgers, later.

Dorothy graduated in 1923, too, and was teaching home economics in Corsica, Pennsylvania. Jim accepted a teaching offer at Iowa State and drove an old Model T Ford to and from Corsica until, its waning lifespan evident, he and a friend drove to Nebraska and sold it for $50 without shedding a tear.

They were married, Dot and Jim, about four years after their respective Penn State graduations. They settled down to a career in biochemistry for Jim, and applied home economics for Dot.

How could anyone with such a history become anything but a humanist? His gentleness, his modesty and humility, his deep respect for the Golden Rule and his good will toward all manner of man and beast, his unwavering commitment to placing human values first in selection of personnel, in the handling of disciplinary problems and in the selection of the young people to follow the careers he respected most of all, *in academe,* were among his finest traits. Employees and colleagues were truly fellow men and responded with intense affection and cooperation that gave the Bureau of Biological Research a special *esprit de corps* in the years of his directorship. The international reputation it gained was a tribute to his genius. Conferences were attended by a representation of the best brains in comparative biological research, especially in protein metabolism study, and their respect for him as a scientist was matched only by their amazement at the Bureau's physical plant, 2800 square feet of converted skating rink over an antiquated garage. *Eureka!*

In 1961 a new physical plant was completed with federal and state funds, supplemented by private and industrial grants and gifts. In November the new Bureau was occupied by a group of scientists still dedicated to its *modus operandi:* unhindered and unquestioned basic biological research of whatever type interested the scientist elected to membership. Membership called for one quarter of research time to be placed at the call of any fellow member in need of one's scientific aid from any of the disciplines represented by the Bureau's roster of research personnel: biochemistry, organic chemistry, clinical and anatomical pathology, endocrinology, parasitology, microbiology, experimental chemotherapy, pharmacology, physiology, enzymology, botany, zoology, immunology, serology (including the unique Serological Museum), statistical analysis, information retrieval and computer technics, to name some. The interdisciplinary scientific roster permitted formation of research programs in nutrition, cardiovascular and tropical diseases, cancer, muscular dystrophy, dysfunctions of reproduction, parasitic diseases, all characterized by a special flavor of social significance, studies aimed at bettering the welfare of mankind.

Jim Allison was a member of the University President's cabinet, deeply involved in developing a modern, excellent Rutgers. He dedicated himself through his humanitarian genius to produce *so* much that was truly constructive of a better world; this was his monumental contribution.

He chose to ignore the warning signs of oncoming relentless disease in himself; he faced the needed but delayed abdominal surgery stoically, but exhausted, his vitality drained by unsuccessful laser therapy of a detached retina that resulted in constant pain and frequent accidents owing to restricted vision. Still, the humility, the friendly kindness, even a faint glimmer of the old mischievousness sputtered through, but less and less apparently in those last few months in 1964.

Jim had tripped over his briefcase a few weeks before, in the spring. His sling, supporting the dislocated elbow that followed, was chafing his neck. We changed it one day for a softer dressing and sprinkled some corn starch on his irritated neck as he stood under the fluorescent light of the Bureau operating room.

The signs of an old enemy with which we had done battle without success for fifteen years were shockingly presented, signs of an inner depletion and an increasing physical and nervous exhaustion, broadcast in the low-keyed tiredness of his voice as he spoke of the future and this new surgical obstacle to a fall trip to Central America to observe the results of a United Nations project for correction of childhood disease of nutritional origin.

He hoped to be up and around after the surgery in time to make the Central America trip by late fall, free of this most recent invasion of his scientific privacy, so that his international research projects and the beloved Bureau could receive his refreshed, renewed attention.

Not so.

Index

A

ABO blood group, red cell receptor sites and, 89
Abraham, S., 256
Actidione, enzyme responses to estrogen and, 255
Actinomycin and enzyme activity
in kidney cells, 184
response to estrogen and, 255
thymidine kinase activity and, 163, 171
Actomyosin in muscle contraction, 116–18
Adelstein, S. J., 179
Adenosine triphosphate, ATP-dependent calcium transport system, 112
Adrenal glands
malnutrition and, 38
protein repletion and adrenalectomy, 58–59
Age
basal metabolic rate and, 16
enzyme activity and, 103–4
liver tryphotophan activity and, 104–6
Albumin
amino acid incorporation into, 78–80
malnutrition and turnover rate, 42–43
Alkaline phosphatase, thymidine phosphatase activity and, 191–92
Alkylating agents in neoplasm therapy, 264–72, 275
Alleyne, G. A. O., 38
Allison, J. B., 28, 236, 264
an appreciation, 283–88
early studies, 97
on physiologic states, 97, 108–9
protein reserve concept, 121–22, 135–36
Altman, C., 223
Amethopterin (Methotrexate) in chemotherapy, 274, 279
Amino acid(s)
deficiency: cellular content of tissue proteins and, 125; liver cells and, 133
free: nature of pool, 246–49; variations in Newcastle disease virus infections, 242 ff
metabolism: albumin turnover rates in malnutrition, 42–43; hypophysectomy and, 54; incorporation in mice, 78–80, 81; protein depletion and starvation and, 29, 41, 131–32
plasma, in malnutrition, 40–41
translational phase of protein synthesis and, 125
Anoxia, dietary protein and thyroid hormone and, 64, 65
Anorexia, dietary protein and, 4
Antimetabolites in neoplasm chemotherapy, 274–75
Antiserum pattern against liver antigens, 84, 85
Askew, J., 156, 173
Asofsky, R., 77, 78
L-Asparaginase in neoplasm chemotherapy, 275, 276
Assay systems
for thymidine kinase, 208–11

Assay systems (cont.)
for TMP kinase, 212
Atkinson, B., 159, 171, 174
Avoidance behavior, nutrition and, 9–11, 22–23
Azathymine, conversion to ribonucleosides and deoxyribonucleosides, 204–5

B

Baker, R. B., 149
Bulbrook, R. D., 258
Baldwin, R. L., 256
Bang, F., 87
Bar pressing tests of learning behavior and malnutrition, 19–22
Basak, M. N., 127
Baserga, R., 166
Basilico, C., 187
Bednarik, T., 78
Behavior, protein-calorie malnutrition and abnormal, 3–26
Behki, R. M., 183
Bellow, L. J., 185
Beltz, R. E., 173, 182, 183
Bernard, C., 44
Bessman, M. J., 171, 185, 186
Bianchi, P. A., 148, 151, 152, 155, 160, 182, 191
Birgus latro, hemagglutinins of, 87–93
Bisset, K. A., 87
Biswas, C., 197, 199
Blood
erythrocyte receptor sites, 88–89
evaluation for neoplasm chemotherapy, 277
invertebrate hemagglutinins, 87–93
response to MEPA chemotherapy, 265, 266, 267
See also Plasma
Body composition, protein deficiency and childhood, 31–36
Body weight
in comparison of normal vs. malnourished children, 32
food consumption and, 14, 16
height ratio to, in marasmus vs. kwashiorkor, 33
malnutrition and, 26, 30–31
nursing restrictions and, 7
postnatal changes in infant, 31
See also Growth
Bojarski, T. B., 148
Bollum, F. J., 181
Bone, protein/potassium ratio in, 34–35
Booth, A. B., 150
Boyd, W. C., 92
Brain
contribution to nitrogen excretion in malnutrition and stress, 123
malnutrition and development of, 5
protein and potassium malnutrition and, 29
protein/potassium ratio in, 34, 35
Bras, G., 36
Breast cancer
biochemistry in clinical, 258
chemotherapy of, 252, 256, 258–59
diagnosis, 251
incidence, 251
response to estrogens, 253–55, 256
surgical therapy, 251–52
survival rate, 251, 258
Breitman, T. R., 176, 197, 199, 200
Brennan, M. J., 259
Brent, T. P., 164
Bresnick, E., 146, 148, 150, 151, 152, 154, 155, 166, 167, 183
Brookes, V. J., 157, 179
Broun, R., 92
Bukovsky, J., 148, 153, 159, 189
Burk, D., 252
Butler, J. A. V., 164

C

Caddell, J., 36
Calaprice, N. I., 92
Calcium
ATP-dependent calcium transport system, 112
composition in muscle fibers, 115
muscle contraction regulation and, 111–19
Caloric intake, nitrogen retention and, 97–98
Cambria, A., 102
Campbell, D. H., 87
Campbell, P. N., 80

Cancer
mammary, study of experimental, 251–59
See also Breast cancer; Neoplasms
Canellakis, E. S., 183, 204
Cannon, W. D., Jr., 176, 197
Carbohydrate consumption, anorexia and, 4
Carcinogens, biochemical changes and exposure to, 259
Cardiovascular system in malnutrition, 36, 98
Carter, C. E., 160, 178
Casein requirements, 61
Cells
thymidine kinase localization in, 146
thymidine phosphorylation and control of proliferation, 141–212
See also under specific topic
Centifanto, Y., 187
Chang, L. O., 192
Chattaraj, A., 92
Chemotherapy
available agents for, 274–76, 279–80
of breast cancer, 252, 256, 258–59
factors in successful, 276–79
of neoplasms in dogs, 264–80
Chiancone, F. M., 104, 107
Chloramphenicol, thymidine kinase activity and, 163
Chloride, body, in normal vs. malnourished states, 31–36
Chow, B. F., 16
Circulation in malnutrition, 36
Cleaver, J. E., 142, 199
Cobalt cyanocobalamin tissue incorporation in starvation, 99–102
Cohen, E., 87, 88, 89, 90, 91
Cohen, S., 42
Cole, L. J., 179
Collagen, body, in normal vs. malnourished states, 33
Conditioned responses, diet and, 8–11
Conrad, H., 221
Contraction, calcium and muscle, 111–19
Cooper, R. A., 199, 200
Corticosterone, growth and, 53
Cortisone
growth and, 52, 53
thymidine phosphorylation and, 179
Constantin, T. T., 115
Cowley, J. J., 18
Crabs, hemagglutinins of horseshoe and coconut, 87–93
Crathorn, A. R., 164, 182
Cravioto, J., 30
Creatinine/hydroxyproline ratio in urine in malnutrition, 40, 41
Crossely, M. L., 264
Curiosity, malnutrition and animal, 12
Cushing, J. E., 87, 92
Cycloheximide, thymidine kinase activity and, 171
Cyclophosphamide, in chemotherapy, 275, 279
Cytochrome c reductase system, starvation and, 103, 104
Cytochrome oxidase
action of, 220
importance of, 220
kinetic behavior of soluble, 220–29
starvation and, 102
Cytoplasm, in protein synthesis in liver, 75–85
Cytoxins, for chemotherapy, 275

D

D'Angelo, S. A., 66
Davison, A. J., 228
Dean, R. F. A., 31
Degradation of thymidine phosphates, 189–93
DeOme, K. B., 259
Deoxynucleotidase, activity in rat liver and tumors, 191
Deoxynucleotide, control of synthesis in *E. coli,* 176–77
Deoxyribonucleic acid (DNA)
alkylating agents and hydrolysis of, 264
enzymes in metabolism of, in normal and regenerating liver, 182–83
malignancies and cellular, 270–72
synthesis: chain in plants, 162, 163; irradiation and, 194–95; reactions and feedback controls and, 143; in regenerating liver, 180–81; regulation of, 177–78; thymidine and, 198–99, 201–2; thymidine kinase activity and, 164–66; thy-

Deoxyribonucleic acid (DNA), synthesis (cont.)
midine triphosphate and, 142, 143, 199–201; TMP kinase and, 170–71
tissue: growth and muscle, 122; in Newcastle disease virus infections, 242 ff
Deoxyribonucleosides, conversion of thymine and azathymine to, 204–5
Deoxyribonucleotides, phosphorylation, in normal vs. regenerating liver, 183
Deoxyribose-5′-phosphate aldolase, action of, 203
Deoxyribosides, in animal tissues, 193–94
Deoxythymidine phosphotransferase, action of, 204
Deoxyuridine, toxicity of derivatives, 151
Development
hormones and fetal growth, 52
thymidine kinase and TMP kinase in embryonic tissues, 144, 145
thymidine phosphorylation in, 171–75
Diabetes
growth and: fetal, 52; postnatal, 53
protein metabolism and, 55–56
Diarrhea, in malnutrition, 37
Diet
conditioned responses and, 8–11
factors in food consumption, 12–17
food intake and nitrogen retention, 238–39
supplemental, in neoplasm chemotherapy, 277–78
toxicity of chemotherapeutic agents and, 266
Digestion, malnutrition and, 37
Dju, M. Y., 171
DNA, *See* Deoxyribonucleic acid (DNA)
DNA-viruses, kinase activity and infection with, 189
Dorn, C. R., 273
Drysdale, J. W., 123
Dubbs, D. R., 155, 186, 187
Duckworth, D. H., 186
Duspiva, F., 175

E

E. coli, kinase activity and phage infection, 185
Ebashi, S., 113
Ecdysone, thymidine phosphorylation and, 179
Edelman, G. M., 87, 88, 89, 90, 91
Edema
fatty liver and, 35, 39
of kwashiorkor, etiology of, 39
Eker, P., 166, 190
Electrolytes
intracellular, in normal vs. malnourished children, 31–36
in muscle fibers, 115
Embryonic tissues, activity of thymidine kinase and TMP kinase in, 144, 145
Emptions, malnutrition and exhibition in stress, 7–11
Endocrine system
malnutrition and, 37–38, 63–64
nutrition and, 61–70
See also specific endocrine gland and/or hormone
Enesco, M., 5
Enzymes
activity of, 253; aging and, 103–4; in canine neoplasm hosts, 268–69, 270; in clinical breast cancer, 258; intracellular, in malnutrition, 42
for neoplasm chemotherapy, 275, 276
protein depletion and liver, 123, 125
starvation and metabolic systems of, 102–9
thymidine and, 193
in thymidine triphosphate synthesis, 143–44
See also specific enzyme
Eriksson-Quensel, J. B., 89
Estradiol benzoate, sarcoma 180 and uterotrophic response to, 62–63
Estrogens
biochemical responses to, 253, 255
neoplasms and, 253–59
response to mammary gland and mammary tumors to, 253–55, 256
TMP activity and, 178–79
tryptophan pyrrolase activity and, 108

uterine response to: dietary protein and, 63, 64; sarcoma 180 and, 62–63
See also specific compound
Evans, H. M., 61
Excitability in stress, malnutrition and, 7–11
Exploration, malnutrition and spontaneous animal, 11–12

F

Farber, E., 259
Fat
body: in kwashiorkor vs. marasmus, 35; in normal vs. malnourished children, 31–36
dietary, effects of, 5
Fausto, N., 159, 180, 181
Feedback in inhibition of thymidine kinase by TTP, 152–56
Fetus, hormones and growth of, 52
Fiala, S., 189, 191
Finlayson, J. S., 89
Fisher, H., 127
Fitzpatrick, W. H., 236
Fleming, C., 88, 89
Fletcher, K., 39
p-Fluorophenylalanine, enzyme activity in kidney cells and, 184
Food consumption
body weight and, 14, 16
malnutrition and, 12–17
metabolic rate and, 16
Frajola, W. J., 221
Fraňková, S., 9, 11
Frearson, P. M., 187
Friar, W., 92
Frieden, E., 108
Friedkin, M., 199, 202
Fritts, M., 122
Fritzen, P., 192
Funabiki, R., 129

G

Gaito, R. A., 204
Ganoza, M. C., 81, 85
Garren, I., 54
Garrow, J. S., 42
Gebert, R. A., 184
Gentry, G. A., 198
Gitlin, D., 42
Globins
amino acid incorporation into, 78–80
protein depletion and, 125
Gonadal hormones
expression of influence of, 53
growth and, 52, 53
for neoplasm chemotherapy, 252, 275
neoplasms and, 253–59
thymidine kinase activity and, 178
thymidine phosphorylation and, 178–79
See also specific hormone
Goodford, P. J., 119
Gordon, A. H., 76
Gordon, E. E., 122
Grabar, P., 76
Graham, G. G., 31
Grav, H. J., 147, 152, 160
Gray, E. D., 150, 191
Greulich, R. C., 198
Griffiths, D. E., 223
Growth
cell number and size increase in, 122
control, TMP kinase activity control and, 168–69
feedback inhibition by TTP affecting, 156
H^3 thymidine incorporation in exponential phase, 199
hormones and, 51–59; fetal, 52; postnatal, 52–53
malnutrition and, 26, 30–31, 37
muscle changes with, 122
of neoplasms, 252–53; phosphatase activity and, 190; serum ribonuclease and, 269, 270
prenatal vs. postnatal, 51
TMP kinase activity in synchronous, 170–71

H

Hacker, B., 146, 157, 158, 159, 204
Hadden, D. R., 37
Halliday, D., 34
Hansen, J. D. L., 42
Hansen-Delkeskamp, E., 175
Haynes, G. R., 175

Heart, in malnutrition and starvation, 36, 98
Height
body weight ratio to, in marasmus vs. kwashiorkor, 33
as comparison basis of normal vs. malnourished states, 32
protein malnutrition and, 30–31
Hemagglutinins, nature and significance of invertebrate, 87–93
Hepatomas
growth of, 252–253
triphosphate synthesis in, 176
Herbert, V., 101, 102
Herrmann, H., 172
Herpes viruses, kinase activity and infection with, 187
Hiatt, H. H., 148
HMC mammary tumor, hormone therapy and biochemistry of, 257–58
Hochwald, G. M., 77
Holford, R. M., 199
Hormones
growth and, 51–59; fetal, 52; postnatal, 52–53
HMC mammary tumor biochemistry and, 257–58
for neoplasm chemotherapy, 252, 275
noeplasms and, 253–59
thymidine phosphorylation and, 178–79
See also specific gland and/or hormone
Hosts of canine neoplasms
enzyme activity in, 268–69, 270
protein reserves of, 267–69
tissue composition in, 270–72
Hotta, Y., 162
Huff, C. G., 87
Hunter, F. M., 101
Huntley, G. H., 195
Hurlbert, R. B., 156
Huseby, R. A., 255
Hutchinson, W. C., 234
Hydroxyproline/creatinine ratio in urine in malnutrition, 40, 41
Hypophysectomy
in breast cancer, 252
growth and, 52, 54–55
Hypophysis, *See* Pituitary gland

I

Immunoelectrophoresis
of cytoplasmic protein synthesis in liver, 75–85
described, 76
Immunologic factors
antiserum pattern against liver antigens, 84, 85
antitumor antibodies, 272–73
nitrogen retention and, 237–38, 239
Infantile and childhood malnutrition, 28–44
body composition and, 31–36
body size and shape and, 30–31
clinical study of, 28–29
function disturbances in, 36–38
mortality from, 29
See also Kwashiorkor; Marasmus; and specific topic
Infection
dietary protein/fat ratio and susceptibility to, 5
mouse liver incorporation systems and Staphylococcal, 81–82
Inhibition
of thymidine kinase, 161–62, 166; feedback, by TTP, 152–56; inhibitors other than TTP, 149–50
of TMP kinase, estrogens in, 178–79
Injury, *See* Trauma
Insulin
fetal growth and, 52
functions of, 55
hypophysectomy and metabolic response to, 54–55, 56
Invertebrates, hemagglutinins of, 87–93
Irradiation, *See* Radiation
Ives, D. H., 146, 148, 152, 155, 160, 171
Iwamoto, K., 194

J

Jacquez, J. A., 203
Johnson, H., 92
Johnson, R. A., 160, 171
Joklik, W. K., 152, 185

K

Kandatsee, M., 129
Kaplan, A. S., 186, 188

Karjala, R., 155
Kaufman, H. E., 187
Kidney
  Co^{57} cyanocobalamin incorporation into, 99
  Co^{60} B-12 incorporation into, 100–2
  cytochrome c reductase activity, 103, 104
  function: in malnutrition and starvation, 36–37; in neoplasm chemotherapy, 278
  thymidine kinase and TMP kinase induction in, 179–83, 184
  uninephrectomy, effects of, 179 ff
  vitamin B-12 and starvation, 98, 100–2
Kielley, R. K., 157, 158, 159, 176
Kinetic behavior of soluble cytochrome oxidase preparation, 220–29
King, E. L., 223
Kinases
  phage infection and bacterial, 185–89
  See also specific compound
Kit, S., 155, 186, 187
Klemperer, H. G., 175
Klose, E., 34
Kohn, H. I., 179
Kokolis, P., 160
Kornberg, A., 145, 147
Korner, A., 54
Kowalski, K., 122
Kundel, H. L., 194
Künkel, H. A., 198
Kwashiorkor
  body composition in, 32–36
  body weight/height ratio in, 33
  defined, 32
  described, 28
  development of, 4, 31
  etiology of edema of, 39
  functional disturbances in, 36–38
  heart in, 31
  mental state in, 39–40
  mortality from, 40–43
  skin changes in, 39

L

Lajtha, I. G., 195
Lát, J., 11
Latner, A. L., 101
Learning behavior
  interpretation of tests, 17–18
  malnutrition and animal, 17–23
Leathem, J. H., 64, 66
Leblond, C. P., 5
Lee, C. J., 16
Lemberg, R., 228
Lerner, L. J., 64, 66
Leucocytes
  H^3 thymidine incorporation into, 29
  pyrimidine metabolism in leukemic, 199–200
  thymidine nucleotides in, 200–1
Leukemia
  incidence of, 273
  pyrimidine metabolism in leucocytes in, 199–200
  thymidine nucleotide pools in leucocytes in, 200–1
Li, M. F., 88, 89
Lieberman, I., 184
Limulus polyphemus, hemagglutinins of, 87–93
Littlefield, J. W., 162, 187
Liver
  amino acid deficiency and, 125, 133
  cell free systems of mouse, 80–82
  Co^{57} cyanocobalamin incorporation into, 99
  Co^{60} B-12 incorporation into, 100–2
  contribution to nitrogen excretion in malnutrition, 123–25
  cytochrome c reductase activity in, 103, 104
  cytoplasmic protein synthesis in, 75–85
  deoxynucleotide activity in, 191
  diet and metabolism, 63
  edema and fatty, 35, 39
  free amino acid pool, 246 ff
  function in malnutrition, 37
  intracellular enzyme activity in malnutrition, 42
  in Newcastle disease virus infections, 240, 241–49
  nucleotide hydrolysis in cells in culture, 190–91
  regenerating: DNA synthesis in, 180–81; enzyme related to DNA metabolism in, 182–83; H^3 thymidine incorporation into, 192;

Liver, regenerating (cont.)
phosphorylation of deoxynucleotides in, 183; thymidine kinase and TMP kinase activity in, 144, 145
ribosomes: protein depletion and, 125; protein synthesis on, 85, 125
RNA changes with nitrogen loss, 134–35
subtotal hepatectomy, effects of, 179 ff
thymidine kinase of, 167, 179–83
thymidine phosphate degradation in, 189–90
TMP kinase of, 157, 158, 159, 179–83
tryptophan pyrrolase activity, 104–8
tumors: growth of, 252–53; triphosphate synthesis in, 176
vitamin B-12 and starvation, 98, 100–2
Liver tryptophan pyrrolase, starvation and, 104–8
Logsdon, D. F., 194
Looney, W. B., 192
Lorenson, M. G., 183
Lyman, K., 151
Lymphoma, incidence of, 273
Lyon, V. E., 91

M

Maass, H., 198
McAuslan, B. R., 152, 185
McDade, J. E., 88, 89, 90, 91
MacDonald, I., 251
McFarlane, A. S., 43
McGuinness, E. T., 221
McLaren, D. S., 31, 32
McLean, A. E. M., 37
Magnesium
body, in malnutrition, 35–36
in muscle fibers, 115
thymidine kinase activity and, 148
Main, R. K., 179
Maley, F., 156, 183
Maley, G. F., 156, 183
Malignancies, *See* Breast cancer; Cancer; Neoplasms
Malnutrition
clinical study, 28–29
function disturbances in, 36–38; endocrine system, 63–64, 64–70
growth and body weight and, 26
learning behavior and, 17–23
metabolic systems and, 97–109
nitrogen excretion in, 121–22, 133–36; tissue contribution to, 123–27
protein content of brain in, 29
thyroid response to, 64–70
See also Infantile and childhood malnutrition; Kwashiorkor; Marasmus; Starvation; and under specific topic
Mammary glands
HMC tumors, hormone therapy and, 257–58
metabolic activity of normal, 256
response to estrogens, 254, 256
See also Breast cancer
Mansley, G. E., 228
Marchalonis, J. J., 87, 88, 89, 90, 91
Marasmus
body composition in, 32–36
body weight/height ratio in, 33
defined, 32
development of, 6, 31
etiology of, 28
function disturbances in, 36–38
heart in, 31
incidence of, 28
Mayfield, E. D., Jr., 167
Mental state in kwashiorkor, 39–40
MEPA, 275
blood response to, 265, 266, 267
therapeutic vs. toxic dosage, 266
6-Mercaptopurine in chemotherapy, 274, 279
Metabolic rate
age and basal, 16
determination in children, 38
food consumption and, 16
Metabolism, physiologic state and, 97–109
Metal ions, thymidine kinase and, 148
Metcoff, J., 40, 41, 42
Methionine, thyroid response to thiouracil and, 69–70
Methotrexate (Amethopterin) in chemotherapy, 274, 279
Metz, C. B., 87
Microorganisms

control of deoxynucleotide synthesis in *E. coli,* 176–77
thymidine kinase and TMP kinase in, 145
Microsomes, liver
serum protein synthesis on, 82–85
staphylococcal infection and incorporation systems, 81–82
Migailo, J. A., 91
Mighorst, J. C. A., 65
Miller, O. N., 101
Miller, O. T., Jr., 146, 199
Minnaert, K., 222
Mitochondria, thymidine kinase and TMP kinase of, 146, 159
Monckeberg, F., 37
Montgomery, R. D., 35
Moore, E. C., 156
Morgan, W. S., 183
Morse, P. A., Jr., 145, 146, 148, 152, 160
Mortality
in infant malnutrition, 29
in kwashiorkor, 40–43
Mueller, G. C., 199
Mulligan, R. M., 263
Munro, H. N., 123, 125, 127, 133, 234
Muscle(s)
changes with growth, 122
composition in malnutrition, 41
contraction: calcium and, 111–19; muscle type and calcium influx in, 118; parameters related to excitation-contraction coupling, 116–18; spacial relations in, 111, 112
contribution to nitrogen excretion in malnutrition and stress, 123, 126–27
contribution of protein fraction to body protein loss, 135
electrolyte composition of fibers, 115
intracellular enzyme activity, 42
postnatal growth, 52
protein composition, 128–31
protein metabolism in: protein reserve concept and, 121–36; in starvation and protein depletion, 128–32
protein/potassium ratio in, 34, 35
RNA and nitrogen loss, 134–35
Mycostat, effects of addition to diet, 5
Myers, D. K., 182
Myofibrils
in muscle contraction, 112, 113
proteins of, 129–31
Myosin in muscle contraction, 113
Myxoviruses, Newcastle disease virus stress and protein metabolism, 234, 235 ff

N

Nakamuro, H., 203
Nelson, D. J., 160, 178
Nelson, M. M., 61, 66
Neoplasms
antitumor antibody, 272–73
chemotherapy of, 264–80; agents available for, 274–76, 279–80; factors in successful, 276–79
deoxynucleotidase activity in, 191
feedback effect of TTP and dCTP on, 153–54
growth, 252–53; phosphatase activity and, 190; serum ribonuclease and, 269, 270
hormones and, 253–59
host enzyme activities in, 268–69, 270
host protein reserve and, 267–69
incidence: canine, 262–63; human vs. animal, 273
liver, 176, 252–53
need for classification, 263–64
pre-neoplastic state studies, 259
surgical removal, 274
thymidine kinase activity in, 144, 166–67, 207
tissue constituents in, 270–72
TMP kinase in, 144, 157, 207
triphosphate synthesis in, 176
types of canine, 263
uterotrophic response to estradiol benzoate and, 62–63
Neter, E., 89
Neurones, malnutrition and central nervous system, 5
Newcastle disease virus
infections with: diurnal rhythms of tissue constituents in, 242–46; nitrogen retention in, 236–39; susceptibility to, 239–41; symptoms of, 235, 236

Newcastle disease virus (cont.)
viral stress and protein metabolism, 234, 235 ff
Nitrogen
excretion in malnutrition and stress, 121–22, 133–36; tissue contributions to, 123–27
metabolism, caloric intake and, 97–98
retention: food intake and, 238–39; as immunologic mechanism, 237–38, 239; in Newcastle disease virus infections, 236–39
Noble, A., 5
Noguchi, H., 88
Nohara, H., 186
Nolten, W., 108
Novikoff hepatoma, thymidine kinase in cell cultures from, 145
Novikova, M. A., 179
Nucleic acids
Newcastle disease virus vs. starvation and avian liver, 240–49
See also Deoxyribonucleic acid (DNA); Ribonucleic acid (RNA)
Nucleoside diphosphokinases, action of, 203–4
Nucleotides
hydrolysis, in liver cells in culture, 190–91
phosphorylation, temperature and, 178
Nursing
body size and restrictions on, 7
of large vs. small litters in rats, 6–7
Nutrition, endocrines and, 61–70
Nygaard, O., 194, 197, 198

O

O'Brien, J. S., 199
O'Dell, B. L., 102
Oelkers, W., 108
Ohno, S., 188
Okazaki, R., 145
Okunuki, T., 228
Ol'Shaneiskaya, A. D., 179
Ord, M. G., 198
Ovarian hormones, growth and, 52, 53
N-(3-Oxapentamethylene)-N′, N″-diethylenephosphoramide (MEPA), 275
hematologic response to, in dogs, 265, 266, 267
therapeutic vs. toxic dosage, 266
Oxygen utilization, thyroid function and, 64, 65

P

Perry, S., 199, 200
Phase infections, kinase activity and, 185–89
Phosphatases
tumor growth and, 190
See also specific enzyme
Phosphokinase activity, variations in cell cycle, 164–66
Phosphorylation
of nucleotides: deoxyribonucleotides in liver, 183; temperature and, 178
of thymidine: cellular proliferation and, 141–212; in development, 171–75; hormones and, 178–79; mechanism of, 151–52; miscellaneous enzymes related to, 202–4; miscellaneous factors in control of, 176–78; thymidine phosphorylase and, 202–3
of thymidine-3′-pyrophosphate, 204
in viral infected tissue culture cells, 185–86
Picou, D., 42
Piekarski, L. J., 186
Pimstone, B. L., 37
Pitot, H. C., 166, 253
Pituitary glands
function and malnutrition, 37
growth and, 52, 53–55
hypophysectomy: in breast carcinoma, 252; growth and, 54–55
Placental hormones, fetal growth and, 52
Plants, thymidine kinase in, 148–49, 162–64
Plasma
amino acids in malnutrition, 40–41
invertebrate hemagglutinins, 87–93
malnutrition and, 35, 38, 40–41
proteins: amino acid incorporation into, 78–80, 81; of canine neoplasm hosts, 268
ribonuclease, tumor growth and, 269, 270

vitamin B-12 and starvation, 98
volume in malnutrition, 35
Platt, B. S., 4
Podolsky, R. J., 115
Polyoma virus, kinase activity and infection with, 187–88
Potassium
body, in marasmus vs. kwashiorkor, 34–35
brain, in malnutrition, 29
in muscle fibers, 115
protein xxx ratio, organ variations in, 34–35
Potter, R. L., 194, 197, 198
Potter, V. R., 145, 146, 148, 152, 160, 181, 184
Powell, W. F., 177
Prednisone, in chemotherapy, 275, 280
Prescott, D. M., 146, 156
Propylthiouracil, oxygen utilization and, 64
Protein(s)
amino acid deprivation and, 125–27
body: in normal vs. malnourished states, 31–36; potassium ratio to, 34–35
cellular, expression of changes in concentration of, 133–34
composition of skeletal muscle, 128–31
contractile, in muscle contraction, 111, 112, 113, 117, 118
dietary, 5; amino acid metabolism and, 41; anoxia in mice and, 64, 65; estrogen-induced uterotrophic response and, 63, 64; thyroid gland and, 66–70; L-triiodothyronine action and, 66–67
dietary deficiency: behavioral abnormalities in animals and, 3–26; body composition and, 31–36; body size and shape and, 30–31; effects on human infant, 28–44; enzymes and, 123, 125; protein metabolism in skeletal muscle and, 128–32
metabolism: adrenalectomy and, 58–59; diabetes and, 55–56; hypophysectomy and, 54–55, 56; kwashiorkor mortality and, 40–41; in muscle, 121–36; study techniques, 40–43; thyroidectomy and, 56–57; viral stress and, 233–49
reserve, 121–36; in amino acid depletion, 125–27; defined, 122; history of concept, 121–22; metabolism in muscles and, 121–36; in neoplasms, 267–69
synthesis: amino acids and translational phase, 125; cytoplasmic, in liver, 75–85; in Newcastle disease virus infections, 242 ff
variations in nutritional value and utilization of, 68–69
Prusoff, W. H., 204
Puromycin, thymidine kinase activity and, 161
Pyrimidine metabolism in leukemic leucocytes, 199–200
Pyrimidine deoxyribosyltransferase, action of, 203

R

R 3230 AB tumor, estrogens and, 253–55, 256
R 3230 AC tumor, estrogens and, 253–55, 256
Radiation
DNA synthesis and, 194–95
thymidine kinase and TMP kinase synthesis and, 179, 181–82
Radioisotopic tracer technique in protein metabolism study, 42–43
Raine, L. C. D. P., 101
Ratliff, R. L., 203
Ray, G. S., 221, 222
Reisfield, D. R., 64
Rejnek, J., 78
Ribonuclease, tumor growth and plasma, 269, 270
Ribonucleic acid (RNA)
liver: nitrogen loss and, 134–35; protein depletion and, 125
malignancies and cellular, 270–72
messenger: stability for thymidine kinase, 166–68; synthesis, 125
muscle: nitrogen loss and, 134–35; protein depletion and starvation and, 128, 131–32
in Newcastle disease virus infections, 240–41, 242 ff
protein depletion and, 125–26, 128, 131–32

Ribonucleosides, conversion of thymine and azathymine to, 204–5
Ribosomes
protein depletion and, 125–26, 131–32
protein synthesis and, 85, 125
Riman, J., 204
RNA, *See* Ribonucleic acid (RNA)
RNA-viruses, kinase activity and infection with, 188
Roberts, D., 202
Roth, J. S., 12, 148, 153, 156, 159, 166, 167, 172, 173, 174, 189
Rouse sarcoma virus, kinase activity and infection with, 188
Rubini, J. R., 191
Rueckert, R. R., 199
Russel, W. C., 166
Rutishauser, I. H. E., 37

S

Sachsenmaier, W., 171
Sakuma, K., 179
Sanders, S. J., 43
Santi, D. V., 149
Sarcoma 180, uterotrophic response to estrogens and, 62–63
Sarcoplasmic proteins, 129–30
synthesis of, 129
turnover rate, 129–30
Sarcoplasmic reticulum in muscle contraction, 111, 112, 113, 115–16, 118–19
Sartorelli, A. C., 150
Scheidegger, J. J., 76
Schmidt, R. R., 160, 171
Schnieder, W. C., 193
Scholl, A., 172
Scholtissek, C., 178
Schor, I. M., 108
Schwan, T. J., 149
Seidenberg, J., 203
Seifert, J., 204
Sekuzu, I., 228
Serum, *See* Plasma
Sex factor in behavioral abnormalities and malnutrition, 19, 23–24, 24–25
Shen, S. R. C., 171
Shimabayashi, Y., 194
Shimkin, M. B., 258
Shock, N. W., 16
Shooter, K. V., 182
Shou, P. G. D., 156
Shuster, C. N., Jr., 89
Sidransky, H., 131
Simian viruses, kinase activity and infection with, 187
Skin
changes in kwashiorkor, 39
contribution to nitrogen excretion in malnutrition and stress, 124
protein/potassium ratio in, 34, 35
Skinner box bar-pressing tests, 19–22, 23–24
Slater, E. C., 221
Smellie, R. M. S., 147, 152, 160
Smith, L., 221
Smith, R., 37
Sodium
body, in malnourished vs. normal states, 31–36
in muscle fibers, 115
Soska, J., 197
Soskova, I., 197
Spleen
contribution to nitrogen excretion in malnutrition and stress, 125
pool sizes of thymidine derivatives in, 197–98
starvation and vitamin B-12 of, 98
Squibb, R. I., 234
Srebnik, H. H., 66
Staehelin, T., 54
Stannus, H. S., 39
Staphylococci, mouse liver incorporation systems and, 81–82
Starvation and metabolic systems, 97–109
enzyme systems, 102–9
nitrogen retention, 97–98
nucleic and free amino acids in avian liver, 241–42
protein metabolism in skeletal muscle, 128–32
vitamin B-12 metabolism, 98–99, 100–2
See also specific topic
Stephens, J. M. L., 30, 126
Stern, H., 162
Stevens, L., 182
Stewart, P. A., 194, 195, 196

Stirling, G. A., 36
Stocken, L. A., 182, 198
Stone, G. E., 146, 201
Stress
  free amino acids in avian liver and, 247–49
  nitrogen excretion in, 133–36; tissue contribution to, 123–27
  nutrient requirements and, 61
  protein deprivation as, 58
  response to malnutrition and, 7–11
  viral, protein metabolism and, 233–49
Stroup, M., 89
Stubblefield, E., 166
Sugino, Y., 142, 203
Surface membrane in muscle contraction, 111, 112, 113, 116–18, 119
Sussman, O., 273
Svedberg, T., 89

T

Takahashi, M., 188
Terayama, H., 179
Testosterone
  growth and, 53
  thymidine kinase activity and, 178
Tetrahymena, thymidine and DNA synthesis in, 201–2
Thio TEPA in chemotherapy, 275, 279
Thiouracil
  fetal growth and, 52
  thyroid gland and, 65–67
Thomas, L. B., 255
Thompson, U. B., 146, 151, 154
Thorbecker, G. J., 77, 78
Thymidine
  DNA synthesis and, 198–99
  enzymes and, 193
  labeled: incorporation in experimental growth phase, 199; in normal vs. leukemic leucocytes, 201
  Michaelis constants, 150
  nucleotides, in normal vs. leukemic leucocytes, 200–1
  phosphorylation: in development, 171–75; hormones and, 178–79; mechanism of, 151–52; miscellaneous enzymes related to, 202–4; miscellaneous factors in, 176–78; thymidine phosphorylase and, 202–3
  pools, in cells, 193–202; DNA synthesis in Tetrahymena and, 201–2; in normal vs. irradiated animals, 194–202
  toxicity of, 151, 196
  thymidine kinase activity and, 162–64
Thymidine diphosphate (TDP), in thymidine triphosphate formation, 151–52
Thymidine kinase
  aggregation and disaggregation, 151
  assay systems, 208–11
  in development, 171–75
  distribution, 145–46; intracellular, 146; in plants, 148–49
  importance, 143–44
  induction, 179–89
  inhibition, 161–62; 166; by agents other than TTP, 149–50; feedback, by TTP, 152–56
  messenger RNA stability, 166–68
  metal ions and, 148
  molecular weight, 149
  properties, 147–56
  protein-free diets and, 184–85
  purification, 146–47
  radiation and synthesis, 179, 181–82
  regulation of activity, 160–68
  stability, 147–48; in tumors vs. normal cells, 166–67
  substrate specificity, 150–51
  testosterone and, 178
  thymidine induction of activity, 162–64
  thymidine triphosphate synthesis and, 143–44
  turnover rate, in liver, 167
  variations in activity, 160–61
  viral infections and activity, 185–89
Thymidine phosphates, degradation of, 189–93
Thymidine phosphorylase, thymidine phosphorylation and, 202–3
Thymidine-3′-pyrophosphate, phosphorylation of, 204
Thymidine triphosphate (TTP)
  action, 141–42

Thymidine triphosphate (TTP) (cont.)
degradation, 189–93
DNA synthesis and, 199–201
feedback inhibition of thymidine kinase by, 152–56
metabolism and metabolic control actions, 141, 142
phosphorylation mechanism, 151–52
reactions in synthesis, 142–44
regulatory control, 142
Thymidylate kinase (TMP kinase), 156–60
activity of fresh vs. frozen homogenates, 159
assay systems, 212
degradation in animal tissue, 189–91
deoxynucleotide synthesis in *E. coli* and, 176–77
in development, 173–75
distribution, 156–5
DNA synthesis and, 170–71, 198–99
estrogens and, 178–79
growth and: activity in synchronous growth, 170–71; growth control and control of, 168–69
induction, 179–89
properties, 158–60
purification, 160
radiation and synthesis of, 179, 181–82
regulation of, 168–71; growth control and, 168–69
stability, 159–60
thymidine triphosphate synthesis and, 143–44
viral infection and activity, 185–89
viral oncogenesis and synthesis of, 207
Thymine, conversion to ribonucleosides and deoxyribonucleosides, 204–5
Thymus, pool size of thymidine derivatives in rat, 197–98
Thyroid glands
effects of function imbalances, 56–57
growth and, 53
malnutrition and, 38, 64–70
thyroidectomy: oxygen utilization and, 64, 65; protein repletion and, 56–57
Tissue(s)
contribution to nitrogen loss in malnutrition and stress, 123–27
contribution to protein depletion, 134–36
deoxyribosides in, 193–94
in Newcastle disease virus infections, 242–46
normal vs. malignant, 270–72
thymidine kinase in, 144, 145
thymidine phosphate degradation in, 189–92
thymidine pools in cells of, 193–202
TMP kinase in, 157, 158–59
vitamin B-12 and starvation, 98, 100–2
See also specific tissue and/or topic
TMP kinase, *See* Thymidylate kinase (TMP kinase)
Toxicity
of deoxyuridine derivatives, 151
of thymidine derivatives, 151, 196
Trauma
TMP kinase and thymidine kinase activity in, 144
See also Stress
Triad, in muscle contraction, 111–12, 115–16
Triethylene melamine (TEM), in chemotherapy, 275, 279
L-Triiodothyronine, dietary protein and action of, 66–67
Triphosphate synthesis, in liver vs. ascites hematoma, 176
Tripp, M. R., 88, 89, 90, 91
Tropomyosin in muscle contraction, 113–14
Troponin in muscle contraction, 113–14
Trubowitz, S., 190
Trump, G., 92
TTP, *See* Thymidine triphosphate (TTP)
Tuberculosis, free amino acids in liver and, 247
Tumors, *See* Cancer; Neoplasms
Tyler, A., 87

U

Urine
composition in malnutrition, 37, 38, 40, 41
vitamin B-12 excretion and starvation, 98–99
Uterus, response to estrogens, 62–63

V

Vaccinia (pox) virus, kinase activity and infection with, 185–87, 188
Van Bibber, M. J., 185
Van Lancker, J. L., 159, 180, 181
Verney, E., 131
Vestling, C. S., 221
Vinblastine, in chemotherapy, 275, 279
Virus(es)
infections with: nitrogen retention in, 236–39; kinases and, 185–89; thymidine kinase and, 144, 155; TMP kinase and, 144, 157
mutants deficient in thymidine kinase-inducing activities, 186–87
oncogenesis and TMP kinase, 207
thymidine phosphate degradation in, 192–93
viral stress and protein metabolism, 233–49
Viscera
contribution to nitrogen excretion in malnutrition and stress, 124
protein/potassium ratio in, 34, 35
Vitamin B-12
intracellular deoxyribosyl compounds and, 196–97
starvation and metabolism of, 98–99
Vitamin deficiencies, intracellular deoxyribosyl compounds and, 196–97
Vitamin E, protein/fat ratio in diet and, 5
Voit, C., 121, 135
Vollmer, E. P., 258
Von der Decken, A., 80

W

Wainio, W. W., 102, 221, 222
Walwick, E. R., 179
Wanka, F., 147, 148
Wannemacher, R. W., 234
Ward, E. E., 36
Water, body, in normal vs. malnourished children, 31–36
Waterlow, J. C., 30, 37, 42, 126
Watne, A., 91
Weber, G., 252
Weight, *See* Body weight
Weinstock, I. M., 171
Weiss, A., 178
Weissman, S. M., 155
Wemyss, C., 76, 78
Wharton, D. C., 223
Whitehead, R. G., 40
Williams, C. A., 81
Williams, C. D., 39
Williams, C. M., 157, 179
Williams, S. E., 150
Winick, M., 5
Wykes, J. R., 156

Y

Yokoyama, M., 89
Yonetani, T., 221, 222
Young, V. H., 38
Y-shape water maze tests of learning behavior, 18–19

Z

Zimmerman, M., 203

This book was set in Baskerville Linofilm and printed by offset on P & S Wove manufactured by P. H. Glatfelter Co., Spring Grove, Pa. Composed, printed and bound by Quinn & Boden Company, Inc., Rahway, N. J.